Biochemistry

Biochemistry

Khushboo Chaudhary

ARCLER
PRESS

www.arclerpress.com

Biochemistry

Khushboo Chaudhary

Arcler Press
224 Shoreacres Road
Burlington, ON L7L 2H2
Canada
www.arclerpress.com
Email: orders@arclereducation.com

ISBN: 978-1-77469-122-9 (Hardcover)

ABOUT THE AUTHOR

Dr. Khushboo Chaudhary is presently working as a Technical officer-I in Translational Health Science and Technology Institute, Faridabad Haryana, India and having one year of teaching experience and eight years of research experience. Previously, she worked on the PPR virus and New Castle Disease virus and isolates many other viruses from animal outbreak samples. She has published several research papers in international and national journals. She has published five international textbooks. She has got seven best paper and poster presentation awards from the Indian Society of Genetics and Biotechnology Research and Development and received the president appreciation awards also in the International conference. She has got the best poster award by ISSGPU Central Institute Research on Goats, Makhdoom. She has published several gene banks in NCBI Pubmed. She has also published a research article in a virology journal. She is likely to be a co-author in several of the publications and coauthor in J. Virological Methods.

TABLE OF CONTENTS

LIST OF FIGURES

LIST OF TABLES

LIST OF ABBREVIATIONS

AMP	Adenosine monophosphate
ATP	Adenosine triphosphate
CCMB	Center for Cellular and Molecular Biology
CMP	Cytosine monophosphate
dATP	Deoxyadenosine triphosphate
dGTP	Deoxyguanosine triphosphate
DNA	Deoxyribonucleic acid
dTTP	Deoxythymidine triphosphate
EMS	Ehylmethanesulfonate
GMP	Guanosine monophosphate
HW	Hardy Weinberg
NA	Numerical aperture
PKU	Phenylkeptonuria
PCR	Polymerase chain reaction
PrP	Prion protein
RNA	Ribonucleic acid
STEM	Scanning transmission electron microscope
SDS	Sodium dodecylsulfate
TEM	Transmission electron microscope
UMP	Uridine monophosphate

PREFACE

Biochemistry has always been a fundamental area of science. The exponential enhancement in the quantity of scientific information and the rate at which new discoveries are made require much elaborated, interdisciplinary and up-to-date information and their understanding. The book "Biochemistry" has been written to serve as a guide book (notes form) for undergraduate, postgraduate and research scholars of different universities in the world. This book has covered a range of syllabus and the student has an easy way to go through this book content. Some books are available on the subject of Biochemistry but still not up to the syllabus run in university so the students face a problem with notes.

While this book has been prepared for students that they have acquired or gain a fair knowledge of Biochemistry and easy presentation of language has written for a proper understanding of each and every topic. The chapters have been so arranged to give a sequential knowledge of the introductory biological chemistry, biophysical chemistry, cell biology and genetics, biostatistics and computers and molecular biology and genetic engineering in this book with practical approaches. In the preparation of this book, I have been greatly helped by several books and presentations which I express my acknowledgment in the references and if I have forgotten any references in my notice, please ignore them. This book would serve not only undergraduate but postgraduate students of all life science students, and also research scholars of various universities. The numbers of scientists and authors have been quoted or fully acknowledged in the bibliography at the end of the book. I claim no originality for the matter or text presented in this book but the method of presentation and illustration is my own. Last but not least I would be thankful to readers for pointing out errors and most welcome for suggestions in the next book.

PART I

INTRODUCTORY BIOLOGICAL CHEMISTRY

1. GENERAL ACCOUNT OF THE CHEMICAL NATURE OF LIVING CELLS.

All living organisms, from microbes to mammals are composed of chemical substances from both the inorganic and organic world that appear in roughly the same proportions and perform the same general activity. Hydrogen, oxygen, nitrogen, carbon, phosphorus, and sulfur normally make up more than 99% of the mass of living cells and when combined in various ways, form virtually all known organic biomolecules. They are initially utilized in the "synthesis of a small number of building blocks that are, in turn, used in the construction of a vast array of vital macromolecules. There are four general classes of macromolecules within living cells such as nucleic acids, proteins, polysaccharides and lipids. These compounds, which have molecular weights ranging from 1×10^3 to 1×10^6, are created through the polymerization of building blocks that have molecular weights in the range of 50 to 150 MW. Although many differences do exist between cells (such as erythrocyte, liver, muscle or fat cell), they all generally contain a greater variety of proteins than any other type of macromolecules, with about 50% of the solid matter of the cell being protein.

Cells generally contain many more protein molecules than DNA molecules, yet DNA is typically the largest biomolecule in the cell. About 99% of cellular molecules are water molecules, with water normally accounting for approximately 70% of the total wet-weight of the cell. Although water is obviously important to the vitality of all living cells, the bulk of our attention is usually focused on the other 1% of biomolecules. Data regarding the chemical composition of the unicellular *Escherichia coli* (*E. coli*), are not greatly different for multicellular organisms which including mammals. Each *E. coli,* and similar bacterium, contains a single chromosome; therefore, it has only one unique DNA molecule. Mammals contain more chromosomes and thus have different DNA molecules in the nucleus. The elements such as Hydrogen, oxygen, nitrogen, carbon, sulfur, and phosphorus normally make up more than 99% of the mass of living cells. Homogenous polymers are non-informational. All non-essential lipids can be generated from acetyl-CoA. Like certain amino acids and unsaturated fatty acids, various inorganic elements are dietary "essentials". Most diseases in animals are manifestations of abnormalities in biomolecules, chemical reactions, or biochemical pathways. (www.scribd.com., www. slideshare.com., www.coursehero.com., www. biocyclopedia.com., www. molecular-plant-biotechnology.info., www.slideplayer.com., www. documents.mx., www.docplayer.net., www.bmb.psu.edu.,www.archive. org.,www.mafiadoc.com., www.powershow.com., www.bioscience.jbpub. com., www.outhorstream.com., www.docslide.com., www.msluay.com., www.scienceprofonlue.com., www.csus.edu., www.krishikosh.egrouth. ac.in., www.tankonyvtor.hu., www.silo.pub., www.studyblue.com., www. Preservearticles.com., www.biotechuniverse.blogspot.com").

2. CARBOHYDRATES: CLASSIFICATION, CONFIGURATIONS AND CONFORMATIONS, SUGAR DERIVATIVES, STRUCTURAL AND STORAGE POLYSACCHARIDES.

It is polyhydroxy aldehydes or polyhydroxy ketones or compounds that can be hydrolyzed to them. In the majority of carbohydrates, H and O are present in the same ratio as in water hence also called as hydrates of carbon. Carbohydrates are polyhydroxyaldehydes or polyhydroxy-ketones of formula $(CH_2O)_n$, or compounds that can be hydrolyzed to them (aldehyde sugars or saccharides). It is classified into the following classes depending upon whether these undergo hydrolysis and if so on the number of products formed.

Monosaccharides

It consists of a single polyhydroxy aldehyde or ketone unit. It is simple sugars and they have the general formula $C_nH_{2n}O_n$. It is freely soluble in water but insoluble in non-polar solvents. Carbohydrates cannot be hydrolyzed to simpler carbohydrates such as glucose or fructose.

Disaccharides

It is the simplest and most common oligosaccharides. A disaccharide consists of two monosaccharides joined by an O-glycosidic bond. Carbohydrates can be hydrolyzed into two monosaccharide units such as sucrose, which is hydrolyzed into glucose and fructose.

Oligosaccharides

Oligosaccharides containing three or more residues are relatively rare and occurring almost entirely in plants. These carbohydrates can be hydrolyzed into a few monosaccharide units.

Polysaccharides

It is ubiquitous in nature. It is also called glycans. They are two types: structural and storage polysaccharides. Structural polysaccharide provides mechanical stability to cells, organs and organisms. Storage polysaccharides serve as carbohydrate stores that release monosaccharides as required. Carbohydrates are polymeric sugars; e.g. starch or cellulose such as Triose, tetrose, pentose, hexose, etc. carbohydrates that contain three, four, five, six, etc. Hydrates of carbon of general formula is $C_n(H_2O)_n$. In plants: photosynthesis $6\ CO_2 + 6\ H_2O\quad C_6H_{12}O_6\ + 6\ O_2$

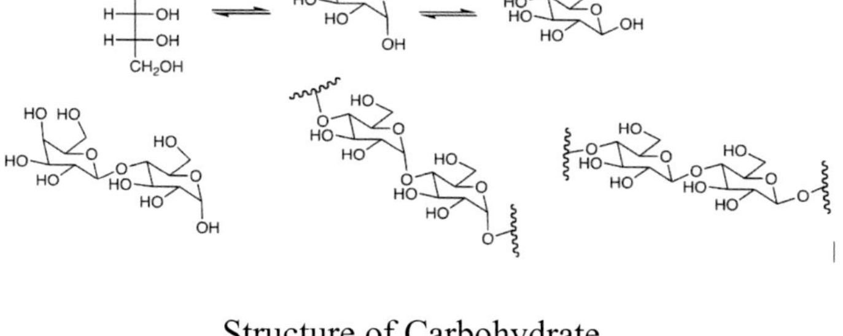

Structure of Carbohydrate

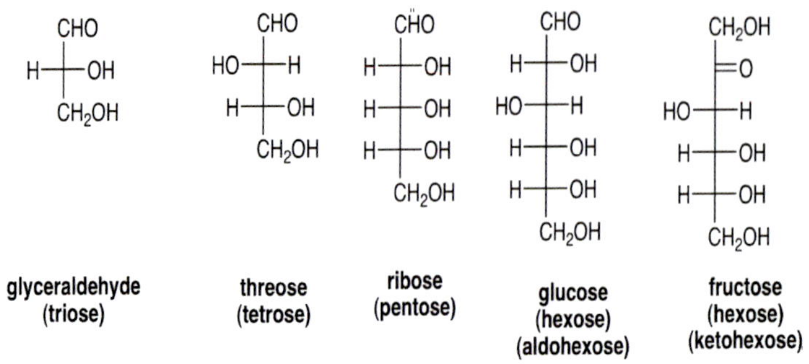

Structure of Polysaccharides

- ***Polymers*:** Large molecules are made up of repeating smaller units (monomer) is called polymers.
- ***Aldose*:** Polyhydroxyaldehyde **example** glucose.
- ***Ketose*:** Polyhydroxyketone **example** fructose.
- Carbons per molecule (usually five or six); eg. Aldohexose and ketopentose etc.
- ***Reducing sugar*:** It is a carbohydrate that is oxidized by Tollen's, Fehling's or Benedict's solution is called reducing sugar.

Tollen's: $Ag^+ \rightarrow Ag$ (silver mirror)

Fehling's or Benedict's: Cu^{3+} (blue) $\rightarrow$ Cu^{2+} (red ppt).

These are reactions of aldehydes and alpha-hydroxyketones. All monosaccharides (both aldoses and ketoses) and most disaccharides are reducing sugars. Sucrose (table sugar) is a disaccharide, is not a reducing sugar.

Glucose (a monosaccharide)

Plants:

photosynthesis

$$6\,CO_2 + 6\,H_2O \xrightarrow[\text{sunlight}]{\text{chlorophyll}} \underset{\text{(+)-glucose}}{C_6H_{12}O_6} + 6\,O_2$$

$$\text{(+)-glucose} \rightleftharpoons \text{starch or cellulose}$$

respiration

$$C_6H_{12}O_6 + 6\,O_2 \longrightarrow 6\,CO_2 + 6\,H_2O + \text{energy}$$

Glucose in Plants

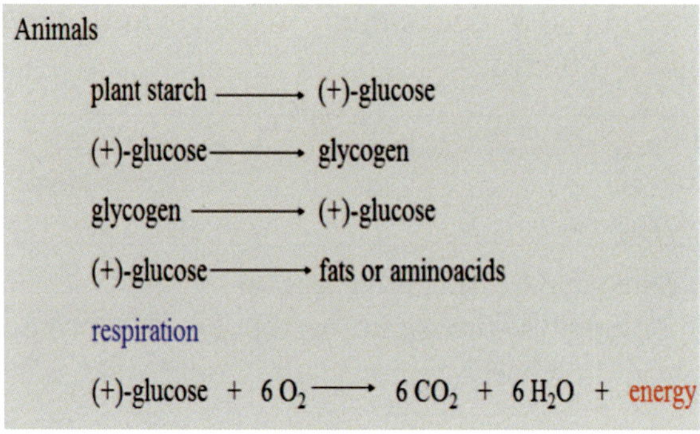

Glucose in Animals

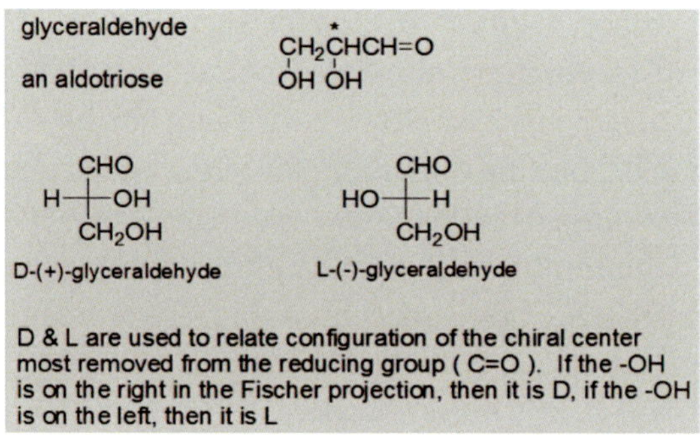

Fischer Projections and the D, L Notation

- Representation of a three-dimensional molecule as a flat structure.
- Tetrahedral carbon is represented by two crossed lines.

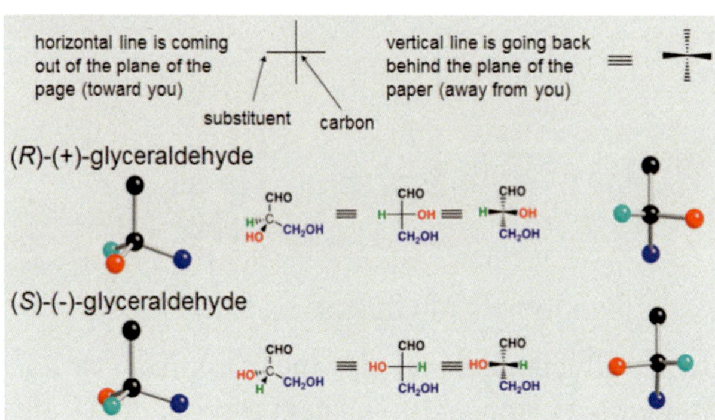

Absolute configuration of Glyceraldehyde is specified by the RS sysem in Fisher projection formulas.

- Before the *R/S* convention, stereochemistry was related to (+)-glyceraldehyde.

D-glyceraldehyde *R*-(+)-glyceraldhyde (+)-rotation = dextrorotatory = d	L-glyceraldehyde *S*-(-)-glyceraldhyde (-)-rotation = levorotatory = l

D-carbohydrates have the -OH group of the highest numbered chiral carbon pointing to the right in the Fischer projection as in R (+)-glyceraldehyde. For carbohydrates, the convention is to arrange the Fischer projection with the carbonyl group at the top for aldoses and closest to the top for ketoses. The carbons are numbered from top to bottom. Carbohydrates are designated as D- or L- according to the stereochemistry of the highest numbered chiral carbon of the Fischer projection. If the hydroxyl group of the highest numbered chiral carbon is pointing to the right, the sugar is designated as D (*Dextro*: Latin for *on the right side*). If the hydroxyl group is pointing to the left, the sugar is designated as **L** (*Levo*: Latin for on *the left side*). Most naturally occurring carbohydrates are of the D-configuration. Chiral molecules such as glucose and arabinose can exit in two configurations that are non-superimposable mirror images of each other. These two configurations are called enantiomers.

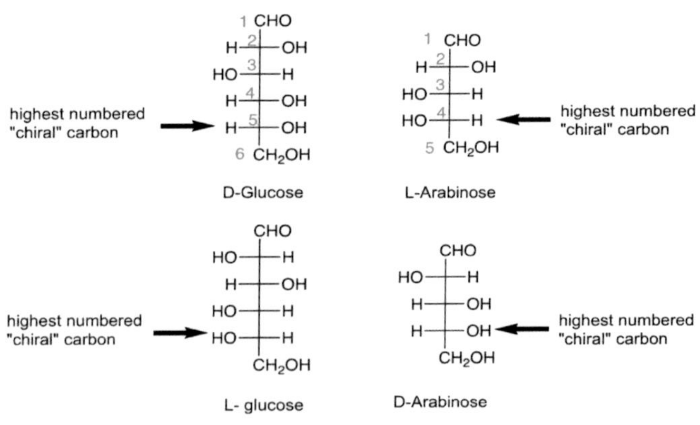

Enantiomers of Glucose and Arabinose

Aldotetroses: Glyceraldehyde is the simplest carbohydrate (C_3, aldotriose, 2,3-dihydroxypropanal). The next carbohydrates are aldotetroses (C_4, 2,3,4-trihydroxybutanal).

aldotriose

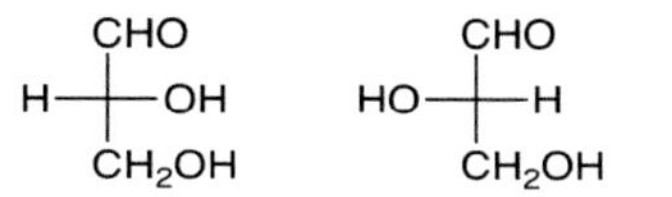

D-glyceraldehyde L-glyceraldehyde

Structure of DL aldoses with three carbons

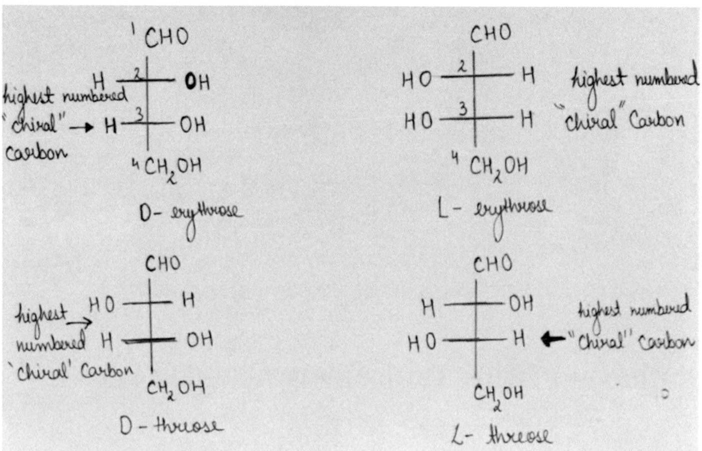

Aldopentoses and Aldohexoses

- **Aldopentoses:** It contains five carbon **C_5**, three chiral carbons and eight stereoisomers.

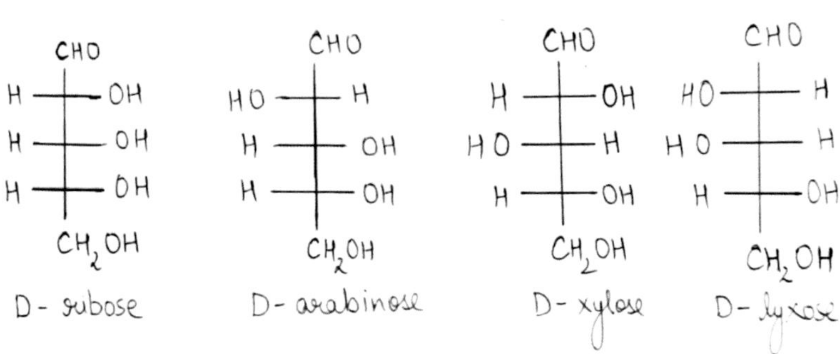

- **Aldohexoses:** It contains six carbon **C_6**, four chiral carbons and sixteen stereoisomers.

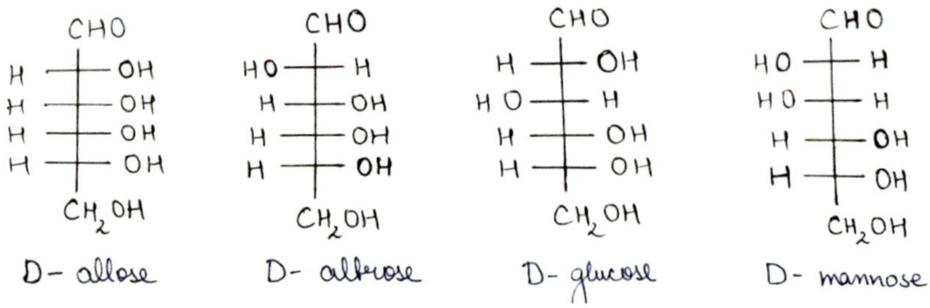

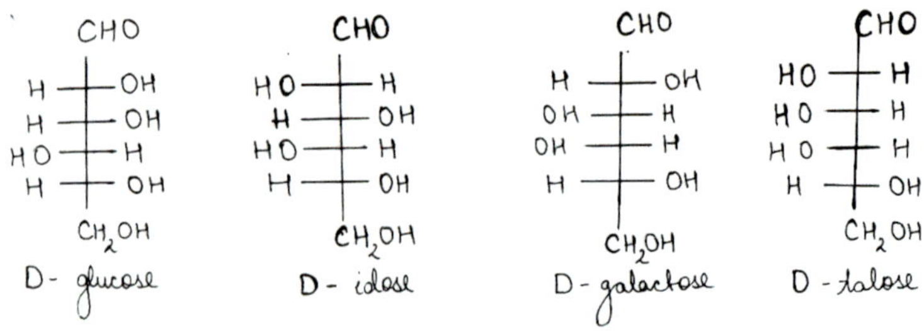

Manipulation of Fischer Projections

Fischer projections can be rotated by 180°.

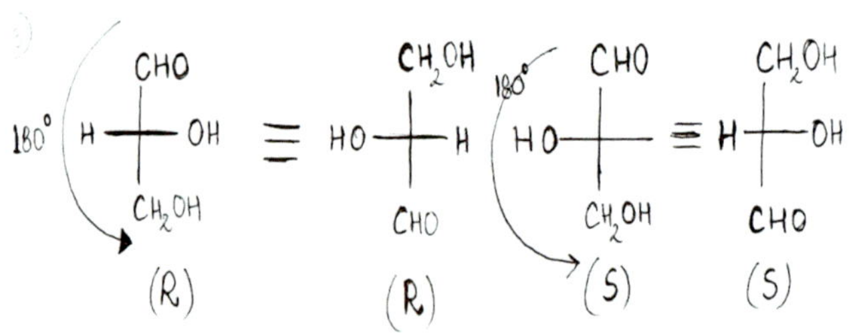

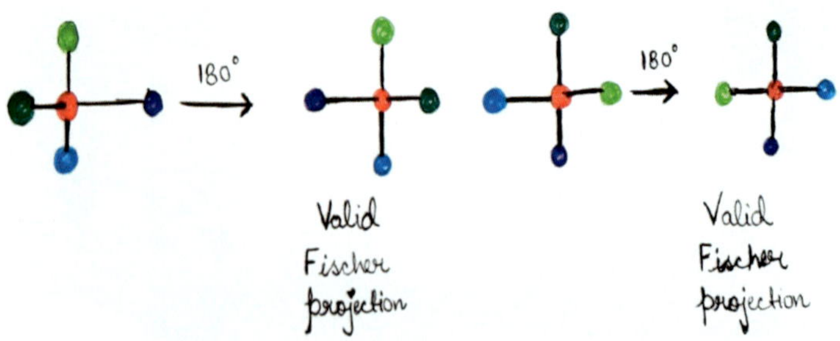

A 90° rotation inverts the stereochemistry and is not right.

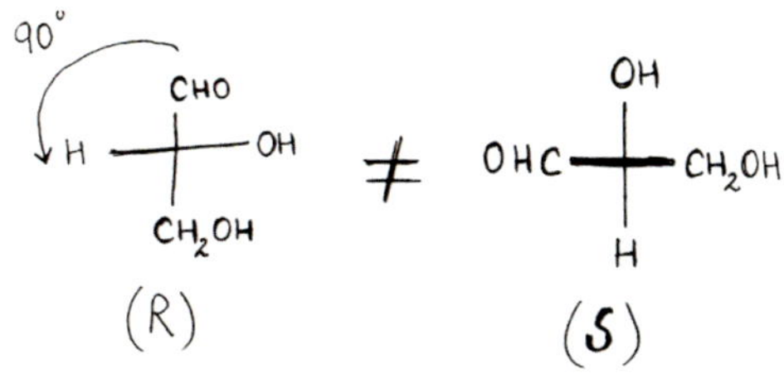

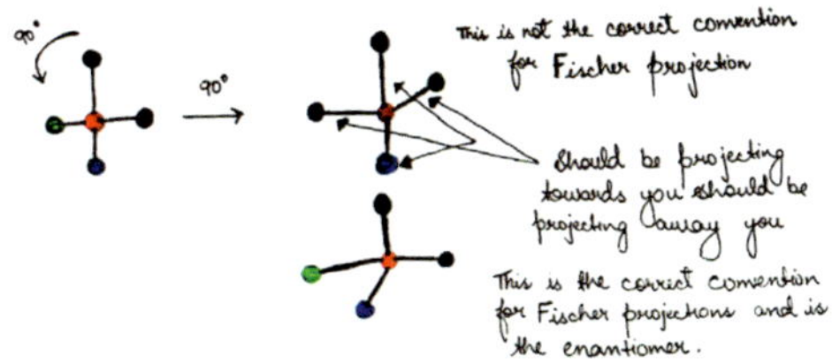

If one group of a Fischer projection is held steady, the other three groups can be rotated clockwise or counterclockwise (data collect from www.scribd.com., www.slideshare.com., www.coursehero.com., www. biocyclopedia.com., www. molecular-plant-biotechnology.info., www.slideplayer.com., www.documents.mx., www.docplayer.net., www.bmb.psu.edu.,www.archive.org.,www.mafiadoc.com., www.powershow.com., www.bioscience.jbpub.com., www.outhorstream.com., www.docslide.com., www.msluay.com., www.scienceprofonlue.com., www.csus.edu., www.krishikosh.egrouth.ac.in., www.tankonyvtor.hu., www.silo.pub., www.studyblue.com., www.Preservearticles.com., www.biotechuniverse.blogspot.com).

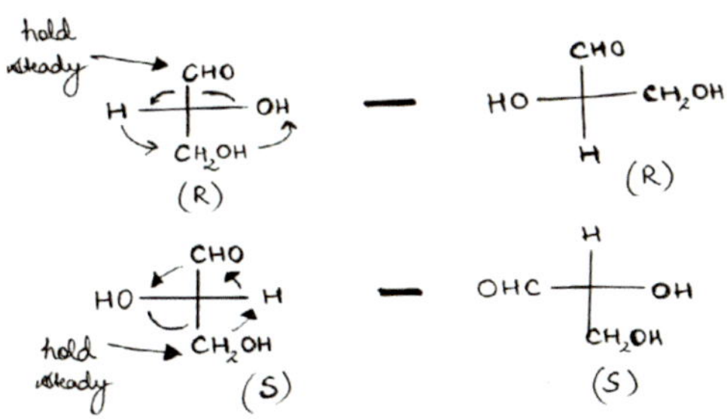

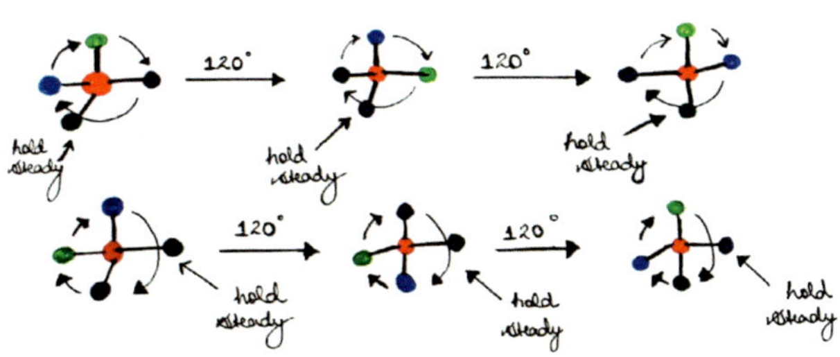

3. AMINO ACIDS: GENERAL PROPERTIES, PEPTIDE BOND, ESSENTIAL AND NON-ESSENTIAL AMINO ACIDS

An amino acid is a compound containing carbon, hydrogen, oxygen and nitrogen and serves as monomers of proteins. It contains both α-Amino acid and carboxylic acids. **Note: except for glycine (R = H), all a-amino acids are chiral molecules.** There are 20 naturally occurring R-groups

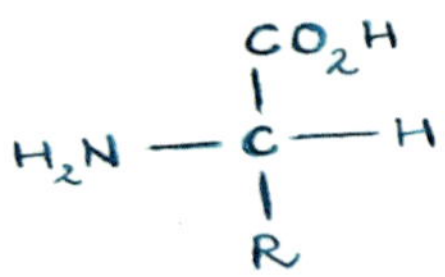

As with carbohydrates, it is traditional to use the D and L nomenclature with amino acids based on the configuration of glyceraldehyde. Naturally occurring amino acids generally have the same configuration as L-glyceraldehyde (*S*-configuration at the α-carbon).

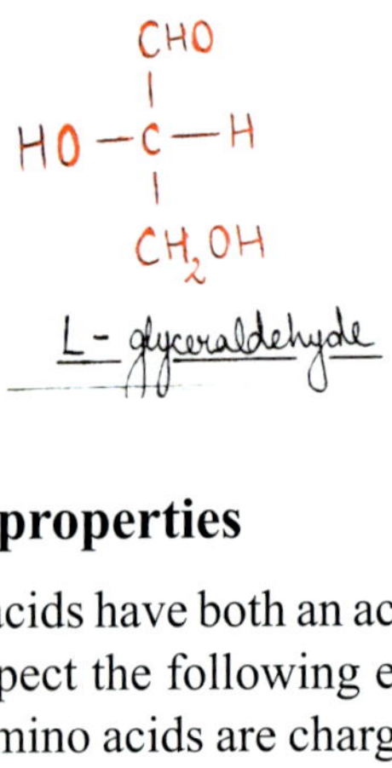

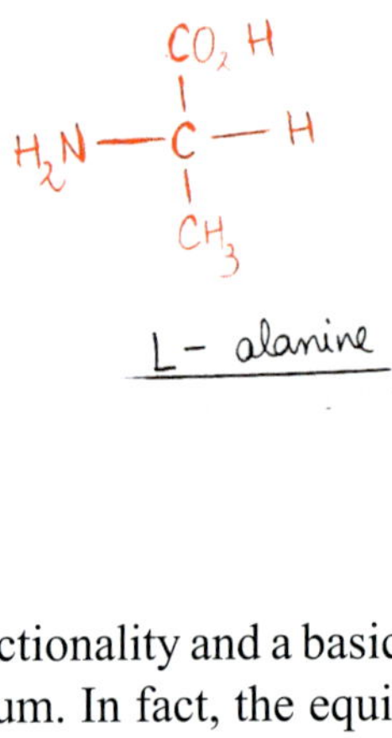

Acid–Base properties

Since amino acids have both an acidic functionality and a basic functionality, we should expect the following equilibrium. In fact, the equilibrium lies to the right all amino acids are charged at any pH. Such species that are overall neutral molecules but contain charged ends are called ***zwitterions.***

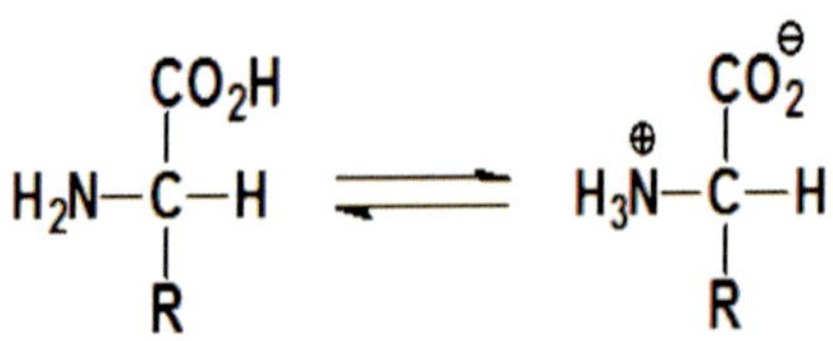

Amino acids can react as either acids or bases.

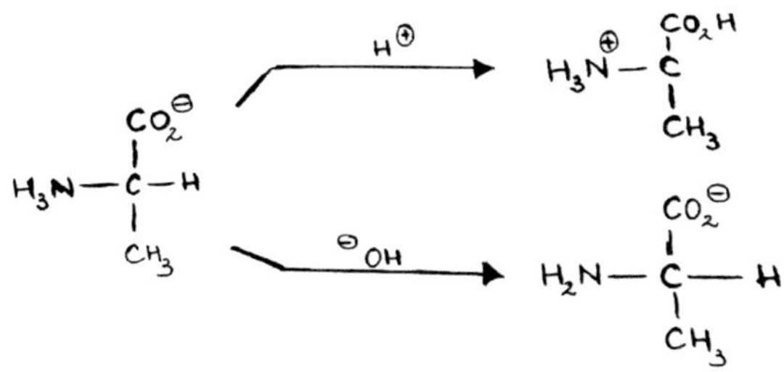

Amino acids can also have side chains containing acidic or basic groups.

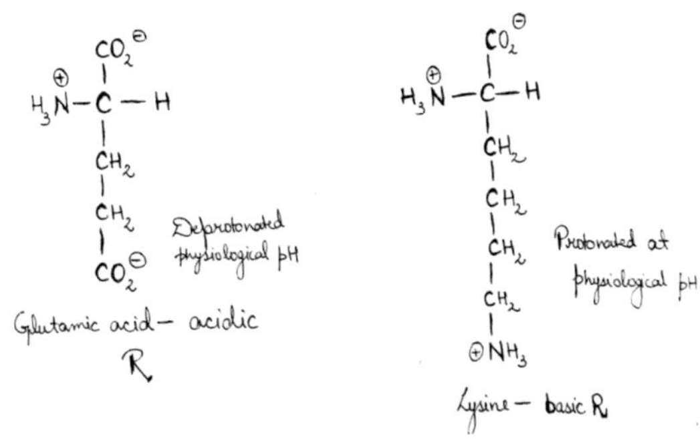

Classification

Amino acids are classified on the basis of the structure of R.

- Aliphatic side chains: Hydrophobic
- Polar side chains: Text classifies as HO-, S-, and amide containing hydrophilic groups
- Acidic – Hydrophilic
- Basic – Hydrophilic
- Heterocyclic/Aromatic: Hydrophilic or hydrophobic
- Aliphatic side chains**:** Hydrophobic

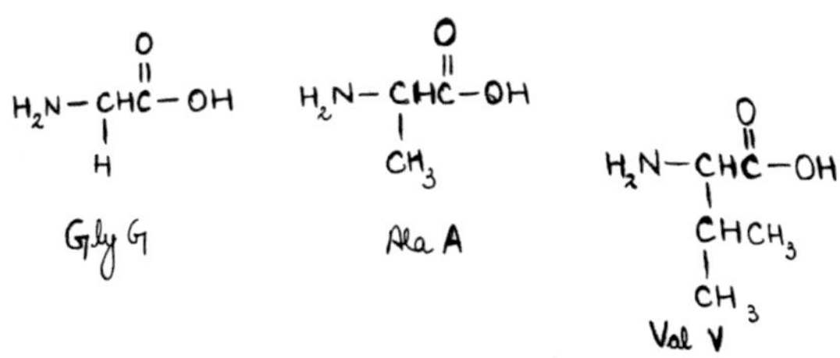

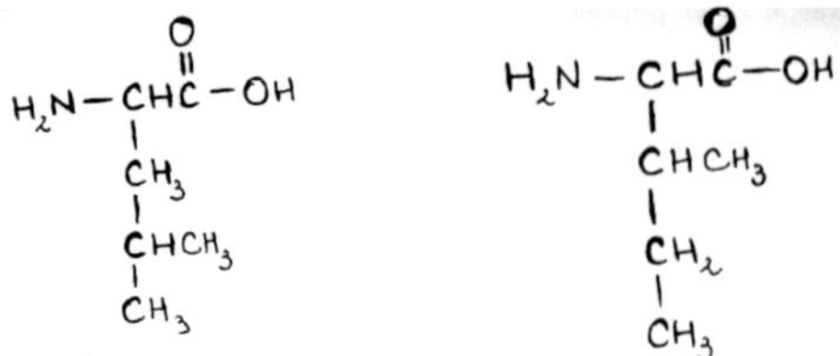

Leu, L   Ile, I

Polar side chains: Text classifies as HO-, S-, and amide containing hydrophilic groups.

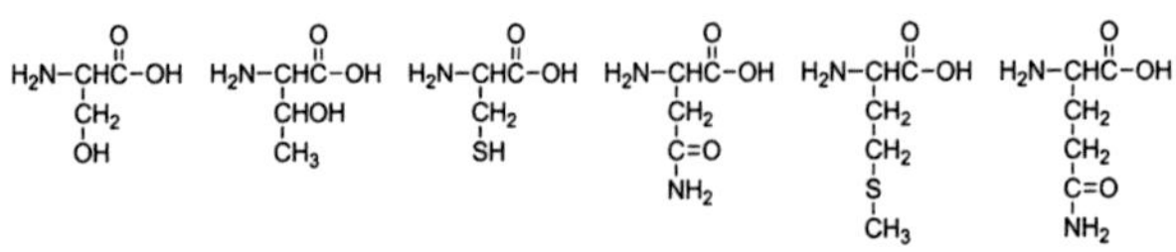

Ser, S   Thr, T   Cys, C   Met, M   Asn, N   Gln, Q

- **Basic** – hydrophilic

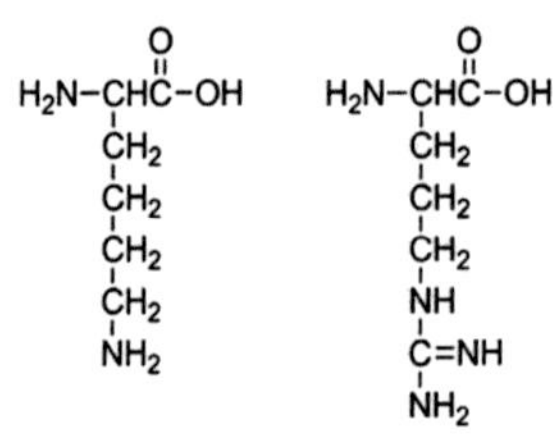

Lys, K   Arg, R

Heterocyclic/Aromatic: Hydrophilic or Hydrophobic

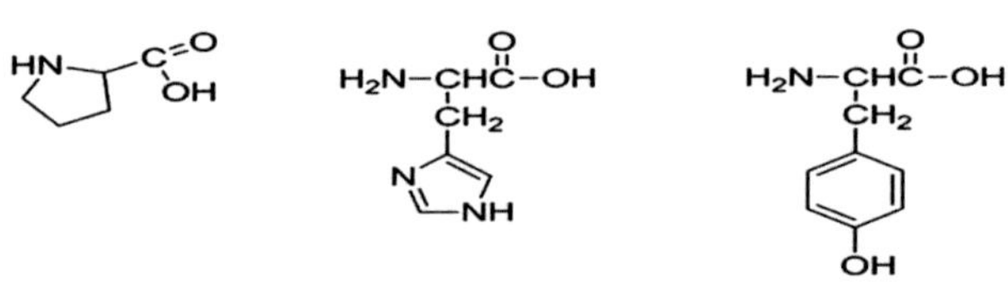

Pro, P   His, H   Phe, F

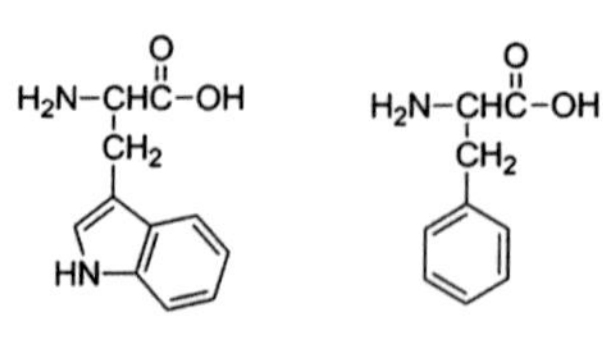

Trp, W   Tyr, Y

Acid-Base properties

Nonpolar & polar side chains	pKa of α–COOH	pKa of $\alpha-NH_3^+$
alanine	2.35	9.87
asparagine	2.02	8.80
glutamine	2.17	9.13
glycine	2.35	9.78
isoleucine	2.32	9.76
leucine	2.33	9.74
methionine	2.28	9.21
phenylalinine	2.58	9.24
proline	2.00	10.60
serine	2.21	9.15
threonine	2.09	9.10
tryptophan	2.38	9.39
Valine	2.29	9.72

Acidic Side Chains	pKa of α- COOH	pKa of $\alpha-NH_3$	pKa of side chain	Side Chain Group
aspartic acid	2.10	9.82	3.86	Carboxyl
glutamic acid	2.10	9.47	4.07	Carboxyl
Cysteine	2.05	10.25	8.00	Sulfhydril
tyrosine	2.20	9.11	10.07	phenolic

Basic Side Chains	pKa of α- COOH	pKa of $\alpha-NH_3^+$	pKa of Side Chain	Side Chain group
arginine	2.01	9.04	12.48	guanidino
histidine	1.77	9.18	6.10	imidazole
lysine	2.18	8.95	10.53	1° amino

Acidity of a-COOH groups

The average pK_a of the a-carboxyl group is 2.19, which makes them considerably stronger acids than acetic acid (pK_a 4.76). The greater acidity is accounted for by the electron-withdrawing inductive effect of the adjacent -NH_3^+ group.

The ammonium group has an electron-withdrawing inductive effect

$$RCH(NH_3^+)COOH + H_2O \rightleftharpoons RCH(NH_3^+)COO^- + H_3O^+ \quad pK_a = 2.19$$

Side chain COOH groups

Due to the electron-withdrawing inductive effect of the α-NH_3^+ group and side-chain -COOH groups are also stronger than acetic acid. The effect decreases with distance from the a-NH_3^+ group compared as:

α-COOH group of alanine (pK_a 2.35)

β-COOH group of aspartic acid (pK_a 3.86)

γ-COOH group of glutamic acid (pK_a 4.07)

Acidity of a-NH_3^+ groups

The average value of pK_a for the α-NH_3^+ group is 9.47as compared with a value of 10.76 for 1° alkylammonium ions.

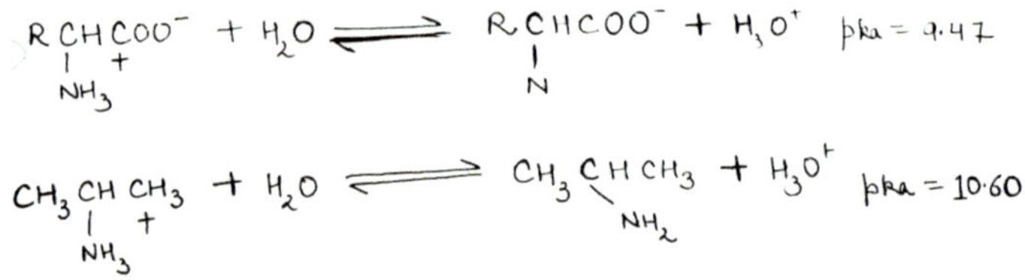

Basicity of guanidine group

The side chain of arginine contains a guanidine group.

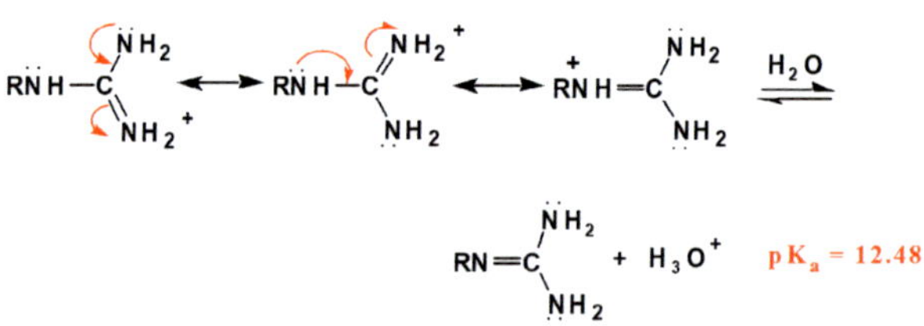

The Imidazole groups of His

The imidazole group is a heterocyclic aromatic amine.

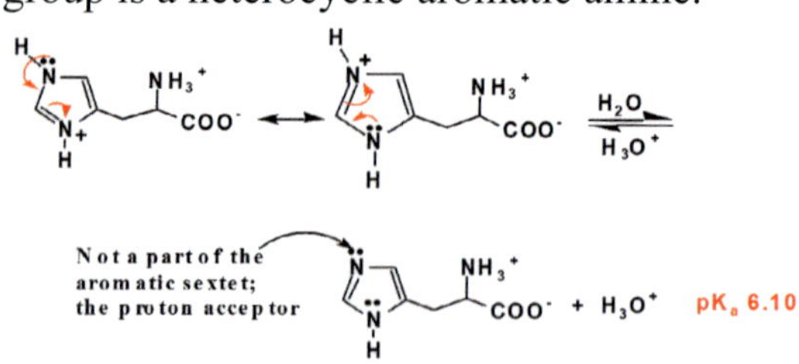

Titration of amino acids

Titration of glycine with NaOH.

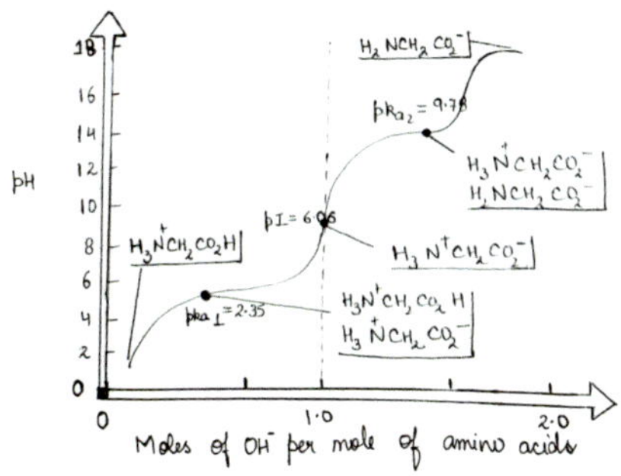

Isoelectric point (pI)

The isoelectric point, pI: the pH at which the majority of amino acid molecules in solution have no net charge. The pI for glycine, for example, falls between the pK_a values for the carboxyl and amino groups. The following tables give isoelectric points for the 20 protein-derived amino acids.

$$pI = \frac{1}{2}\left(pKa\ \alpha\text{-}COOH + pK_a\ \alpha\text{-}NH_3^{+}\right)$$

$$= \frac{1}{2}(2.35 + 9.78) = 6.06$$

Nonpolar & polar side chains	pKa of α-COOH	pKa of α-NH_3^+	pKa of side chain	pI
alanine	2.35	9.87	---	6.11
asparagine	2.02	8.80	---	5.41
glutamine	2.17	9.13	---	5.65
glycine	2.35	9.78	---	6.06
isoleucine	2.32	9.76	---	6.04
leucine	2.33	9.74	---	6.04
methionine	2.28	9.21	---	5.74
phenylalanine	2.58	9.24	---	5.91
proline	2.00	10.60	---	6.30
serine	2.21	9.15	---	5.68
threonine	2.09	9.10	---	5.60
Valine	2.29	9.39	---	6.00
tryptophan	2.38	9.72	---	5.88

Acidic side chain	pKa of α-COOH	pKa of α-NH_3^+	pKa of side chain	pI
aspartic acid	2.10	9.82	3.86	2.98
glutamic acid	2.10	9.47	4.07	3.08
cysteine	2.05	10.25	8.00	5.02
tyrosine	2.20	9.11	10.07	5.63

Basic side chain	pKa of α-COOH	pKa of α-NH_3^+	pKa of side chain	pI
arginine	2.01	9.04	12.48	10.76
histidine	1.77	9.18	6.10	7.64
lysine	2.18	8.95	10.53	9.74

A reagent is commonly used to detect amino acids is ninhydrin.

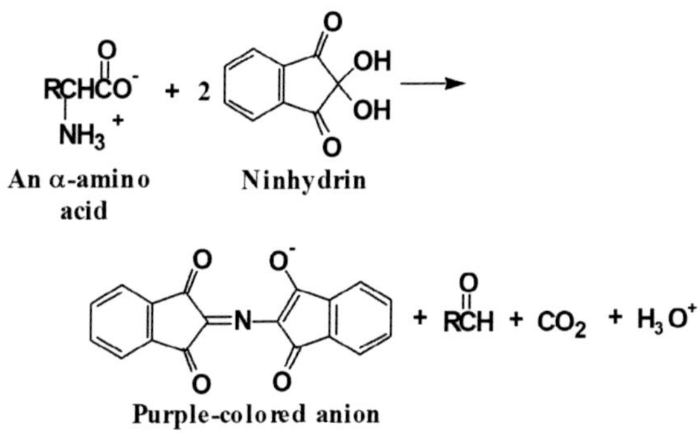

Polypeptides and proteins

In 1902, Emil Fischer proposed that proteins are long chains of amino acids joined by peptide bonds. Peptide bond: amide bond between the α-carboxyl group of one amino acid and the a-amino group of another.

Peptides

- *Peptide*: the short polymer of amino acids are joined by peptide bonds; they are classified by the number of amino acids in the chain.
- *Dipeptide*: a molecule containing two amino acids joined by a peptide bond.
- *Tripeptide*: a molecule containing three amino acids joined by peptide bonds.

- *Polypeptide*: a macromolecule containing many amino acids joined by peptide bonds.
- *Protein*: a biological macromolecule of molecular weight 5000 g/mol or greater, consisting of one or more polypeptide chains.

Peptides are amino acid polymers containing 2–50 individual units. Peptides with >50 units are called proteins. By convention, peptide structures are written with the *N*-terminal amino acid on the left and the *C*-terminal amino acid on the right.

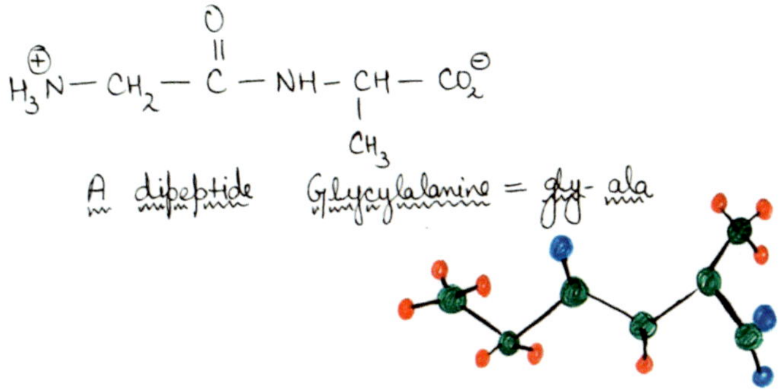

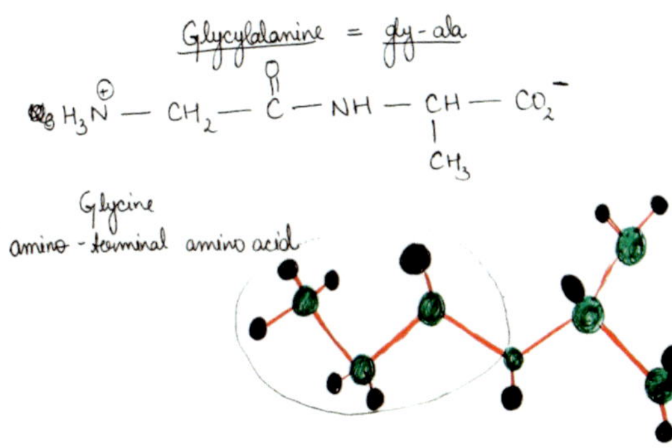

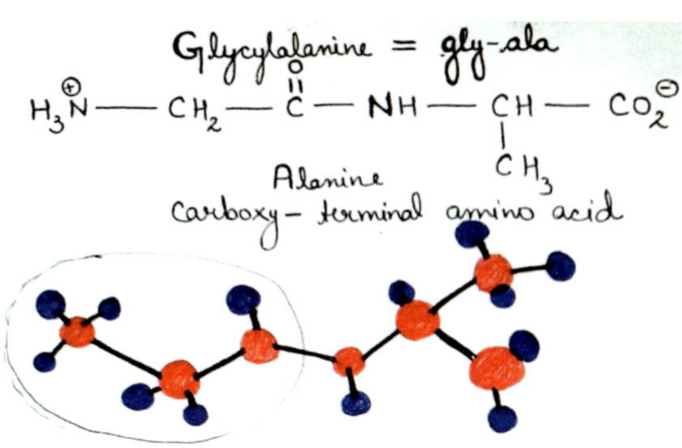

glycylalanine = gly-ala

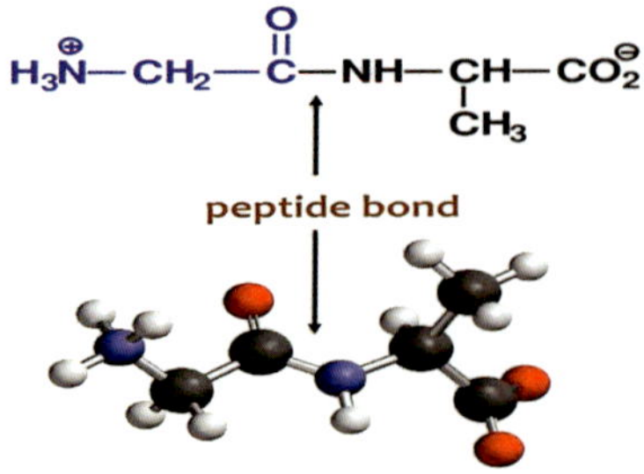

• A tetrapeptide:

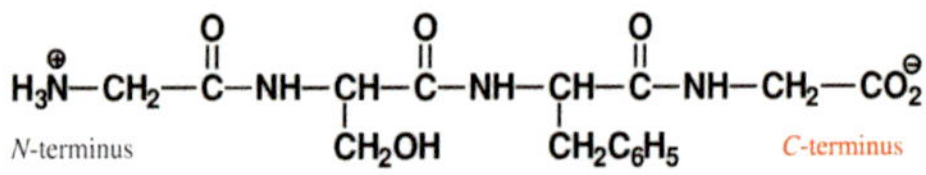

glycylserylphenylalanylglycine
gly-ser-phe-gly

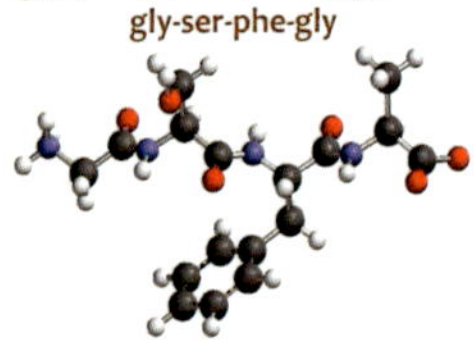

The Peptide (amide) bond

The amide nitrogen is sp^2 hybridized and the lone pair is conjugated with the carbonyl group. There is considerable C–N double-bond character. Rotation about the C–N bond is difficult.

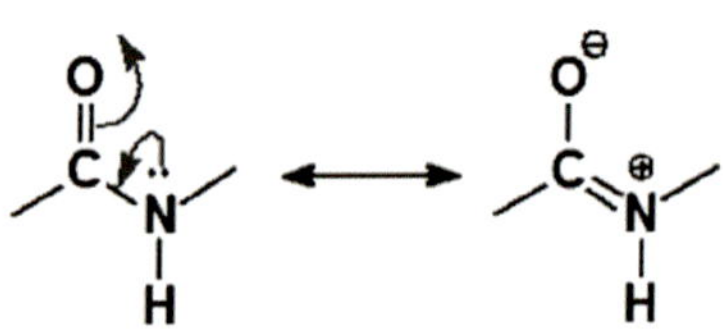

The amide groups are *planar*. The R-groups are on opposite sides of the plane.

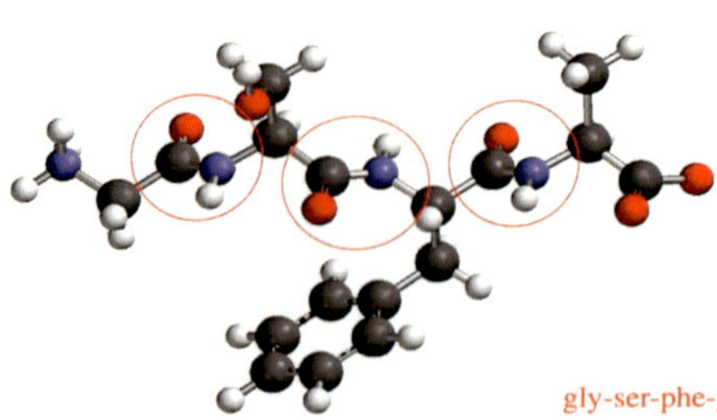

gly-ser-phe-gly

The only other type of covalent bond between amino acids in proteins and peptides is the disulfide linkage between two cysteine units.

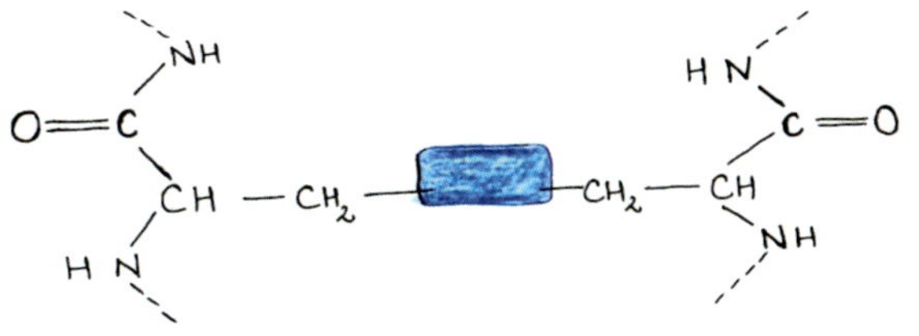

Note:

- Thiols are readily oxidized to disulfides.
- Disulfides are readily reduced to thiols.

Disulfide links serve to connect polypeptide chains.

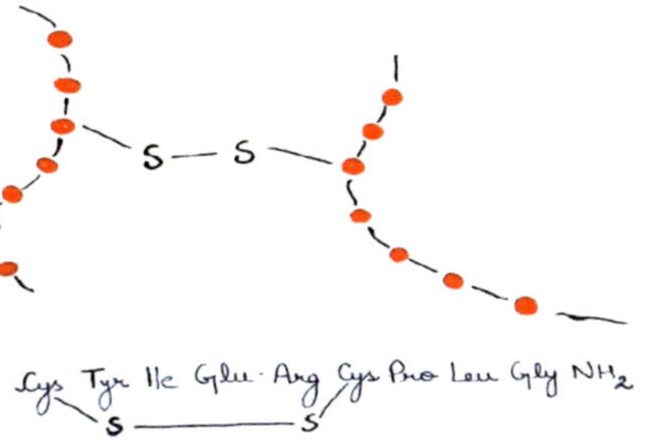

It can form a macrocycle.

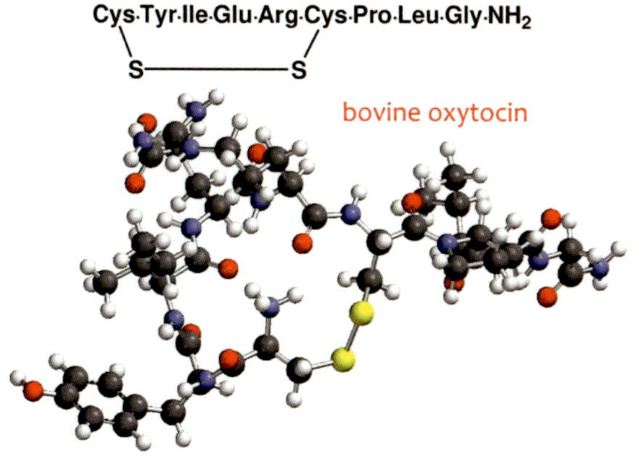

Structure determination of peptides

It contains are:

- Amino Acid Analysis.
- Terminal Residue Analysis.
- N-Terminal Analysis (Edman degradation).
- C-Terminal Analysis (Carboxypeptidase)
- Partial Hydrolysis (enzymatic).
- Hydrolyze the peptide into smaller fragments which are
  - Trypsin - Cleaves at lys and arg.
  - Chymotrypsin - Cleaves at phe, tyr, and trp.
  - Pepsin - Cleaves at phe, tyr, trp, leu, asp, glu.
  - Cyanogen bromide (not enzymatic) - Cleaves at met.
  - Determine the sequence of the fragments.
  - Successive Edman degradations (www.scribd.com., www.slideshare.com., www.coursehero.com., www. biocyclopedia.com., www. molecular-plant-biotechnology. info., www.slideplayer.com., www.documents.mx., www.docplayer.net., www.bmb.psu.edu.,www.archive. org.,www.mafiadoc.com., www.powershow.com., www. bioscience.jbpub.com., www.outhorstream.com., www. docslide.com., www.msluay.com., www.scienceprofonlue. com., www.csus.edu., www.krishikosh.egrouth.ac.in., www.tankonyvtor.hu., www.silo.pub., www.studyblue. com., www.Preservearticles.com., www.biotechuniverse. blogspot.com).

4. LIPIDS: CLASSIFICATION, PROPERTIES OF LIPID AGGREGATES, BIOLOGICAL SIGNIFICANCE

A lipid is an organic substance found in living systems that is insoluble in water but is soluble in organic solvents. Lipids are organic molecules essential for life that are composed mostly of C, H, O. They serve as a storage form of metabolic fuel. They serve as a transport form of metabolic fuel. They provide the structural components of membranes. They have protective functions in bacteria, plants, insects, and vertebrates serving as a part of the outer coating between the body of the organism and the environment.

There are four types of lipids

- Fats (triglycerides)
- Phospholipids
- Steroids
- Waxes

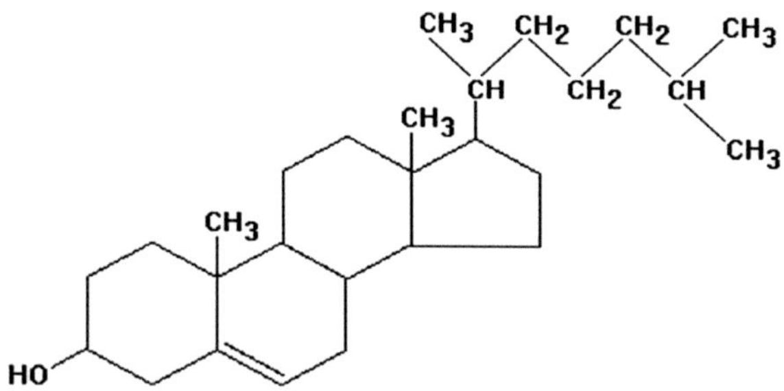

Structure of Lipid

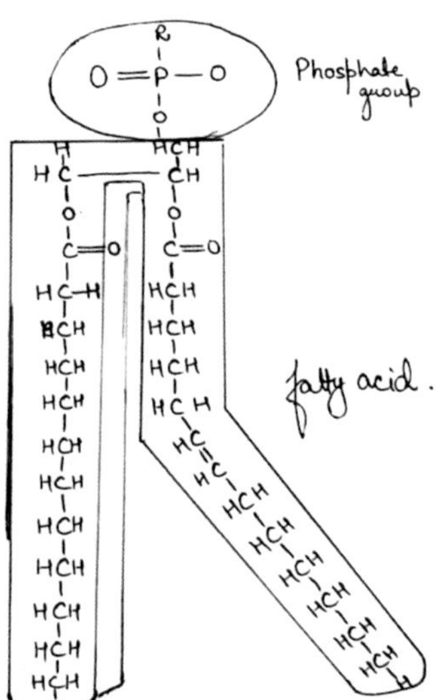

Fats & oils make up 95% of the nutritional lipids; the other 5% are steroids. Waxes are functional only. Fats are solid triglycerides and oils are liquid triglycerides.

Representative Fatty Acids

C12-C18 (also C20) Dietary fats and Oils

$CH_3-(CH_2)_{12}-COOH$ (1)

$CH_3-(CH_2)_{14}-COOH$ (2)

$CH_3-(CH_2)_{16}-CO_2H$ (3)

$CH_3-(CH_2)_7-CH=CH-(CH_2)_7-CO_2H$ (4)

$CH_3-(CH_2)_4-CH=CH-CH_2-CH=CH-(CH_2)_7-CO_2H$ (5)

$CH_3-CH_2-CH=CH-CH_2-CH=CH-CH_2-CH=CH-(CH_2)_7-CO_2H$ (6)

	Name	Mp (°C)	class
(1)	Myristic acid	58	Saturated C-14
(2)	Palmitic acid	63	Saturated C-16
(3)	Stearic acid	71	Saturated C-18
(4)	Oleic acid	4	Monounsaturated C-18
(5)	Linoleic acid	-5	Polyunsaturated C-18
(6)	Linolenic acid	-11	Polyunsaturated C-18

Triglycerides

There are length and the number of carbons in the side chains. The number of carbon-carbon double bonds in the side chains (the degree of unsaturation).

Fat molecules

- Structure = 3 fatty acid chains linked to glycerol.
- Functions = Long term energy storage and insulation for animals.
- It can be saturated or unsaturated.

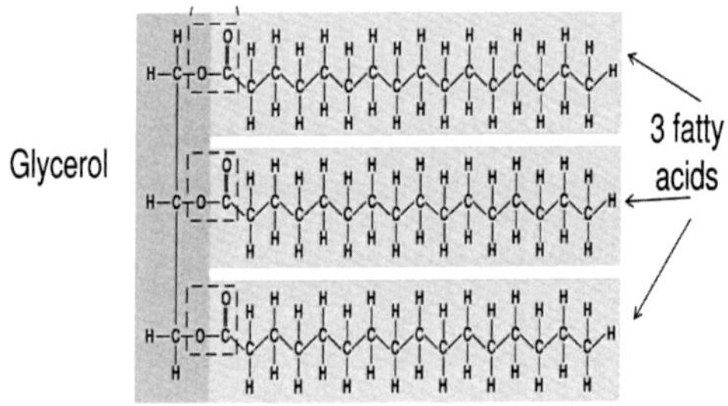

Body fat is Stored Energy

The body converts the unused carbohydrates, proteins and triglycerides that make up our macronutrients into small globules of fat that end up in the specialized cells of adipose tissue and the fatty tissue of the body. One pound of adipose tissue stores, and provides when needed, ~3500 Cal of energy. The high energy density of fat and its ability to store energy (9 Cal/

gram) compactly in relatively little space and with relatively little weight, compared with carbohydrates and proteins (4 Cal/gram) which allows us to carry stores of energy with us.

Fat

Most of our long-term energy supplies operate via the formation, storage and metabolism of body fat (triglycerides). Short term energy storage, from one meal to another, occurs through a starch-like substance called glycogen (a carbohydrate).

Adipose Tissue and Fat

Adipose tissue forms cushioning shields around our major organs, protecting them against damage from physical shock and provides insulation to our bodies, guarding against a rapid loss of body heat to the external environment. Fats carry the flavors and vitamins of many of our foods although fats have no flavors of their own such as Carrying vitamins A, D, E and K from our foods to our tissues. Fatty acids form not only the triglycerides but other compounds as well, including such vital classes as prostaglandins and phospholipids.

Saturated fats

All carbon bonded to hydrogen in saturated fats. There are no C=C double bonds. It has a long, straight chain. It includes most animal fats and butter. It is solid at room temperature. It contributes to cardiovascular disease.

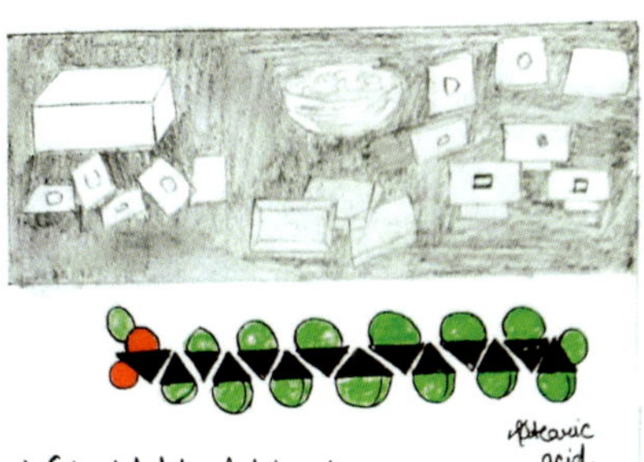

a) Saturated fat and fatty acid.

Unsaturated fats

At least one C=C double bond in the fatty acids such as plant & fish fats, vegetable oils and it is a liquid at room temperature.

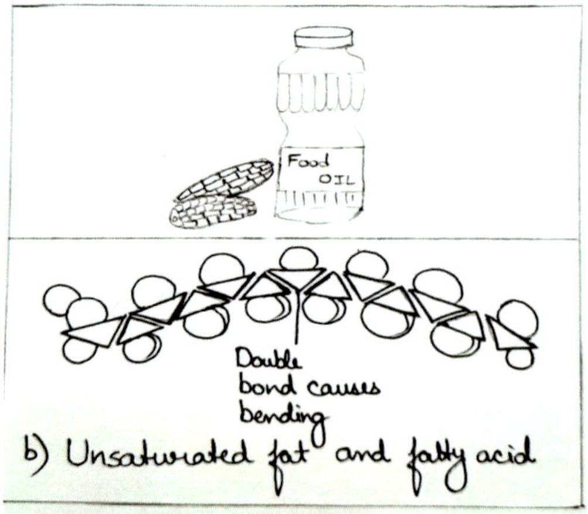

b) Unsaturated fat and fatty acid

Reactions ("Synthetic")

Iodine # (↓ satt ↑ Unsat.)

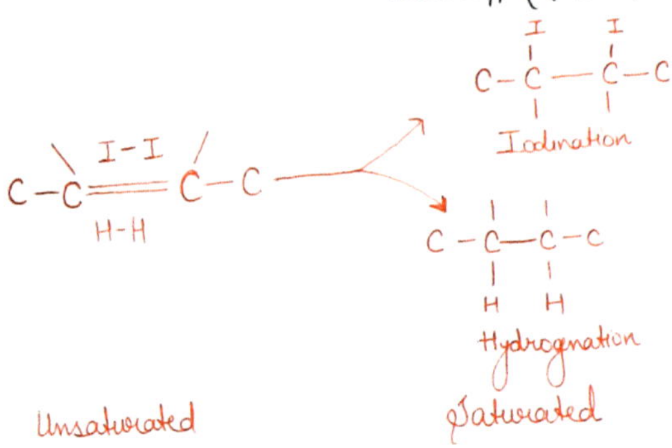

Saponification | Hydrolysis of a Triglyceride

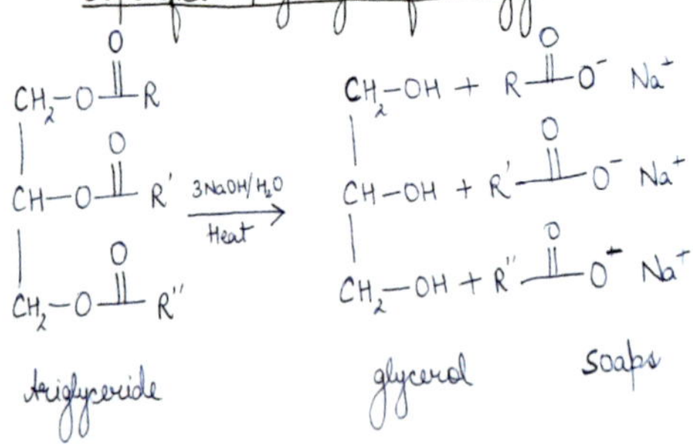

Omega (3 Fatty Acids, Especially in Fish Oils)

It contains mainly 20 & 22 C's, polyunsaturated and 1st 'ene' is 3 from Ω-end. It is Eicosapentaenoic acid (EPA): 20 (5, 8, 11, 14, and 17) mp -50°C. It is Docosahexaenoic acid (DHA): 22 (5,8,11,14,17, 20). It includes α-linolenic acid: 18 (9, 12, 15).

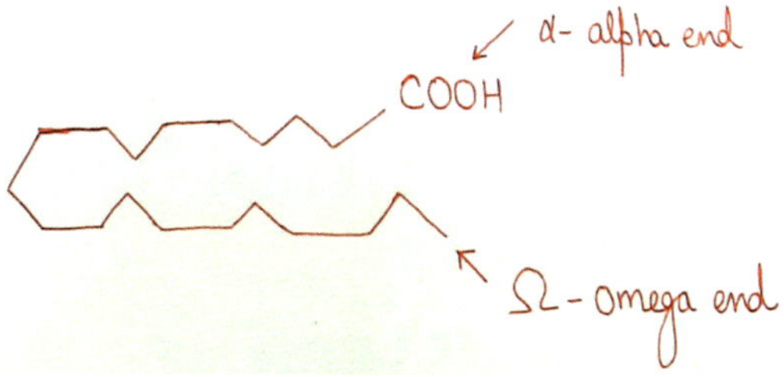

Phospholipids

It has two fatty acid chains with a positive phosphate head. The fatty acid chains are non-polar hydrophobic (water-hating) and also a phosphate head with polar Hydrophilic (water-loving).

- *Function*: It makes up cell membranes (phospholipid bilayer).

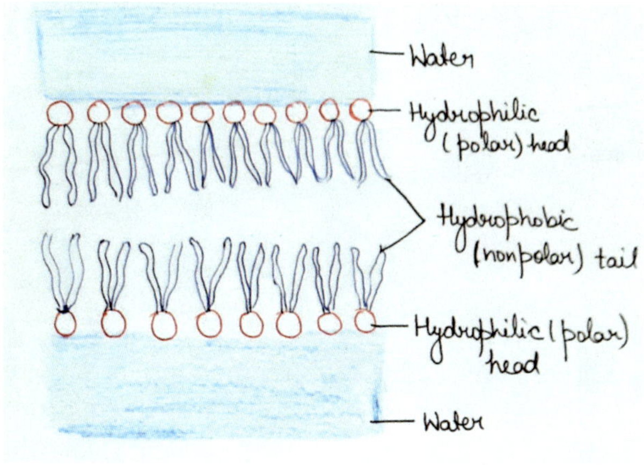

Steroids

It contains four fused carbon rings.

- ***Examples***:
- Cholesterol: It controls the fluidity of the cell membrane.

- Hormones: It regulates processes in the body (such as pregnancy).
- Vitamins (A, B, D): It supports metabolism and cell processes.

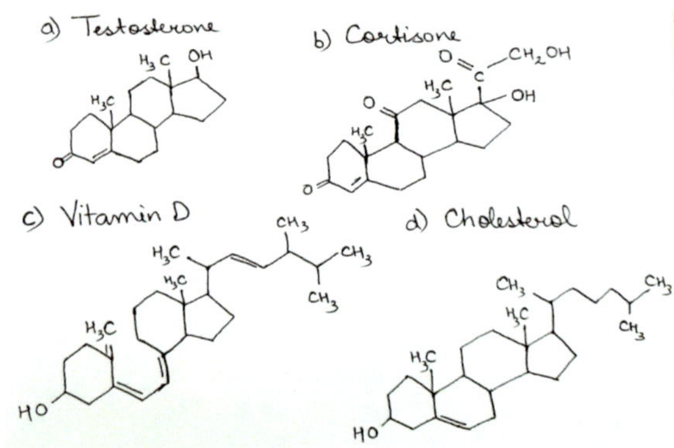

Waxes

It is solid at room temp and insoluble in water.

Function

- It is used protective covering for animals and plants.
- ***Examples*:** Beeswax, earwax, the cuticle of a leaf (keeps water in).

Lipids may be broadly defined as hydrophobic or amphiphilic small molecules that originate entirely or in part from two distinct types of biochemical subunits: *ketoacyl* and *isoprene* groups. Using this approach, lipids may be divided into eight categories: fatty acyls, glycerolipids, glycerophospholipids, sphingolipids, saccharolipids and polyketides (derived from condensation of ketoacyl subunits); and sterol lipids and prenol lipids (derived from condensation of isoprene subunits). (Fahy, E., et al, Journal of Lipid Research, Vol. 46, 839-862, May 2005 and www.slideshare.net), (www.scribd.com., www.slideshare.com., www.coursehero.com., www. biocyclopedia.com., www. molecular-plant-biotechnology.

info., www.slideplayer.com., www.documents.mx., www.docplayer.net., www.bmb.psu.edu.,www.archive.org.,www.mafiadoc.com., www.powershow.com., www.bioscience.jbpub.com., www.outhorstream.com., www.docslide.com., www.msluay.com., www.scienceprofonlue.com., www.csus.edu., www.krishikosh.egrouth.ac.in., www.tankonyvtor.hu., www.silo.pub., www.studyblue.com., www.Preservearticles.com., www.biotechuniverse.blogspot.com).

Fundamental biosynthetic units of lipids

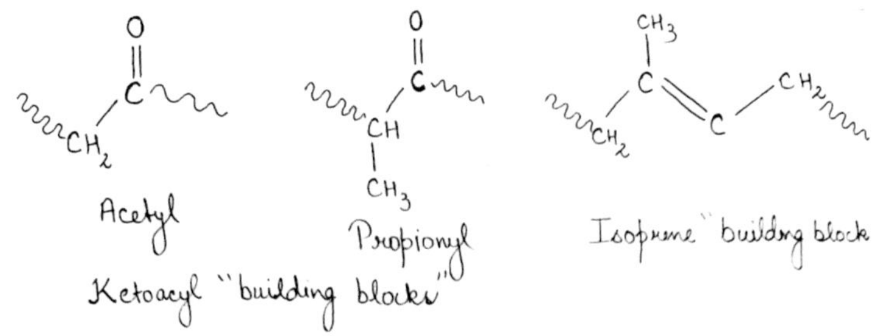

Table 4.1 List of Lipids

Category	Abbreviation	Example
Fatty acyls	FA	Dodecanoic acid
Glycerolipids	GL	1-hexadecanoyl-2-(9Z-octadecenoyl)-sn-glycerol
Glycerophospholipids	GP	1-hexadecanoyl-2-(9Z-octadecenoyl)-sn-glycero-3-phosphocholine
Sphingolipids	SP	N-tetradecanoyl-sphing-4-enine
Sterol lipids	ST	Cholest-5-en-3ß-ol
Prenol lipids	PR	2E, 6E-farnesol
Saccharolipids	SL	UDP-3-O-3R-hydroxy-tetradecanyoyl- Ad-n-acetylglucosamine
Polyketides	PK	Aflatoxin B_1

Consistent Structure Representation Across Classes of Lipids

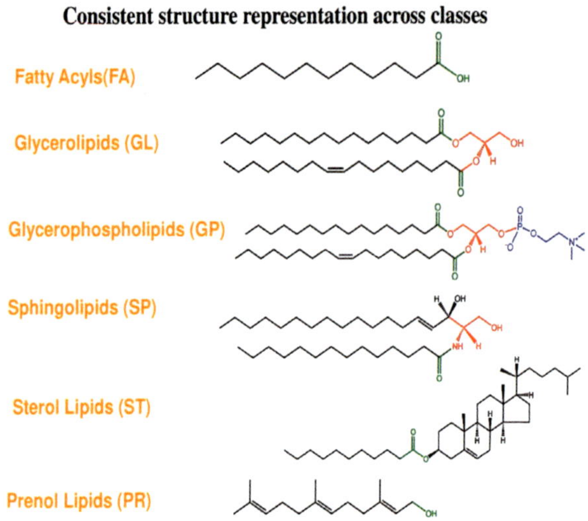

Structural comparison of Sphingomyelin and Phosphatidyl-choline

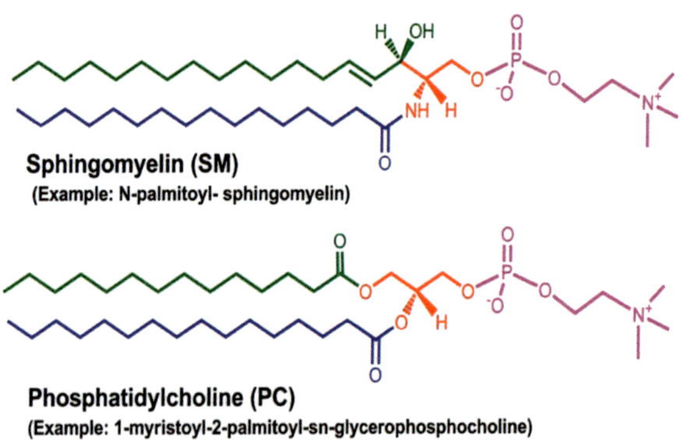

5. NUCLEIC ACID: CHEMICAL STRUCTURE AND BASE COMPOSITION, DOUBLE-HELICAL STRUCTURES, TM, SUPER-COILED DNA

It was first discovered by **Friedrich Miescher** from the nuclei of the pus cells (Leukocytes) from discarded surgical bandages and was called nuclein. "Nuclein was chemically a nucleoprotein. Later, Richard Altman in 1889 obtained the first protein free material, to which he gave the name nucleic acid. It is the polymer of nucleotides. There are two types of nucleic acids: ribonucleic acid (RNA) and deoxyribonucleic acid (DNA).

Genetic material

The cell is the basic unit of life. All living organisms including bacteria, plants and animals are built from cells. Generally, cells are very small, with diameters much less than 1 mm, so they are invisible to the naked eye. In the simplest cells, bacteria, a cell wall surrounds a very thin fatty acid containing outer plasma membrane, which in turn surrounds a superficially unstructured inner region located in the bacterial DNA that carries the genomic information. These nuclei free bacteria and their close relatives are known as prokaryotes. In prokaryotes, the nuclear material is not separated from the cytoplasm by a discrete membrane. These also lack an extensive cellular architecture: membranous organelles are absent and the genetic material is not enclosed in a distinct structure. In all the cells other than bacteria, the inner cellular mass is partitioned into a membrane-bound spherical body called the nucleus and an outer surrounding cytoplasm.

In the nucleus is located the cellular DNA in the form of coiled rods known as chromosomes. Cells that contain a nucleus and with other distinctive membranous organelles such as mitochondria, vesicles and Golgi bodies are referred to as eukaryotes. In the cells of eukaryotes, a nuclear membrane separates the genetic material from the cytoplasm, which is then further subdivided by other distinct membranous structures. Eukaryotes can be unicellular and multicellular and comprise all macroscopic forms of life such as plants, animals and fungi. Plant cells differ from animal cells in the presence of outer cell walls surrounding the cell membrane and chloroplasts, whereas animal cells contain centrosomes, which are involved in cell division and lysosomes. The essence of cells is the ability to grow and divide to produce progeny, which are capable of generating new cellular molecules and undergoing replicating themselves. Thus, cells are in effect tiny factories that grow by taking up simples molecular building blocks. Considering the complexity of an organism, one is struck by the relatively small number of organic components used in the formation of substances **Figure 5.1.** Cells contain molecules of great variety, but living cells are composed of restricted sets of elements six of which (C, H, N, O, P, and S) make up more than 99% of its weight. Certain simple combination of atoms such as methyl (-CH3), hydroxyl (OH), carboxyl (-COOH), and amino groups (-NH2) groups occur repeatedly in biological molecules. Each such group has distinct chemical and physical properties. Molecules can be broadly grouped into two categories.

Small organic molecules

These are compounds with a molecular weight in the 100-1000 range containing up to 30 or so carbon atoms. Molecules of this kind are usually found in free solution in the cytoplasm where they form a pool of intermediates from which large molecules are made. Some of the small organic molecules are sugars, amino acids and fatty acids. At least 750 different types of small molecules are found.

Macromolecules

These are invariably polymeric molecules that are formed by the joining together of specific types of small molecules. A large polymeric molecule or more than one polymer associated into large molecules can be referred to as a macromolecule. Some of the macromolecules such as carbohydrates, proteins and nucleic acids" are the most important. Three simple molecules such as sugar, amino acids and nucleotides that form macromolecules have been briefly described.

Sugars

There are three types of sugar molecules that are involved in different roles in organisms.

Monosaccharide $(CH_2O)n$
Disaccharide $(C_{12}H_{22}O_{11})$
Polysaccharide (C6H10O5)n

Nucleotides

Nucleotides have 3 basic parts:

1. A 5-carbon sugar.
2. A phosphate group.
3. A nitrogenous (nitrogen-containing) base **Figure 5.2.**

A. Nucleic acids

Nucleic acids are containing instructions that cells need to carry out all the functions of life. Nucleic acid: It is a very large organic molecule made up of carbon, oxygen, hydrogen, nitrogen, and phosphorus.

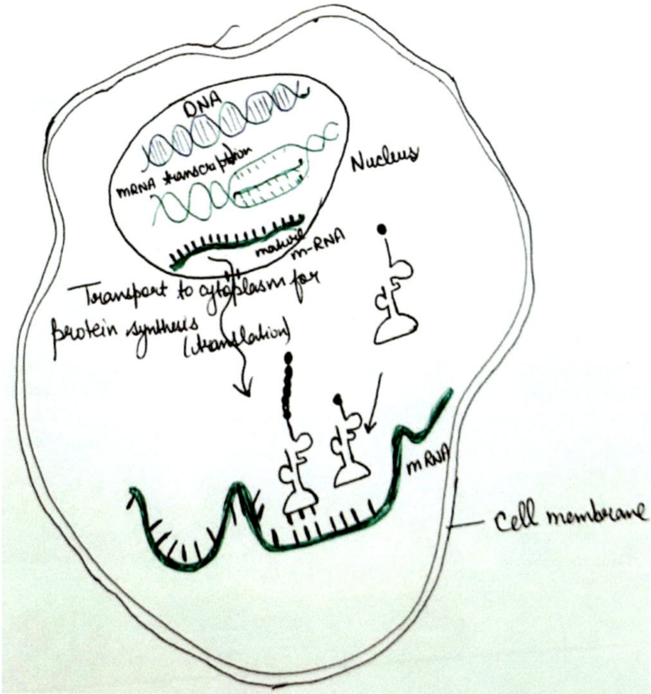

Figure 5.1 Cell structure

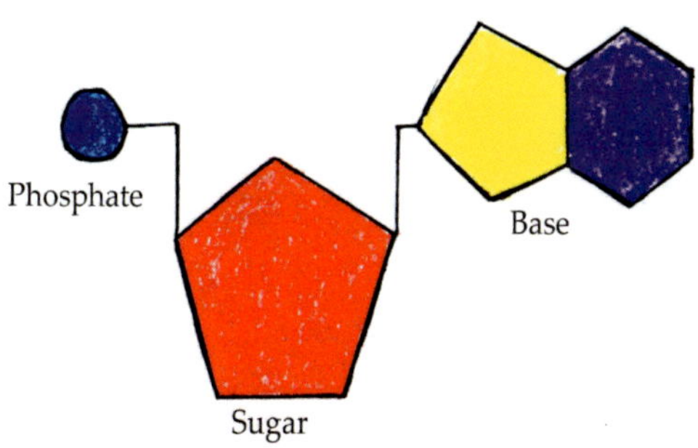

Figure 5.2 Nucleotides

Types of Nucleic acid

- DNA (deoxyribonucleic acid) - the genetic material that carries information about an organism. It is passed from the parent to the offspring.
- DNA directs all cellular activities and is only found in the nucleus of a eukaryotic cell.
- RNA (ribonucleic acid) – a nucleic acid found in the nucleus and cytoplasm of cells that plays an important role in the production of proteins **Figure 5.3.**

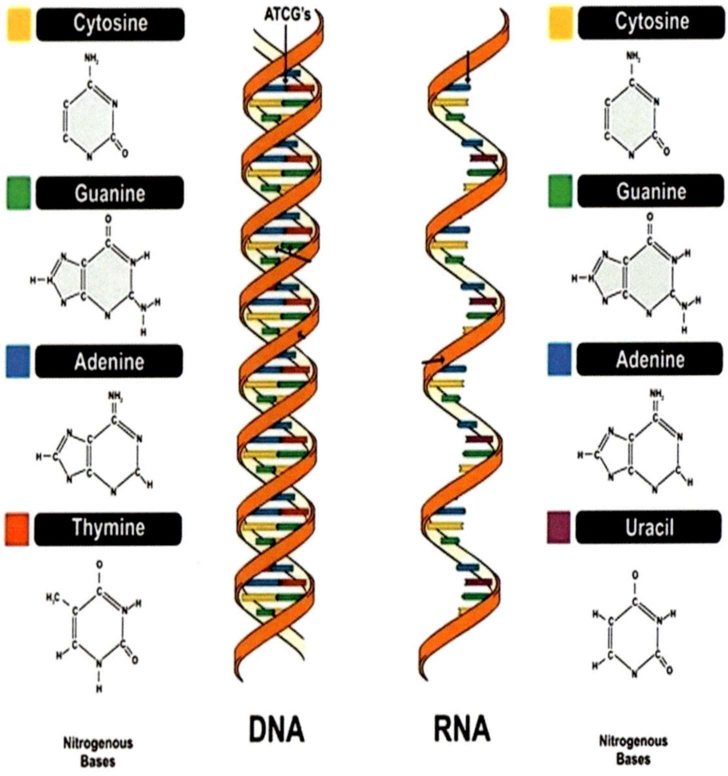

Figure 5.3 DNA and RNA sugars

Nucleotides and genetic organization

Nucleic acids are made up of nucleotides.

Nucleotides have 3 basic parts

1. A 5-carbon sugar.
2. A phosphate group.
3. A nitrogenous (nitrogen-containing) base.

Levels of genetic organization

Nucleotide	Gene	DNA	Chromosomes	Genome
The smallest building block of gene	Code for traits	Made of many genes	Tightly coiled strands of DNA	The entire genetic make-up of an organism

Composition of DNA

- Major parts of a DNA nucleotide:
- A 5-carbon sugar called deoxyribose.
- Phosphates
- Four different nitrogenous bases (rungs of the ladder) **Figure 5.4.**

  a. Adenine (A)

  b. Thymine (T)

  c. Guanine (G)

  d. Cytosine (C)

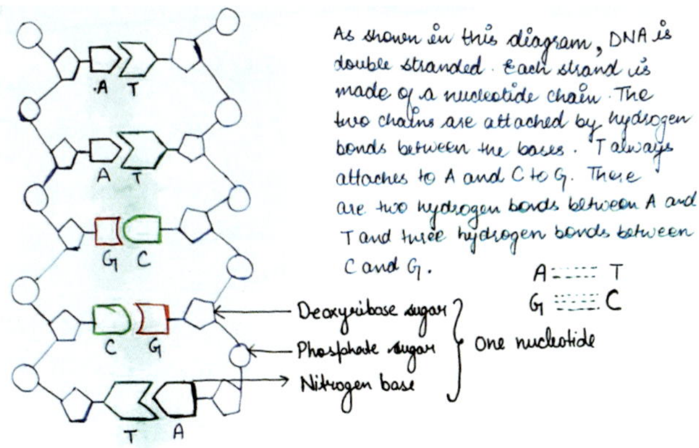

Figure 5.4 Nucleotide chain

Pairing the Nitrogenous Bases in DNA

Adenine is always paired (matches up) with thymine and guanine always pairs (is complementary to) with cytosine. These nitrogenous bases make up four different possible nucleotides of DNA. Many nucleotides (basic unit of nucleic acids) are placed together to make one DNA molecule Figure 5.5.

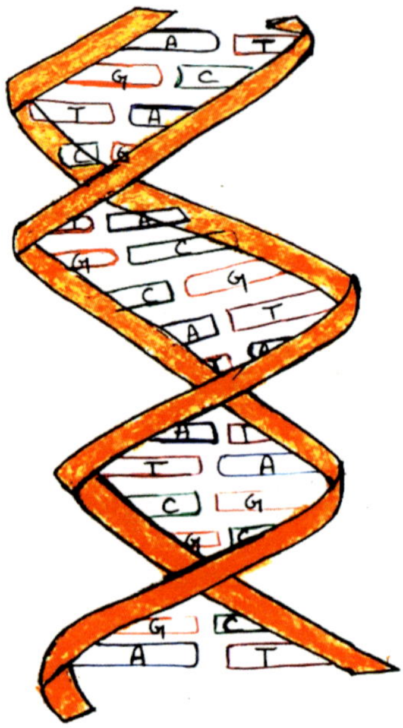

Figure 5.5 DNA structure

Structure of DNA

- **James Watson & Francis Crick** – an American biologist & a British physicist who built the first accurate structural model of DNA (Won the Nobel Prize in 1962).
- **Watson & Crick's** model of DNA is a double helix (twisted ladder), in which 2 strands are wound around each other.
- The outside/sides of the ladder are made of deoxyribose sugar and phosphates.
- The stages of the ladder are the nitrogenous bases (connect to deoxyribose sugar in the outside/side of the ladder).

Base pairs

DNA strands are complementary to one another (match up with each other). The bases are A & T and G & C. The following sequences of DNA are given, and also the complementary strands.

AGT TCA	CAG GTC
CTT GAA	AGC TCG

DNA replication

They make copies of their DNA before cells divide. DNA Replication ensures that each daughter cell will have all of the genetic information and it needs to carry out its activities. DNA replication is semiconservative, meaning that when DNA copies, each new strand has ½ of the original strand.

- DNA unwinds (untwists) exposing the nitrogenous bases.
- DNA unzips – weak hydrogen bonds between nitrogenous bases break.
- Hydrogen bonds are the weakest type of bond that occurs in molecules.
- New complementary strands form nitrogenous bases floating in the nucleus of the cell pair up with the bases on each half of the DNA molecule.
- Two new DNA molecules that are exactly alike are formed.
- Both new strands are identical to the original strand of DNA.

Genes in DNA

Proteins determine the size, shape, and many other traits of an organism, e.g. Eye color, hair color, type of hairline, etc. Each gene in a DNA molecule codes for the production of a particular protein. A Gene is a small portion of the DNA molecule that codes for a particular trait or protein.

Nitrogenous bases

The order of the nitrogenous bases along a gene forms a genetic code that specifies what type of protein will be produced. The order of the nitrogenous bases tells the cell which amino acids are to be put together to make a protein. If one amino acid is changed or out of place, a different protein is made.

Since DNA cannot leave the nucleus to make proteins at the ribosomes, RNA directs protein synthesis; see Figures 5.6 and 5.7.

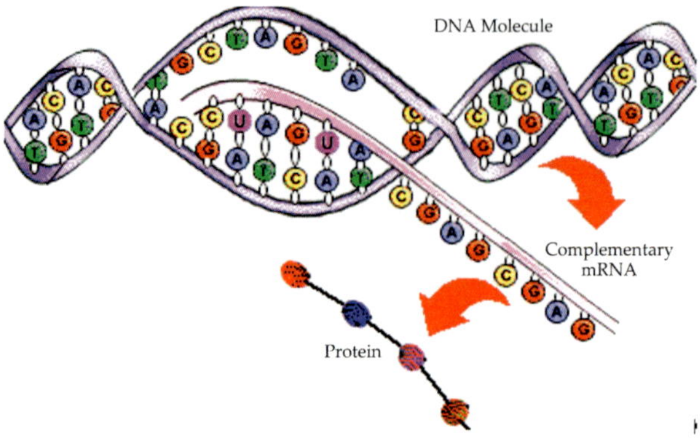

Figure 5.6. Nitrogenous bases involved to make protein.

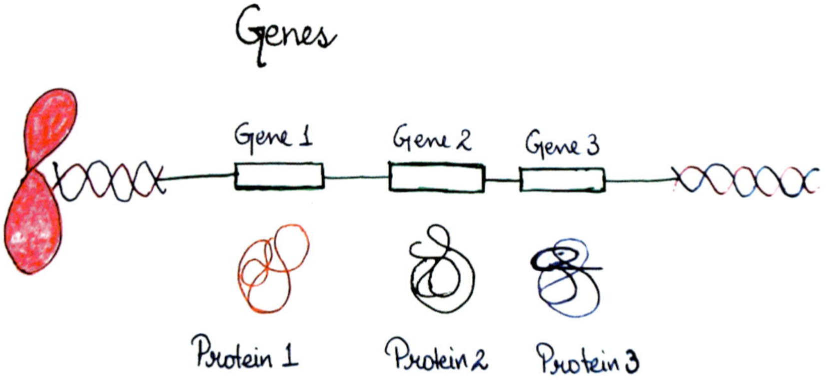

Figure 5.7. Sequence of genes

It all started with Mendel

The beginning of our thinking of the possibility of genetic material begins with Mendel. He described the genetics of pea crosses, although he did not know what was genetic material, he referred to "factors" as the things that gave his pea plants their physical characteristics. While Mendel's work gave a logical description of heredity it was largely ignored for many decades. His work was finally reproduced in the early 1900s and was thereafter regarded as facts read from genetics (www.scribd.com., www.slideshare.com., www.coursehero.com., www. biocyclopedia.com., www. molecular-plant-biotechnology.info., www.slideplayer.com., www.documents.mx., www.

docplayer.net., www.bmb.psu.edu.,www.archive.org.,www.mafiadoc.com., www.powershow.com., www.bioscience.jbpub.com., www.outhorstream.com., www.docslide.com., www.msluay.com., www.scienceprofonlue.com., www.csus.edu., www.krishikosh.egrouth.ac.in., www.tankonyvtor.hu., www.silo.pub., www.studyblue.com., www.Preservearticles.com., www.biotechuniverse.blogspot.com).

6. PROTEIN CHEMISTRY: CLASSIFICATION, DIFFERENT LEVELS OF PROTEIN STRUCTURE, FORCES, STABILIZING PROTEIN STRUCTURE, PROTEIN FOLDING, PROTEIN MODIFICATION

Proteins are including a diversity of structures, resulting in a wide range of functions. Proteins are account for more than 50% of the dry mass of most cells. "Protein functions include structural support, storage, transport, cellular communications, movement, and defense against foreign substances. At first, all we have is a string of amino acids bound with peptide bonds. Once the string of amino acid interacts with itself and its environment (often aqueous), then we have a functional protein that consists of one or more polypeptides precisely twisted, folded, and coiled into a unique shape. The sequence of amino acids determines a protein's three-dimensional structure. A protein's structure determines its function.

- A **protein** is a biologically functional molecule that consists of one or more polypeptides
- Proteins are unbranched polymers constructed from 22 standard α-amino acids.
- They have four levels of structural organization.
- **Primary structure:** the amino acid sequence is specified by genetic information.
- As the polypeptide chain folds and it forms certain localized arrangements of adjacent amino acids that constitute the **secondary structure**.
- The overall three-dimensional shape that a polypeptide assumes is called the **tertiary structure**.
- Proteins that consist of two or more polypeptide chains are said to have a **quaternary structure**.

Amino acids

Amino acids are chemically diverse but they all contain a carboxyl acid group and an amino group both linked to a single carbon atom. Proteins are long linear polymers of amino acid joined head to tail by a peptide bond between the carboxyl acid group of one amino acid and the amino group of the next amino acid. There are twenty amino acids in proteins each with a different side chain attached to the α-carbon atom. All amino acids contain nitrogen atoms; therefore the protein represents a class of nitrogenous compounds. The properties of amino acid side chains in aggregates determine the properties of proteins they constitute. Such amino acids are called standard amino acids. Amino acids that occur naturally in cells but are not incorporated ribosomally into proteins are called nonstandard amino acids. Some nonstandard amino acids are constituents of ribosomally synthesized proteins, but they are generated by post-translational modification of standard amino acids which are incorporated ribosomally into proteins.

- Selenocysteine is the 21st standard amino acid.
- Pyrrolysine is the 22nd standard amino acid.

Primary structure

The sequence of amino acids linked by peptide bonds along the polypeptide chain is the primary structure. Each polypeptide has its own unique sequence of amino acids. The primary structure of polypeptide determines to a large extent the native secondary and tertiary structure.

Secondary Structure

The structure of a protein in which the polypeptide chain is folded in a regular manner with the helical or pleated structure is called secondary structure. Much of the folding is the result of linking carbonyl and amide groups of the peptide chain backbone by hydrogen bonds. In many proteins, the hydrogen bonding produces a regular coiled arrangement called the α-helix and the ß-sheet.

Tertiary Structure

A final structure of a protein in three-dimensional from a single polypeptide chain is formed. The final confirmation of a protein from a single polypeptide that is bent or tightly folded and compact is determined by a variety of interactions including hydrogen bonds, salt bonds, hydrophobic or nonpolar bonds and van der Waals forces. It is by virtue of their tertiary structure that proteins adopt a globular shape.

Quaternary Structure

It contains more than a single polypeptide chain to indicate the spatial relationships among the separate chains or subunits. It means how two or more chains are arranged in relation to each other. Most large proteins contain two or more polypeptide chains, and in between, there are no covalent linkages. Proteins with two or more polypeptide chains are known as oligomeric proteins. A well-known example of an oligomeric protein is hemoglobin.

Enzymatic Proteins

It is used for the acceleration of chemical reactions.

- ***Example***: Digestive enzymes catalyze the hydrolysis of bonds in food molecules.

Storage Proteins

The storage of amino acids

- ***Examples***: Casein, the protein of milk, is the major source of amino acids for baby mammals. Plants have storage proteins in their seeds. Ovalbumin is the protein of egg white, used as an amino acid source for the developing embryo.

Hormonal proteins

It allows coordination of an organism's activities.

- ***Example***: Insulin, a hormone secreted by the pancreas, causes other tissues to take up glucose, thus regulating blood sugar concentration

Defensive proteins

It is used for protection against disease.

- ***Example***: Antibodies inactivate and help destroy viruses and bacteria.

Transport proteins

It is used for the transport of substances.

- ***Examples*:** Hemoglobin, the iron-containing protein of vertebrate blood, transports oxygen from the lungs to other parts of the body. Other proteins transport molecules across cell membranes.

Receptor proteins

It is the responses of a cell to chemical stimuli.

- ***Example*:** Receptors built into the membrane of a nerve cell detect signaling molecules released by other nerve cells.

Structural proteins

It is used for support.

- ***Examples*:** Keratin is the protein of hair, horns, feathers, and other skin appendages. Insects and spiders use silk fibers to make their cocoons and webs, respectively. Collagen and elastin proteins provide a fibrous framework in animal connective tissues.

The primary structure consists of its unique sequence of amino acids. Secondary structure, found in most proteins, consists of coils and folds in the polypeptide chain. Tertiary structure is determined by interactions among various side chains (R groups). Quaternary structure results when a protein consists of multiple polypeptide chains. **Primary structure**, the sequence of amino acids in a protein, is like the order of letters in a long word. The primary structure is determined by inherited genetic information.

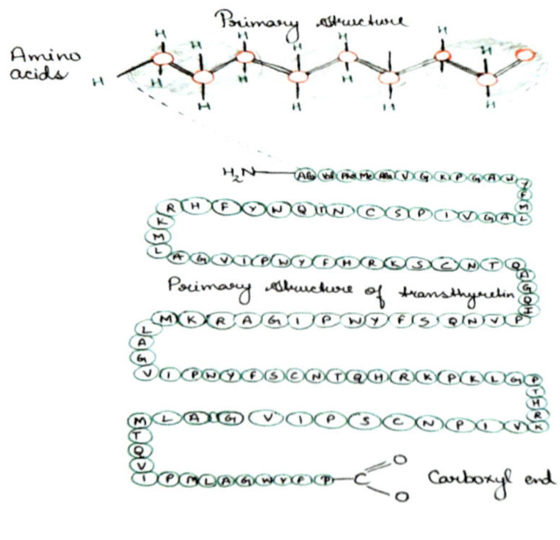

Hydrogen bonding arrangement in the α-helix.

The coils and folds of the **secondary structure** result from hydrogen bonds between repeating constituents of the polypeptide backbone. Typical secondary structures are a coil called an **α helix** and a folded structure called a **β pleated sheet.**

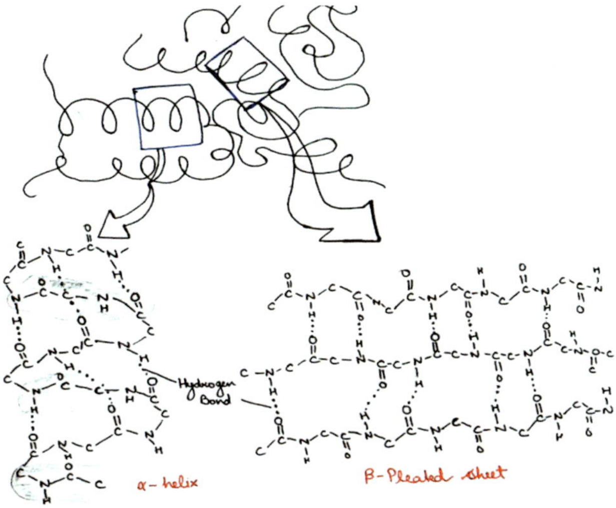

Describing the geometry of α-helix and ß-sheet.

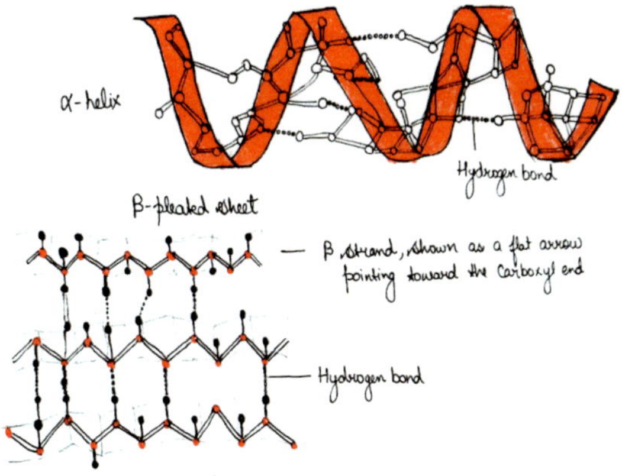

Hydrogen bonding arrangement in the α-helix and ß-sheet. The helix is defined by the pitch and rise per residue

Tertiary structure is determined by interactions between R groups, rather than interactions between backbone constituents. These interactions between R groups include actual ionic bonds and strong covalent bonds called disulfide bridges that may reinforce the protein's structure. IMFs such as London dispersion forces (LDFs a.k.a. and van der Waals interactions), hydrogen bonds (IMFs), and hydrophobic interactions (IMFs) may affect the protein's structure.

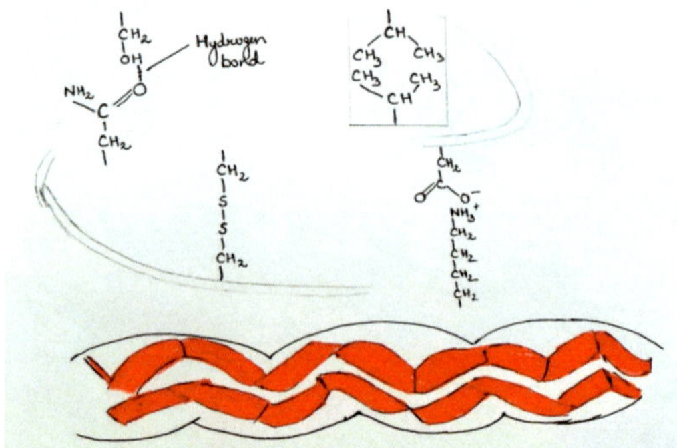

Quaternary structure results when two or more polypeptide chains form one macromolecule. Collagen is a fibrous protein consisting of three polypeptides coiled like a rope.

Collagen fiber

Hemoglobin is a globular protein consisting of four polypeptides: two alpha and two beta chains.

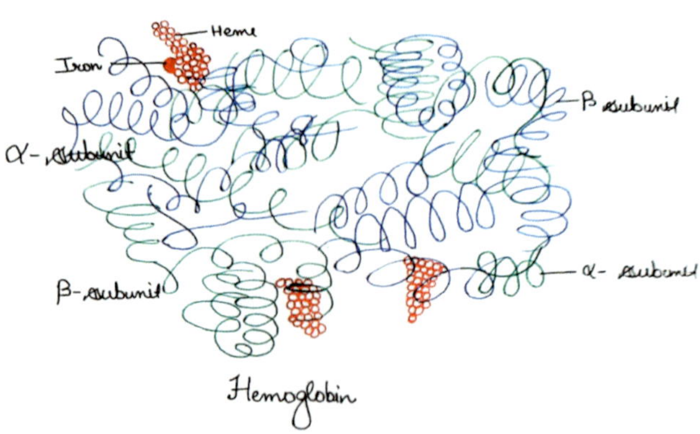

Structure of Hemoglobin

Sickle-cell disease

A slight change in the primary structure can affect a protein's structure and ability to function. Sickle-cell disease, an inherited blood disorder, results from a single amino acid substitution in the protein hemoglobin. A slight change in primary structure can affect a protein's structure and ability to function. Sickle-cell disease, an inherited blood disorder, results from a single amino acid substitution in the protein hemoglobin. In addition to the primary structure, physical and chemical conditions can affect the structure. Alterations in pH, salt concentration, temperature, or other environmental factors can cause a protein to unfold. This loss of a protein's native structure is called denaturation. A denatured protein is biologically inactive.

Protein Subunits and Domains

- A single polypeptide may have regions along its length that fold up separately - **domains**
- More than one polypeptide coming together to form a protein - **subunit**

Protein shape

- Globular proteins – The polypeptide folds tightly together on itself – **most of the enzymes**
- Fibrous proteins – It is an elongated structure that spans great distances - **hair - keratin**
- Actin & collagen
- Strong & rigid (Collagen - ligaments)
- Strong & flexible (elastin - skin)

Proteins linked together

- Domain to domain links (same P')
- Protein to protein links (between P's)
- Covalent bonds used to stabilize
- The disulfide bond is the most common
  - Cysteine AA
  - No impact on the final shape, but stabilizes it.

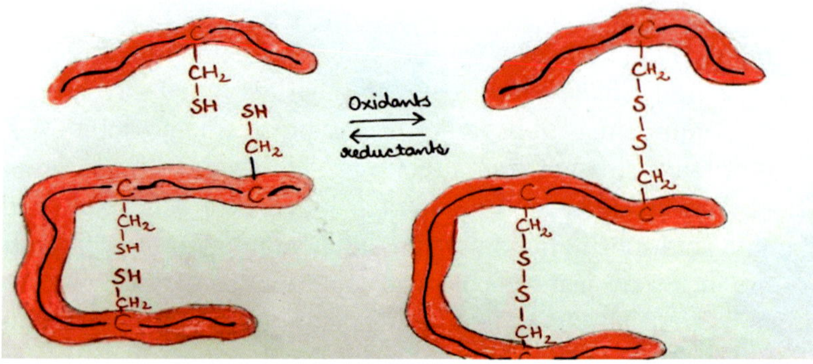

Proteins bind

- The biological properties of proteins result from interactions with other molecules.
  - Antibodies, enzymes, structure, etc.
- Binding is always very specific.
- Ligand - Anything that is bound by a protein.
- Ligand binding is by ionic bonds only.
- Many ionic bonds required to stabilize the link - matching configurations needed specificity.

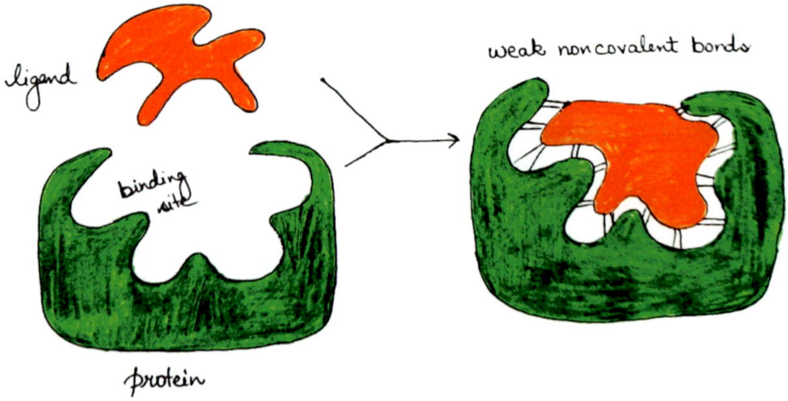

Modes of Ligand binding with protein

- A ligand binds to a protein at a binding site.
- A protein may have more than one binding site.

- It may bind the same ligand many times or it may bind different ligands.
- Ligands can bind to regulate the activity of the protein.
- Interior amino acids have a say in the conformation of the molecule too.

Proteins pair with other molecules

Some biologically active molecules are a result of a very convenient relationship.

- Rhodopsin & Opsin.
- Retinal (a light sensing pigment) + opsin protein.
- Haem (Fe containing molecule) + globin protein.

Protein folding and Protein Modification

Protein folding is the physical process by which a "polypeptide folds into its characteristics and functional three-dimensional conformations. The correct three-dimensional structures are essential to protein function. Failure to fold into native structure produces inactive proteins. A protein molecule folds spontaneously during or after biosynthesis. However, the process also depends on the nature of the solvent and the concentration of salts the temperature and the presence of molecular chaperones. To understand the complete three-dimensional structure of a protein, it is important to analyze its folding pattern. Two terms are used to describe a folding pattern-motif and domain. A motif is a recognizable folding pattern involving two or more elements of secondary structure and the connections between them. Motifs can be simple, such as the ß-α-ß loop motif; motifs also can be elaborate, such the ß barrel motif. In some cases, a single large motif may comprise the entire protein. Motif and fold can be used interchangeably; however, the term fold is more commonly applied to describe more complex folding patterns". **For example**, the distinctive arrangement of eight α helices in myoglobin is found in all globins and is called the globin fold.

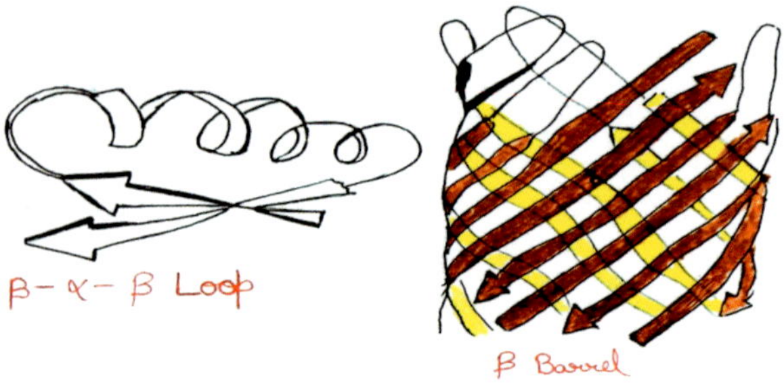

A domain is a part of a "polypeptide chain that is independently stable or could undergo movements as a single entity with respect to the entire protein. Each domain can appear as a distinct globular lobe region as is true for the calcium-binding protein of muscle, troponin C. However, extensive contacts can occur between domains and make individual domains hard to discern. The two or more domains of a protein often have different functions, which are coordinated in the overall function of the protein. Small proteins such as myoglobin usually only have one domain".

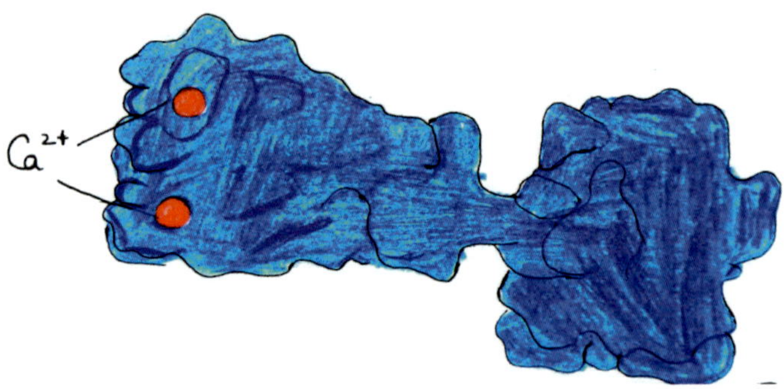

Several rules have emerged from studies of common protein folding patterns

The "hydrophobic interactions are important for protein stability. At least two layers of secondary structure are required for shielding hydrophobic residues from water. The α helices and ß sheets usually are found in different structural layers of a protein. Segments of adjacent to each other in the primary structure are usually stacked next to each other in the folded structure". The ß conformation is most stable when the individual strands are twisted slightly in a right-handed sense.

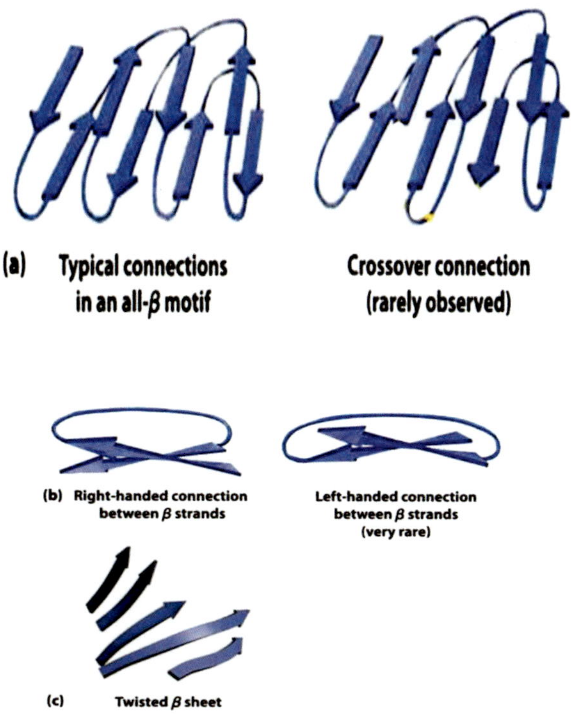

Protein denaturation

The conformation of a “native protein is only marginally stable. The conditions are different from those in a cell, therefore can cause protein structural changes resulting in the loss of three-dimensional structure needed for function (i.e., denaturation). In the denatured state, the conformation of the protein need not be completely randomized. A number of physical and chemical agents can cause protein denaturation. Strong acids and bases, organic solvents (ethanol), detergents, reducing agents (ß-mercaptoethanol) and heavy metals (mercury and lead) denature the proteins. A classic agent is heat, which has complex effects on many weak interactions in a protein (primarily on the hydrogen bonds). On heating, a protein’s conformation generally remains intact until an abrupt loss of structure occurs over a relatively narrow temperature range (the T_m). The unexpectedness of the loss of structure suggests a cooperative process in which loss of structure in one or more parts of the protein rapidly destabilizes the structure” of other parts.

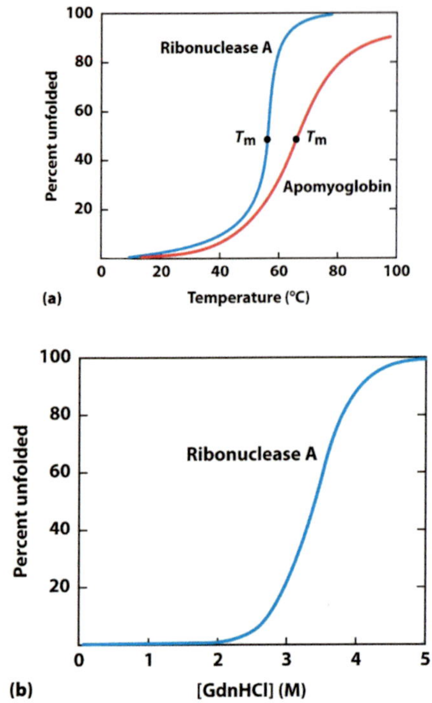

Renaturation of ribonuclease

The early studies are examining the "renaturation (refolding) of the enzyme ribonuclease which provided experimental proof that the tertiary structure of a protein is determined by its primary structure. In this experiment, purified ribonuclease was denatured to its unfolded inactive state by exposure to a concentrated solution of urea and the reducing agent, mercaptoethanol. This unfolded the protein and reduced all disulfide cross-links to cysteine residues. When urea and mercaptoethanol were removed, the randomly-coiled, denatured ribonuclease spontaneously refolded to its catalytically active state. Even the protein's four unique disulfide bonds reformed correctly. This experiment provided the first evidence that the amino acid sequence of a polypeptide chain contains all the information required to fold the chain into its native" three-dimensional structure.

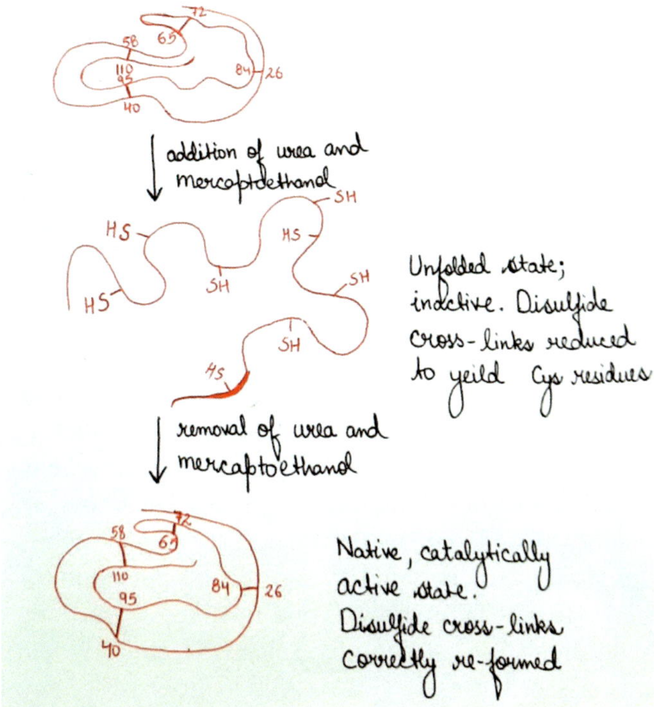

Renaturation of Ribonuclease

Protein folding

Protein folding is "supposed to occur via a hierarchical pathway. The certain amino acid sequences are folded first into local secondary structures. Assemblies of local structures are followed by the formation of longer-range interactions between these secondary structure elements to obtain folding motifs. Throughout the process, hydrophobic interactions play a significant role in aggregating non-polar amino acid side chains in the core of folding intermediates and the final tertiary structure. The computer programs have now been developed that can often predict the structures of small proteins on the basis of their amino acid sequences".

Chaperones

Many proteins do not fold spontaneously as they are synthesized by a cell. The folding of these "proteins require the general class of proteins called chaperones. Chaperones interact with partially folded or improperly folded polypeptides, facilitating correct folding pathways or providing microenvironments in which folding can occur. Several types of chaperones are found in organisms ranging from bacteria to humans. The two major classes of chaperones are the molecular chaperones (e.g., the Hsp70 family of proteins) and chaperonins (e.g., the GroEL/Hsp60 family of proteins) and (Hsp stands for heat shock protein). These proteins are induced in bacterial cells by exposure to high temperatures". Molecular chaperones act as monomers, whereas chaperonins are large multi-subunit complexes.

Chaperone-assisted protein folding

Molecular chaperones such as "Hsp70, bind to regions of unfolded polypeptides that are rich in hydrophobic residues. They protect both proteins denatured by heat and newly synthesized and not yet folded proteins from aggregation. They also block the folding of certain proteins that must remain unfolded until they have been translocated across a membrane. Some chaperones also facilitate the quaternary assembly of oligomeric proteins. The Hsp70 proteins bind to and release polypeptides in a cycle that uses energy from ATP hydrolysis and involves several other proteins, such as Hsp40 and NEF (nucleotide-exchange factor). When the bound polypeptide is released after a cycle of ATP hydrolysis, it has a chance of folding properly. If not, the protein may be bound again, and the process repeated. Hsp70 can also deliver unfolded polypeptides to chaperonins. Chaperonins are elaborate protein complexes required for the folding of many cellular proteins that do not fold spontaneously. In *E. coli*, 10-to-15% of newly translated proteins require the GroEL/GroES chaperonin system for folding under normal conditions. This increases to 30% when cells are subjected to heat stress.

The mechanism of protein folding by the GroEL/GroES chaperonin is shown in the next slide. Each GroEL complex consists of two large chambers formed by two heptameric rings. GroES (also a heptamer) blocks one of the GroEL chambers after an unfolded protein is bound inside. The chamber with the unfolded protein is referred to as cis; the opposite one is trans. Folding occurs within the cis chamber during the time it takes to hydrolyze the 7 ATPs that are bound to subunits in the heptameric GroEL ring. The GroES and the ADP molecules then dissociate and the protein is released. The two chambers of the GroEL system are alternate in the binding and facilitated folding of client proteins. Inside the chamber, a protein has about 10 seconds to fold. The constraining the protein within the chamber prevents inappropriate protein aggregation and also restricts the conformational space that a polypeptide chain can explore as it folds".

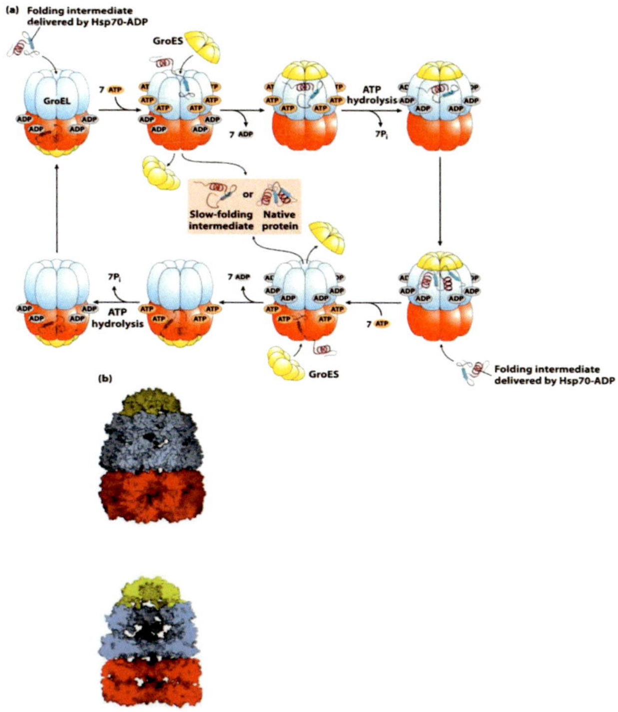

Structure and function of the chaperonin, GroEL

Protein folding disorders: amyloidoses

In a number of human disorders such “amyloidoses; including Alzheimer’s disease, a soluble protein that is normally produced by a cell adopts a misfolded state and converts into an insoluble extracellular amyloid fiber. Commonly, fiber formation is promoted by the aggregation of regions of proteins that have a ß conformation. In **Alzheimer’s disease**, proteolytic cleavage of a neuronal membrane protein (amyloid-ß precursor protein, APP) produces an α-helical membrane-spanning peptide (amyloid-ß peptide) that converts to the ß conformation and aggregates into amyloid fibrils. The extracellular deposition of amyloid fibrils is associated with plaque formation and ultimately death of the nearby neurons”.

Protein folding disorders: prion diseases

The “neurodegenerative diseases are known as spongiform encephalopathies and a misfolded form of a normal neuronal protein PrP is responsible for the disease. Spongiform encephalopathies occur in many species of animals. In humans, the disorders are known as kuru and Creutzfeld-Jacob disease. In cows, the disorder is known as mad cow disease. In sheep, it is called scrapie, and in deer and elk, it is called chronic wasting disease. The diseased brain becomes riddled with holes. Progressive deterioration of the brain leads to a spectrum of neurological symptoms and is always fatal. Prion protein (PrP) is a normal constituent of brain tissue in all mammals. The protein is thought to possibly function in cell signaling. Illness occurs when the normal cellular PrP (PrP^C) folds into an altered conformation called PrP^{Sc} (Sc denotes scrapie). In PrP^{Sc}, part of the α-helical region of PrP^C is converted to the ß conformation. PrP^{Sc} monomers then associate forming amyloid-like ß sheets. Importantly, the interaction of PrP^{Sc} with PrP^C converts the latter to PrP^{Sc}, initiating a domino effect in which more and more of the brain protein converts to the disease-causing form. For this reason, PrP^{Sc} is infectious. The mechanism by which PrP^{Sc} leads to spongiform encephalopathy is not understood". In inherited forms of prion diseases, a mutation in the gene encoding PrP produces a change in one amino acid residue that is believed to make the conversion of PrP^C to PrP^{Sc} more likely.

Protein modification

Proteins undergo a variety of modifications that are critical for function. There are numerous amino acid modifications such as collagen. Thirty percent of the residues are glycine. The ~ 30 of residues are Proline or hydroxyproline (HyPro). The 5-hydroxylysine (HyLys) also occurs; a site for glycosylation. The Hydroxylation of Pro, Lys is a post-translational modification, requires vitamin C as a reactant. The sequence of collagen bears long stretches of Gly--Pro/HyPro-X repeats.

Structure of collagen

Collagen makes up 25 to 35% of the total protein of mammals. It is found in all forms of connective tissue. "Collagen itself is an insoluble protein because of extensive cross- linking. The structural unit of collagen is a tropocollagen, a supercoil made up of 3 helices, with a molecular mass of ~285 kdal. Each collagen helix consists of ~ 1000 amino acid residues. The helix is left-handed. It is not an α-helix. The helix contains 3 amino acids per turn, with a pitch of 0.94 nm. Each unit of tropocollagen is about 1.5 nm wide and 300 nm long. The bundles of these 3-stranded supercoils can be seen as collagen fibers in the electron microscope. Mature collagen has extensive covalent cross-linkages between individual collagen molecules.

There is no intra-helical H-bonding in collagen helices. Rather, H-bonding occurs between the amide N of glycine residues in the central axis and the carbonyls of other residues in the adjacent chains. Often proline and hydroxyproline" are involved.

Genetic disorders of collagen are numerous

One such group of 10 different collagen deficiency diseases is the Ehlers-Danlos (E-D) syndrome. The "India-rubber man" of circus fame had an E-D syndrome. Osteogenesis imperfecta: abnormal (fragile) bone formation is seen in human babies" (www.scribd.com., www.slideshare.com., www. coursehero.com., www. biocyclopedia.com., www. molecular-plant-biotechnology.info., www.slideplayer.com., www.documents.mx., www. docplayer.net., www.bmb.psu.edu.,www.archive.org.,www.mafiadoc.com., www.powershow.com., www.bioscience.jbpub.com., www.outhorstream. com., www.docslide.com., www.msluay.com., www.scienceprofonlue. com., www.csus.edu., www.krishikosh.egrouth.ac.in., www.tankonyvtor. hu., www.silo.pub., www.studyblue.com., www.Preservearticles.com., www.biotechuniverse.blogspot.com).

7. ENZYMES: NOMENCLATURE, APOENZYME AND HOLOENZYME, SUBSTRATE SPECIFICITY, COENZYMES, FACTORS AFFECTING ENZYME ACTIVITY, REGULATION OF ENZYME ACTIVITY, ENZYME INHIBITION, ISOZYMES, RIBOZYMES, ABZYMES.

An enzyme is a biocatalyst that increases the rate of a chemical reaction without itself being changed in the overall process. Virtually all cellular reactions or processes are mediated by enzymes. Enzymes have several properties that make them unique. An enzyme is a globular protein that functions as a biological catalyst, speeding up reaction rate by lowering activation energy without being affected by the reaction it catalyzes.

Enzymes are Protein in Nature

It is a globular protein. Ribozymes are RNA molecules with enzymatic activity. The catalytic behavior of any enzyme depends upon its primary, secondary, tertiary or quaternary structure. Enzymes of the digestive tract and those found in blood are present in an inactive form called zymogen or proenzymes.

Active site

Enzymes are composed of long chains of amino acids that have folded into a very specific three-dimensional shape that contains an active site. An active site is a region on the surface of an enzyme to which substrates bind to catalyze a chemical reaction.

Enzymes are highly specific for the type of reaction they catalyze and for their substrate. Enzymes with the highest specificity and accuracy are involved in the copying and expression of the genome. These enzymes have a proof-reading function e.g. DNA polymerase.

Enzymes make reactions easier to occur at reasonable temperature by lowering the Activation energy E_A of the reaction.

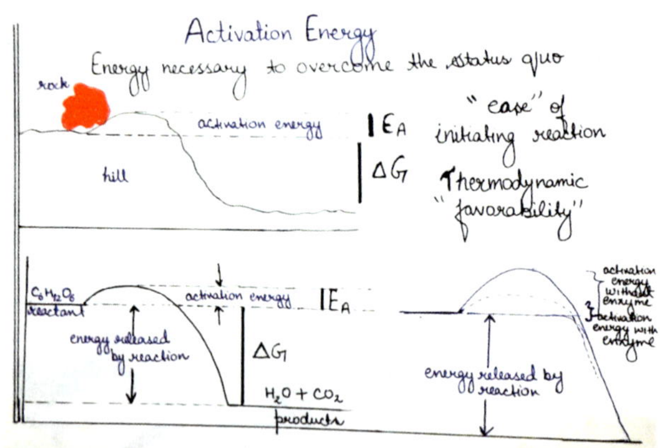

Some enzyme requires a non-protein molecule for their activity. So enzymes are conjugated proteins known as Holoenzymes.

- Holoenzyme = Apoenzyme +cofactor/coenzyme.

Enzyme without non-protein part is inactive and called apoenzyme. Cofactors can be either inorganic or organic compounds.

Mechanism of enzyme action

The enzymatic reactions take place by binding of the substrate with the active site of the enzyme molecule by several weak bonds.

- E + S ‹--------› ES --------› E + P

Formation of the ES complex is the first step in the enzyme-catalyzed reaction then the ES complex is subsequently converted to the product and free enzyme.

Lock and key model

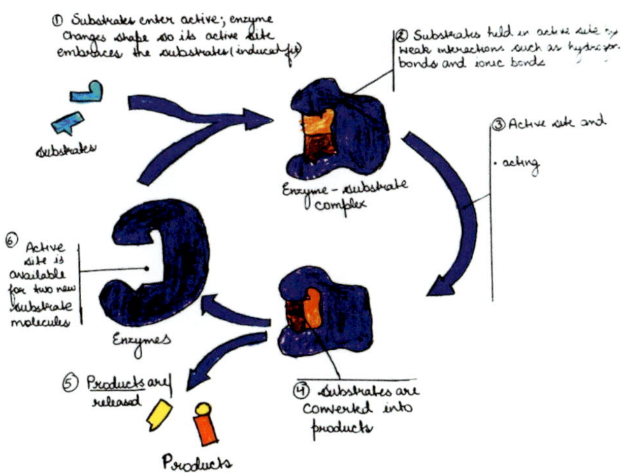

Induced-fit model

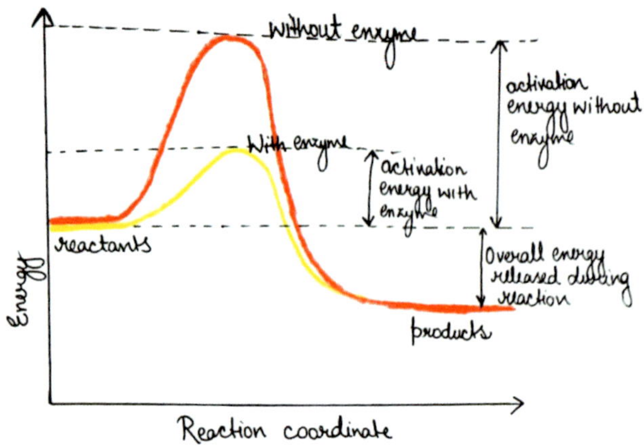

Nomenclature Enzyme Classification

The **IUBMB** has recommended system of nomenclature for enzymes & according to them each enzyme is allotted with two names.

- **Trivial name** (common name, recommended name).
- **Systemic name** (official name).

Systemic name

Each enzyme is characterized by a code number called enzyme Code number or **EC number** and it contains four digits separated by a dot.

- E.g. **EC M.N.O.P**

First digit represents the class. Second digit stands for sub-class. Third digit stands for the sub-sub class or subgroup. Fourth digit gives the serial number of the particular enzyme in the list.

- E.g. **EC 2.7.1.1 for hexokinase**.

According to the IUBMB system of enzyme nomenclature, enzymes are grouped into 6 major classes.

1. EC 1 Oxidoreductases
2. EC 2 Transferases
3. EC 3 Hydrolases
4. EC 4 Lyases
5. EC 5 Isomerases
6. EC 6 Ligases

Factors Affecting Reaction Velocity

- Temperature
- Hydrogen ion concentration (pH)
- Substrate concentration
- Enzyme concentration
- Products of the reaction
- Presence of activator/inhibitor
- Allosteric effects
- Time

Effect of Temperature

The rate of an enzyme catalyzed reaction increases as the temperature increases, enzyme rate of reaction is decreased at high temperature due to denaturation .

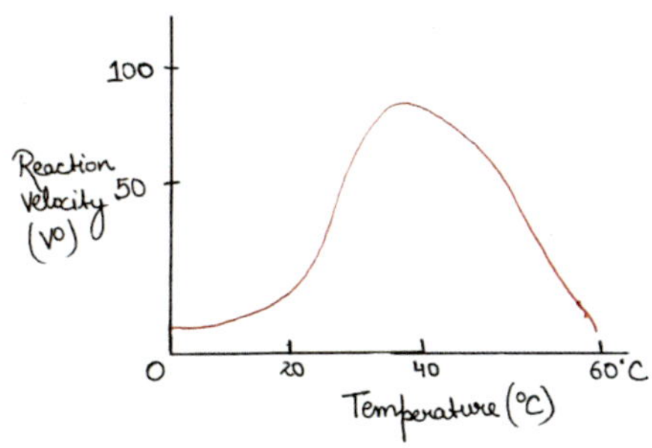

Effect of pH

Every enzyme has an optimum range of pH, slow activity of rate of reaction due to changing the outside range of pH, the enzyme is denatured at extremely high value of pH.

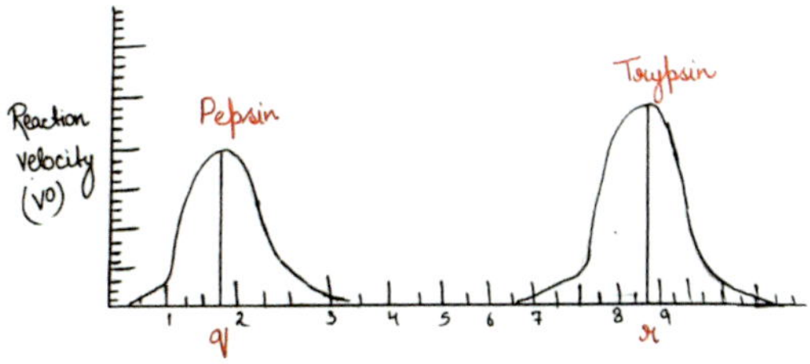

The rate of the reaction or velocity is directly proportional to the enzyme concentration when sufficient substrate is present. Accumulation of product in a reaction causes inhibition of the enzyme activity.

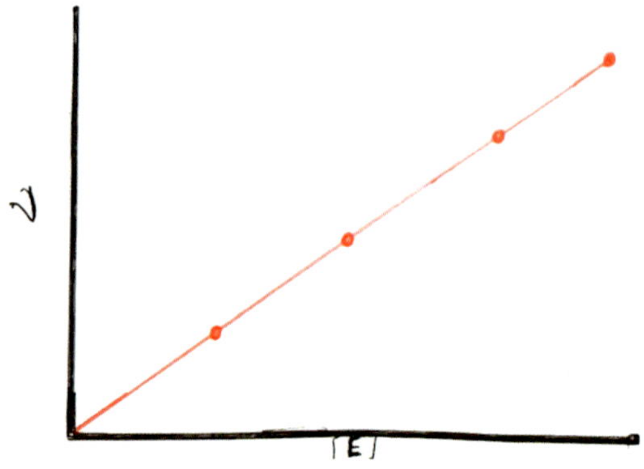

Effect of Substrate Concentration

The rate of reaction increases as enhance of substrate concentration at a specific point. As the amount of substrate increases have no effect on the rate of reaction once all the enzymes have bound. The available enzymes will be saturated and it will work at a maximum rate.

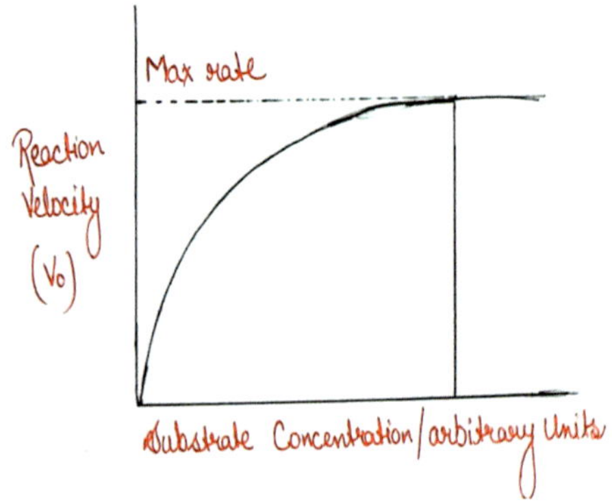

Enzyme kinetics

The study of the rate of reaction and how it changes in response to change in experimental parameters is known as kinetics. The amount of substrate present is one of the key factors affecting the rate of the reaction catalyzed by an enzyme *in vitro*.

Effect of Substrate Concentration On Reaction Velocity

The initial velocity V_0 progressively increases with increasing concentration of the substrate during enzyme-substrate rate of reaction. The increase in V_0 will not depend on the substrate. The substrate concentration on the X-axis and reaction velocity on Y-axis are plotted on a graph.

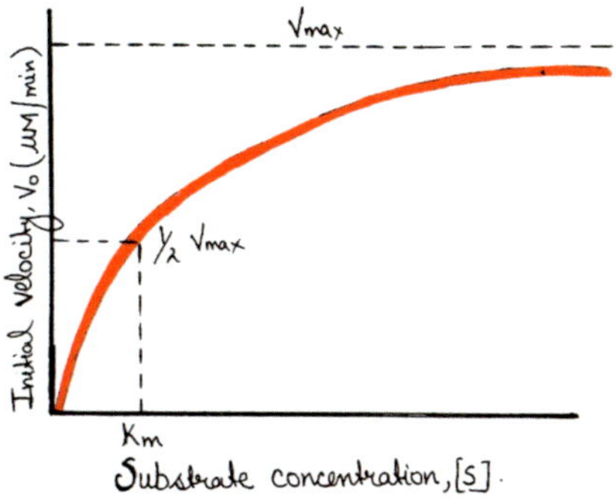

Michaelis- Menten Kinetics

The model involves one substrate molecule.

$$E + S \underset{k_{-1}}{\overset{k_1}{\rightleftharpoons}} ES \xrightarrow{k_2} E + P$$

Where

- S is the substrate
- E is the enzyme
- K_1, k_{-1} and k_2 are the rate constants
- The mathematical equation that defines the quantitative relationship between the rate of an enzyme reaction and the substrate concentration is the *Michaelis-Menten equation*:

$$V_0 = \frac{V_{max}[S]}{K_m + [S]}$$

- V_0 is the observed velocity at the given [S]
- K_m is the Michaelis-Menten constant
- $K_m = (K_{-1} + K_2) / K_1$

- V_{max} is the maximum velocity at saturating [S] conc.

Lineweaver-Burk (double reciprocal) plot

A linear representation is more accurate and convenient for determining V_{max} and K_m. This equation is obtained by taking reciprocal of both the side of the Michelis-Menton equation.

1/[S] vs. $1/V_o$

Lineweaver-burk (Double Reciprocal) plot

$$\frac{1}{v} = \frac{K_m}{V_{max}} \frac{1}{[S]} + \frac{1}{V_{max}}$$

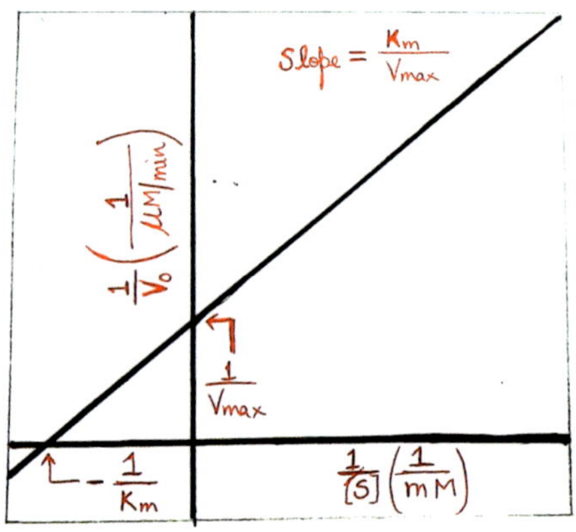

Enzyme Inhibition

Any substance that can reduce the velocity of an enzyme catalyzed-reaction is an inhibitor. These include drugs, antibiotics, poisons, and anti-metabolites. The sequence of enzyme-catalyzed reactions has useful in understanding, metabolic regulation, studying the mechanism of cell toxicity produced by toxicants.

Types of enzyme inhibition

- Reversible inhibitors
- Irreversible inhibitors

Reversible Inhibitors can be classified as

- Competitive
- Non-competitive
- Un-competitive

Competitive Inhibition

The structure of a competitive inhibitor closely resembles that of the enzyme's normal substrate. The competitive inhibitor binds reversibly to the enzyme's active site. The inhibitors form an enzyme-inhibitor complex (EI) that is equivalent to the ES complex.

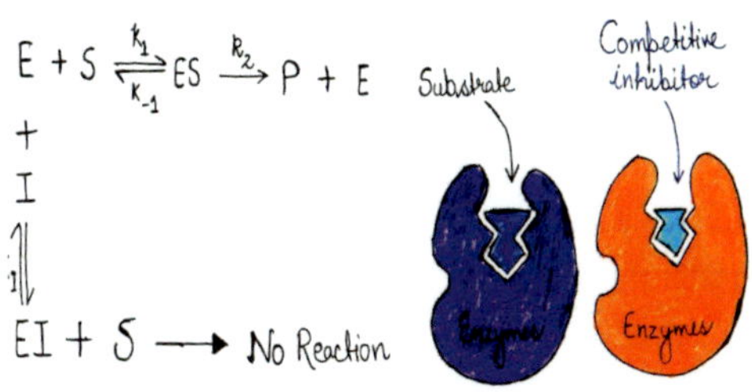

Non-competitive inhibition

The inhibitor binds to the enzyme at a site other than the active site. Inhibitor binding alters the enzyme's three-dimensional configuration and blocks the reaction.

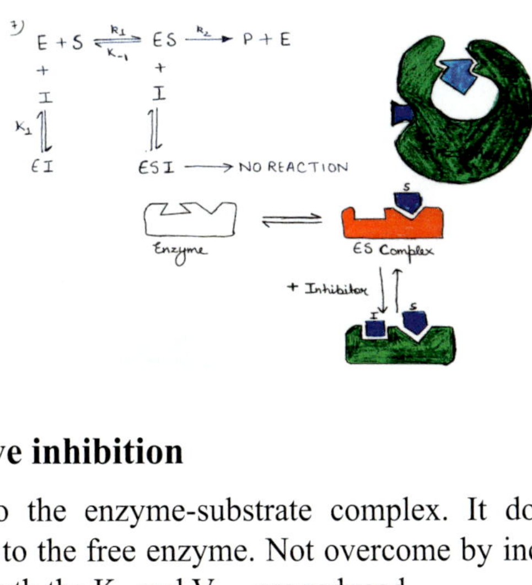

Un-competitive inhibition

It binds only to the enzyme-substrate complex. It does not have the capacity to bind to the free enzyme. Not overcome by increasing substrate concentration. Both the K_m and V_{max} are reduced.

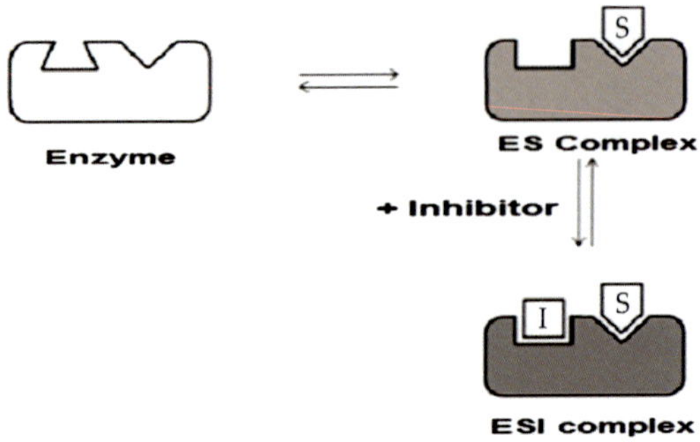

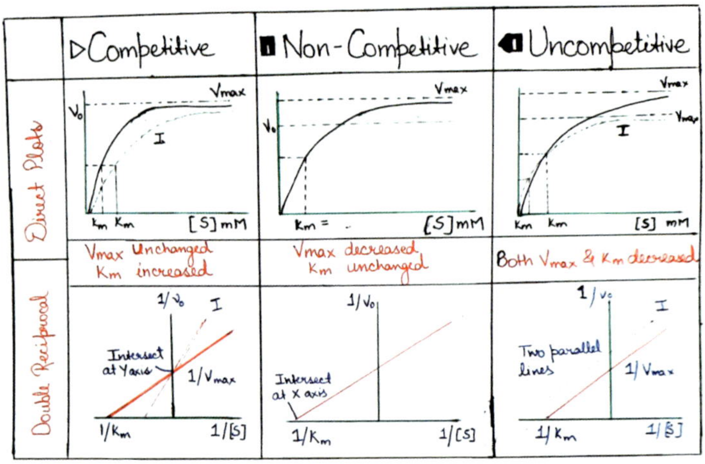

Isoenzymes

Isoenzymes **(**Isozymes**):** Some metabolic processes are regulated by enzymes that exist in different molecular and multiple forms. All forms demonstrate enzymatic activity and catalyze the same biochemical reaction, but they may differ in kinetics (different K, M and Vmax) and regulatory properties (different effectors). The form of coenzyme is prefer and their cellular distribution.

Ribozymes

Catalytic RNA: Some forms of RNA can serve as a biological catalyst was acknowledged by the awarding of the 1989 Nobel Prize in Chemistry Sidney Altman Thomas Cech and Ribozyme**:** RNA enzymes Ribonuclease P: it is used as self-splicing RNA introns.

Significance of ribozymes

Ribozymes are used as tools in specific cleavage of RNA and have been designed to catalyze the *in vitro* cleavage of the RNA of HIV.

Abzymes: catalytic antibodies

It was reported in 1986 by Peter Schultz and Richard Lerner. It has catalyzed the hydrolysis of esters and carbonates.

(www.scribd.com., www.slideshare.com., www.coursehero.com., www. biocyclopedia.com., www. molecular-plant-biotechnology.info., www. slideplayer.com., www.documents.mx., www.docplayer.net., www.bmb. psu.edu.,www.archive.org.,www.mafiadoc.com., www.powershow.com., www.bioscience.jbpub.com., www.outhorstream.com., www.docslide. com., www.msluay.com., www.scienceprofonlue.com., www.csus.edu., www.krishikosh.egrouth.ac.in., www.tankonyvtor.hu., www.silo.pub., www.studyblue.com., www.Preservearticles.com., www.biotechuniverse. blogspot.com).

8. VITAMINS

It is an organic compound required by an organism in trace amounts to perform specific cellular functions. These compounds serve the "same roles in nearly all forms of life, but higher animals have lost the capacity to synthesize them in the course of evolution. Vitamins are not synthesized by humans. Therefore, they must be supplied by the diet. They can be classified according to the solubility and their functions in metabolism. The requirement for any given vitamin depends on the organisms. Not all vitamins are required by all organisms. It may be water-soluble or fat-soluble. There are nine vitamins such as thiamine, riboflavin, niacin, biotin, pantothenic acid, folic acid, cobalamin, pyridoxine, and ascorbic acid that are classified as water-soluble, whereas four vitamins (Vitamins A, D, E and K) are termed as fat-soluble. The water-soluble vitamins are all precursors" coenzymes, except vitamin C. Certain vitamins and minerals are needed for the body to function.

- 13 vitamins
- 22 minerals

Two types of vitamins

- Water-soluble
- Fat-soluble

Fat-soluble vitamins

Fat-soluble vitamins are A, D, E and K and It has excess stored in the liver and in body fat. It is possible to build up to a toxic level.

Vitamin A (Retinol)

Beta-carotene is converted into vitamin A. It promotes good vision. It helps healthy skin. It helps with the growth and maintenance of bones, teeth, and cell structure

- RDA: 900 micrograms for males; 700 micrograms for females.

Too much vitamin A

It may turn your skin orange. It may cause fatigue, weakness, severe headache, blurred vision, hair loss and joint pain.

Toxicity:

- May cause severe liver or brain damage
- Birth defects

Too little vitamin A

It may cause night blindness and there is a lowered immune system

Foods rich in vitamin A

- **Foods**
- Only animal products
  - Liver
  - Eggs
  - Milk, butter and cheese
- Carotenoids
- Orange/Yellow fruits and vegetables
- Cantaloupes, carrots, sweet potatoes, winter squash

- Leafy green vegetables
- Spinach, broccoli

Vitamin D

It is essential for building and maintaining bones and teeth. It is responsible for the absorption and utilization of calcium. It may boost the immune system. It may also help decrease certain cancers.

- RDA: 5 micrograms until age 50
- 10 micrograms / day until 70; 15 mcg for 70+ age

Vitamin D deficiency has been in the news a lot lately. Deficiency may occur from the:

- Inadequate diet
- Vegetarianism, lactose intolerance and milk allergy.
- The body is unable to absorb needed vitamin D. Limited exposure to sunlight.

It may lead to osteomalacia and/or osteoporosis.

Getting vitamin D from sun exposure for 10 minutes a day

- Foods
  - Fortified milk
  - Tuna
  - Salmon
  - It may need a supplement

Vitamin E

It is important in red blood cells, muscles and other tissues. Deficiency is rare.

- Toxicity is rare.
- But, Vitamin E acts as a blood thinner.
- Foods
- Vegetable oils, salad dressings, whole grain cereals, green leafy vegetables, nuts, seeds, peanut butter and wheat germ.

Vitamin K

It is important for blood clotting and also has a role in bone health. It is mostly made in the intestines.

- Foods
  - Turnip greens, cauliflower, spinach, liver, broccoli, kale and cabbage.

Water-soluble Vitamins

Water soluble vitamins are B and C

- There are eight vitamins B such as
- Thiamin (B-1)
- Riboflavin (B-2)
- Niacin (B-3)
- Pyridoxine (B-4)
- Cobalamin (B-12)
- Folic acid
- Pantothenic acid
- Biotin

Thiamin or B-1

It helps to convert carbohydrates into energy.

- Deficiency
  - Fatigue, nausea, depression and nerve damage.
  - Foods: Pork, beef, liver, peas, seeds, legumes, whole-grain products, and oatmeal

Riboflavin or B-2

It is important to metabolism and red blood cells.

- Deficiency
- Dry, scaly skin
- Foods: Milk, yogurt, cheese, whole-grain bread, green leafy vegetables, meat, and eggs

Niacin or B-3

It is involved in energy production. It is also helps the skin, nerves and digestive system.

- Deficiency:
  - Rare but causes: diarrhea, dermatitis, dementia and death
- Foods: Meat, poultry, liver, eggs, brown rice, baked potatoes, fish, milk, and whole-grain foods.

Pyridoxine or B-6

It is involved in chemical reactions of proteins and amino acids.

- **Deficiency:** Skin changes, dementia, nervous system disorders and anemia.
- **Foods:** Lean meats, fish, legumes, green leafy vegetables, raisins, corn, bananas and mangos.

Cobalamin or B-12

It helps the nervous system, red blood cells and DNA synthesis.

- Deficiency
  - Nervous system disorders and pernicious anemia
  - Foods: Only found in animal products
    - Meat, fish, poultry, eggs, milk products and clams

Folic acid (Folacin, Folate)

It has ans important role in red blood cell formation and cell division.

- **Deficiency:**
- Anemia, digestive disorders
- Foods: Leafy, dark green vegetables and also found in liver, beans, peas, asparagus, oranges, avocados

Pantothenic Acid and Biotin

It helps the metabolism and formation of some hormones.

- Deficiencies are rare
- Foods: Almost any food, plant-based or animal-based

Vitamin C

It has important in bone health, blood vessel health, cell structure and absorption of iron. Too much vitamin C is found in Foods: Melons, berries, tomatoes, potatoes, broccoli, fortified juices, kiwi, mangos, yellow peppers and citrus fruits.

Table 8.1 Roles of vitamins

Vitamin	**Role of action**
Water soluble	
Thiamine (vit B1)	Cofactor in decarboxylations Deficiency disease: Beri-beri
Riboflavin (vit B2)	Constituent of flavoproteins
Niacin (vit B3)	Constituent of NAD^+ and $NADP^+$ Deficiency disease: Pellagra
Pantothenic acid (vit B5)	Constituent of CoA and acyl-group transfer
Pyridoxine (vit B6)	Group transfer to or from amino acids
Biotin (vit H)	ATP-dependent carboxylation and carboxyl group transfer
Folate (folic acid)	Coenzymes for 1-carbon transfer Deficiency disease: Megaloblastic anemia
Cyanocobalamin (vit B12)	Coenzyme in amino acid metabolism Deficiency disease: Pernicious anemia
Ascorbic acid (Vit C)	Antioxidant Deficiency disease: Scurvy
Fat Soluble	
Vitamin A	Role in vision, growth, reproduction. Deficiency disease: Night blindness
Vitamin D	Regulation of calcium and phosphate metabolism Deficiency disease: Rickets (children), osteomalacia (adults).
Vitamin E	Antioxidants; cofactors in electron transport in cyto-chrome chain
Vitamin K	Blood coagulation, Post-translational modification of blood clotting factors.

Minerals

- Calcium
- Chromium
- Folate
- Iron

- Magnesium
- Selenium
- Zinc
- Selenium
- Phosphorus
- Sodium
- Potassium
- Chloride
- Sulfur
- Copper
- Iodine
- Cobalt
- Fluoride
- Our body is needed 22 minerals.
- There are two categories such as
- Major
- Include calcium, chloride, magnesium, phosphorus, potassium, sodium, and sulfur
- Trace
- Include iron, zinc, iodine, selenium, copper, manganese, fluoride, chromium, molybdenum, arsenic, nickel, silicon, boron and cobalt.

Sodium

- What does sodium do for you?
  - It helps maintain fluid balance.
  - It helps transmit nerve impulses.
  - It influences the contraction and relaxation of muscles.

Sodium & health

High amount of sodium

It causes high blood pressure and it may lead to fluid retention.

Sodium savvy

The human body requires about 500 mg of sodium per day, while the average American usually ingests between 2,300-6,900 mg each day. It is recommended to stay in a range of 1,500 to 2,400 mg / day.

Where are you getting sodium?

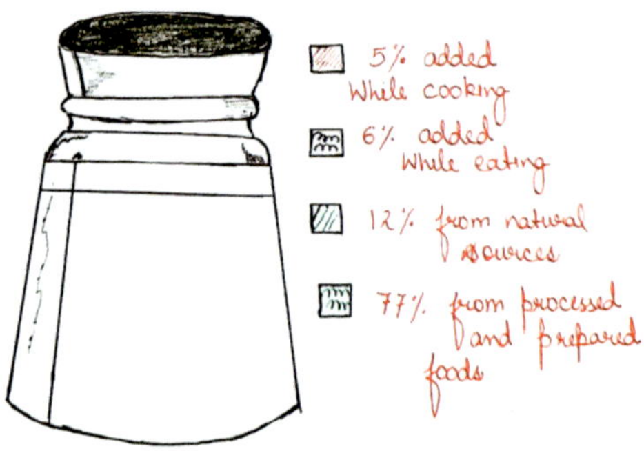

Sodium & Food

- Monosodium glutamate (MSG)
- Baking soda
- Baking powder
- Disodium phosphate
- Sodium alginate
- Sodium nitrate or nitrite

Reducing sodium in your diet

- Eat more fresh foods.
- Eat less processed foods.
- Look for low-sodium products.
- Limit the salt you add to foods.
- Experiment with other seasonings.
- Use salt substitutes with caution.

Calcium

The most abundant mineral in your body and 99% is stored in the bones. It is known for bone health.

- Males 19-50 years old: 1,000 mg / day
- Females 19-50 years old: 1,000 mg / day

- We can take calcium from dairy products, fortified juices and sardines.

Food	**Calcium**
Yogurt, plain (low-fat) **Yogurt, flavored (low-fat)**	**1 cup - 415 mg** **1 cup – 345 mg**
Milk, skim **Milk, 1-2%**	**1 cup – 302 mg** **1 cup – 300 mg**
Ice cream	**½ cup – 88 mg**
Broccoli, cooked	**½ cup – 68 mg**
Salmon, canned	**3 oz – 165 mg**
Fortified orange juice	**8 oz – 300 mg**

Iron

Iron deficiency is the most widespread vitamin or mineral deficiency in the world.

- 70% of your body's iron is in your hemoglobin
- Too little iron = too little oxygen

Iron & foods

It found in animal products

- Red meats, liver, poultry and eggs

Non-heme iron: It found in plant products

- Beans, nuts, seeds, dried fruits, fortified breads and cereals

Iron supplements

- Strict vegetarians
- Those who do not eat a balanced diet
- Those who are over 60

- Smokers and those who regularly drink alcohol
- Chronic dieters
  - Those who suffer from food allergies, intolerances

Potassium

It keeps muscles and nervous system working properly and an electrolyte that conducts electricity in the body.

- Potassium rich foods: bananas, broccoli, tomatoes, potatoes with skins, leafy green vegetables, citrus fruits, dried fruits, and legumes.

Zinc

It supports normal growth and development during pregnancy, childhood, and adolescence. It is required for the proper sense of taste and smell. It is involved in numerous aspects of cellular metabolism.

- Foods rich in zinc: beef, pork, lamb, & legumes.

Magnesium

Approximately 50% of the total body magnesium is found in bone. The other half is found predominantly in cells of body tissue and organs. It helps to maintain muscle and nerve function, keeps heart rhythm steady, supports a healthy immune system, keeps bones strong. It helps to regulate blood sugar levels, promotes normal blood pressure. Eating a wide variety of legumes, nuts, whole grains, and vegetables will help meet your daily need for magnesium.

Trace minerals

There are several trace minerals such as Iron, Manganese, Copper, Iodine, Zinc, Fluoride, and Selenium. Manganese is important for the production of enzymes and antioxidants. Copper is necessary for your body to make connective tissue and the metabolism of iron. Iodine is used to synthesize thyroid hormones. Fluoride is used for strong bones and teeth. It reduces the number of cavities in teeth. Selenium combines with proteins to make antioxidants and is also essential for normal thyroid function.

Simple salmon supper

- 1 each zucchini, yellow squash, and red pepper, chopped
- 4 skinless salmon fillets
- 1 can diced tomatoes, drained
- 1/2c Kraft Sun-Dried Tomato Vinaigrette Dressing
- 2-2/3 cups hot cooked long-grain brown rice

- **Heat** oven to 375° F.
- **Combine** zucchini, squash and peppers in a 13x9-inch baking dish; top with fish. Mix tomatoes and dressing; spoon over fish.
- **Bake** 20 to 25 min. or until the fish flakes easily with a fork.
- **Serve** with rice.

Water

It is essential for life. It makes up about 45-75% of your body weight. It is important for transport. It is used for mechanical functions. It helps to break substances down. It helps to maintain body temperature/pH

For men: 3.7 L/ day

For women: 2.7 L / day

Perfectly 80% of water should come from drinking fluids.

The balance 20% of water should come from food

(www.scribd.com., www.slideshare.com., www.coursehero.com., www.biocyclopedia.com., www. molecular-plant-biotechnology.info., www.slideplayer.com., www.documents.mx., www.docplayer.net., www.bmb.psu.edu.,www.archive.org.,www.mafiadoc.com., www.powershow.com., www.bioscience.jbpub.com., www.outhorstream.com., www.docslide.com., www.msluay.com., www.scienceprofonlue.com., www.csus.edu., www.krishikosh.egrouth.ac.in., www.tankonyvtor.hu., www.silo.pub., www.studyblue.com., www.Preservearticles.com., www.biotechuniverse.blogspot.com).

PART II

BIOPHYSICAL CHEMISTRY

9. WATER: STRUCTURE AND INTERACTIONS, WATER AS SOLVENT, PROTONMOBILITY, ACID- BASE REACTIONS, PH AND BUFFERS, ISOELECTRIC PH

Water is essential for life. The most abundant substance of the living cell is water. It accounts up to 70% or more of cell's weight. It provides the medium in which most substances remain dissolved. Water is colorless, odorless and tasteless in nature. Water is formed when two hydrogen atoms bond to one oxygen atom. It is not symmetrical. Electrons spend more time near the (O) and less time near the (H). The water molecule is polar.

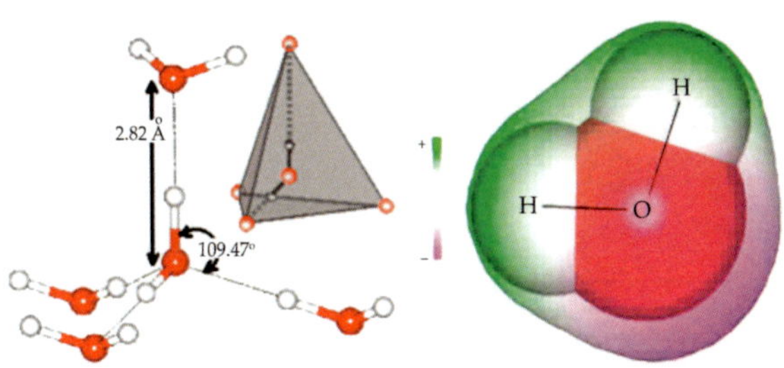

Figure 9.1 Water molecules are held together by a hydrogen bond.

There is a greater concentration of electrons around the nucleus of the oxygen than around the hydrogen. Therefore, the hydrogen end is slightly positive and the oxygen end is slightly negative.

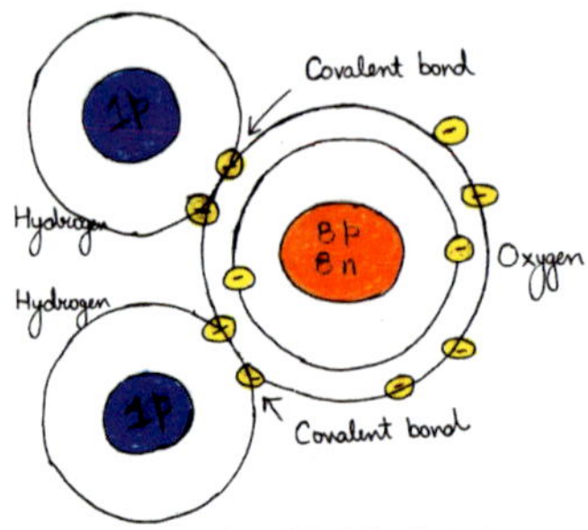

Bohr Model of H_2O

It is made up of two "hydrogen atoms and one oxygen atom, with a total atomic mass of 18 Daltons. The H-O-H bond angle is 104.5 that is smaller than the typical tetrahedral angle of 109. Since oxygen is more electronegative than hydrogen, the oxygen nucleus attracts electrons more strongly than does the hydrogen nucleus. It means that the shared electrons are more shifted towards the oxygen atom than of the hydrogen. The result of this unequal electron sharing is the formation of two electric dipoles in the water molecule, one along each of the H-O bonds. Each hydrogen atom bears a partial positive (δ+) and the oxygen atom carries a partial negative charge equal in magnitude to the sum of the two partial positives (δ-)". When water molecules attract to one another, the hydrogen end of one molecule is attracted to the oxygen end of another.

States of Water

Water is the only substance found on Earth in all three states (phases):

1. Liquid
2. Solid (Ice)
3. Gas (Steam or Vapor)

Physical states

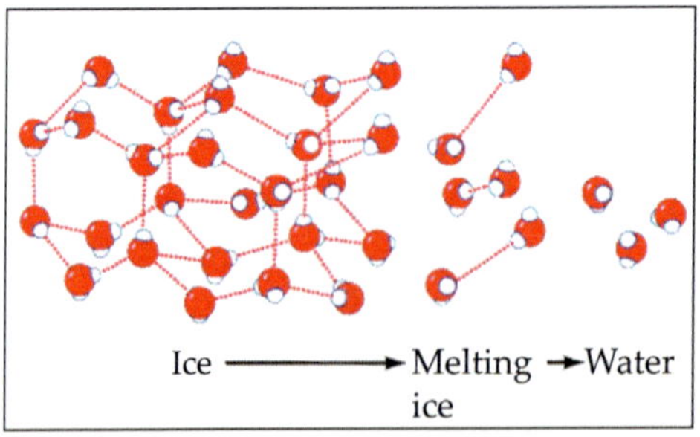

Water molecules are constantly moving. Temperature increases with an increase in movement. "When water molecules move faster, they tend to break their hydrogen bonds. This is called evaporation. When gas or vapor molecules slow down, they clump or join together. This is called condensation. As water becomes cooler, it becomes less dense: 1 gram per cubic cm at 39.2° C and 0.95865 gram per cubic cm at 212° C. As water changes from a liquid to a solid, molecules form crystals. In ice crystals, molecules are spaced further apart. Since molecules are spaced further apart, ice is less dense than water (it floats). Helpful for aquatic organisms; forms a blanket" If ice was denser than water, lakes would freeze from the bottom upwards.

Thermal Properties of Water

Water has a high specific heat, or heat capacity (the amount of heat needed to raise 1g of a substance 1° C.). This quality of water is important for several reasons:

- It regulates the rate at which air temperature changes.
- It affects the global climate (milder temperatures in coastal regions).
- So, aquatic organisms face less temperature variation than terrestrial organisms.
- Therefore, water is used in cooling systems found in cars and industrial plants.

Surface Tension

Next to mercury, water has the highest surface tension of all liquids. This is a result of the tendency of water molecules to attract to one another, or cohere, at the surface of any accumulation of water. The surface tension allows insects to walk on water. Interaction between hydrogen bonding and the earth's gravitational pull

- **Cohesion**- the attraction between water molecules to each other through hydrogen bonds (H).
- **Adhesion**- the attraction of water molecules to another substance.
- **Capillarity**- the movement of water within the spaces of a porous material due to the forces of surface tension, adhesion, and cohesion.

pH: Alkalinity/Acidity

The measurement of the "H^+ ions found in that particular substance. The scale goes from 0 to 14. pH 7 is neutral. Below pH 7 is acidic. Above pH 7 is alkaline (or basic). One pH unit represents a ten-fold change in H^+ concentration. Since water can dissolve more things than any other natural substance, it is known as the "Universal Solvent". Interacts with other polar compounds. It is repelled by non-polar compounds. The small size allows it to saturate areas. It can convey other substances in solutions. Water is especially good at dissolving salts. The salts formed from the combination of particles with opposite electrical charges (or ions) Example $Na^+ + Cl^- =$ NaCl. When salt is placed in water, the strongly charged salt ions attract to the weaker charged water molecules. Water molecules surround each ion. Salt crystals fall apart or dissociates, and the salt dissolves". The chemical/ Mechanical weathering of rocks on land.

  – Carried to the sea by rivers.

- It comes from the Earth's interior

  – Released into oceans by hydrothermal vents.

- The seawater contains almost all the elements found on Earth. Most of the solutes are made of a small group of common ions.
- Solute- the substance being dissolved by the solvent.

The "salinity- total amount of salt dissolved in seawater. It is expressed as the number of grams of salt left when 1000g of seawater evaporates **example:** 35 grams left over = salinity of 35 parts per thousand (35 ppt or $35^0/_{00}$). The salinity varies by the addition or deletion of pure water (**not salts**). The pure water is added by precipitation. The pure water is removed by evaporation or freezing. The pH of a precisely neutral solution is 7. Solutions having a pH greater than 7 are alkaline or basic; the concentration of OH- is greater than that of H+. Conversely, solutions having a pH less than 7 are acidic. Keep in mind that the pH scale is logarithmic, not arithmetic. To say that two solutions differ in pH by 1 pH unit means that one solution has ten times the H^+ concentration of the other" but it does not tell us the absolute magnitude of the difference.

10. PHOTOMETRY: BASIC PRINCIPLES OF UV-VISIBLE SPECTROPHOTOMETRY AND COLORIMETRY, INSTRUMENTATION AND APPLICATION.

Photometry refers to the science of measurement light in terms of the "light's brightness that the human eye can observe. Photometers examine how light interacts with reflective materials are the basic principle. The arrangement of photometers allows for photometry to be applied in different fields of study. Some instruments are used to observe how light is absorbed and/or reflects wavelengths. Other instruments measure light by converting light into an electrical current and then measuring the electrical current produced by the light". Photometry is often used in the study of liquids and solutions in chemistry. Photometers can help measure the masses of organic or inorganic materials in a solution or liquid.

Colorimetry

The technique in which the concentration is measured by its ability to produce or change the color of a solution is called colorimetry.

- Changes the solution's ability to absorb light.

Spectrophotometry

The technique uses light to measure l chemical concentrations. A colorimetric method is where an instrument is used to determine the amount of analyte in a sample by the sample's ability or inability to absorb light at a certain wavelength.

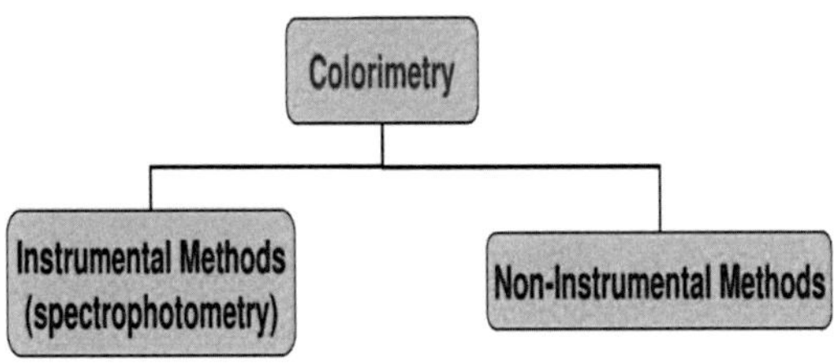

Measurement of ozone (O_3) above South Pole

- O_3 provides protection from ultraviolet radiation
- Seasonal depletion due to chlorofluorocarbons

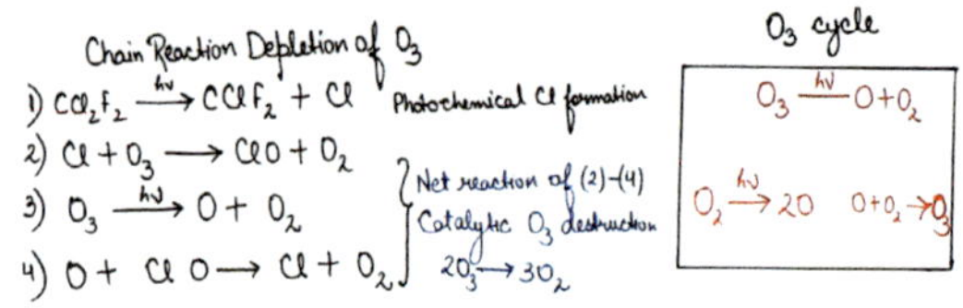

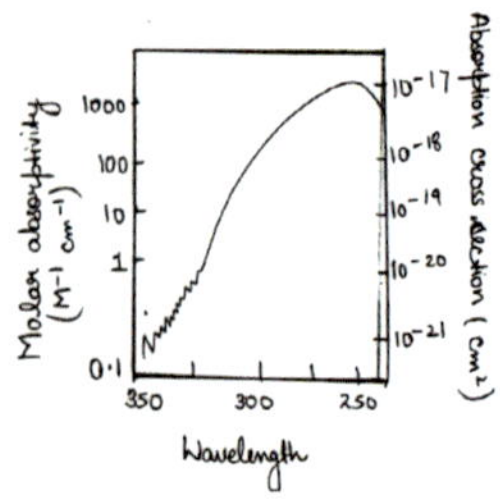

Properties of Light

Particles and Waves

Light waves are consisting of perpendicular, oscillating electric and magnetic fields. Parameters used to describe light.

- Amplitude (A): height of the wave's electric vector.
- Wavelength (l): distance (nm, cm, m) from peak to peak

Frequency (n): number of complete oscillations that the waves make each second

Hertz (Hz): unit of frequency, second^{-1} (s^{-1})

1 megahertz (MHz) = $10^6 s^{-1} = 10^6$Hz

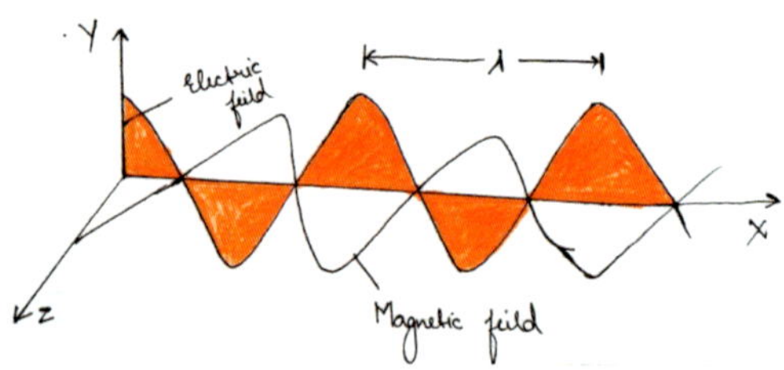

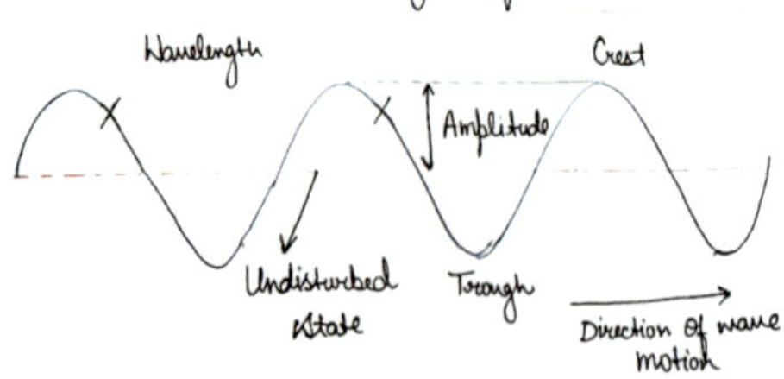

Properties of light

Energy (E): the energy of one particle of light (photon) is proportional to its frequency.

$E = hv$

where: E = photon energy (Joules)

n = frequency (sec^{-1})

h = Planck's constant ($6.626x10^{-34}$J-s)

The frequency(n) is increases as also energy (E) of light increases.

Relationship between frequency and wavelength

$\lambda v = c = V$

$V = c/\lambda$

where: c = speed of light ($3.0x10^{8}$ m/s in vacuum))

n = frequency (sec^{-1})

l = wavelength (m)

Relationship between energy and wavelength

$E = hc/\lambda = hcv$

where: = (1/l) = wavenumber

As frequency (l) is decreases, energy (E) of light increases.

Types of light: The electromagnetic spectrum

Again, energy (E) of light increases as frequency (n) increases or wavelength (l) decreases.

Absorption of Light

1.) *Colors of visible light*

Many types of chemicals absorb various forms of light. The color of light absorbed and observed passing through the compound are complimentary.

Wavelength of maximum absorption (nm)	Color absorbed	Color observed
380—420	Violet	Green-yellow
420—440	Violet-blue	Yellow
440—470	Blue	Orange
470—500	Blue-green	Red
500—520	Green	Purple
520—550	Yellow-green	Violet
550—580	Yellow	Violet-blue
580—620	Orange	Blue
620—680	Red	Blue-green
680—780	Red	Green

2.) Ground and excited state

When a chemical absorbs light, it goes from a "low energy state (ground state) to a higher energy state (excited state). Only photons with energies exactly equal to the energy difference between the two electron states will be absorbed. Since different chemicals have different electron shells that are filled, they will each absorb their own particular type of light. Different electron ground states and excited states".

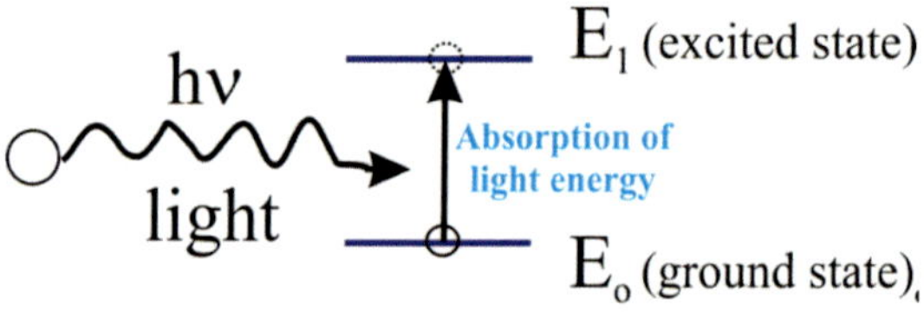

The energy required for a photon to give this transition:

$DE = E_1 - E_0$

3.) Beer's Law

The relative amount of a certain wavelength of light absorbed (*A*) that passes through a sample is dependent on the distance the light must pass through the sample (cell path length - b), the amount of absorbing chemicals in the sample (analyte concentration – c), the ability of the sample to absorb light (molar absorptivity - e), the relative amount of light making it through the sample (P/P_o) is known as the transmittance (T).

$T = P/P_0$

Percent transmittance $\%T = 100 * P/P_0$

T has a range of 0 to 1, %T has a range of 0 to 100%

Absorbance (A) is the relative amount of light absorbed by the sample and is related to transmittance (T). absorbance is sometimes called optical density (OD).

$$A = -\log\left[\frac{P}{P_0}\right] = -\log(T) = -\log(\%T/100)$$

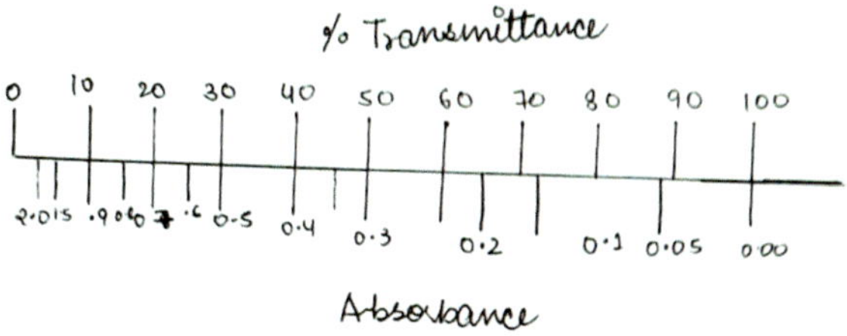

A has a range of 0 to infinity

Absorbance is useful since it is directly related to the analyte concentration, cell pathlength and molar absorptivity. This relationship is known as Beer's Law.

$$A = \varepsilon bc$$

where:
- A = absorbance (no units)
- e = molar absorptivity (L/mol-cm)
- b = cell pathlength (cm)
- c = concentration of analyte (mol/L)

Beer's law allows compounds to be quantified by their ability to absorb light, Related directly to concentration (c)

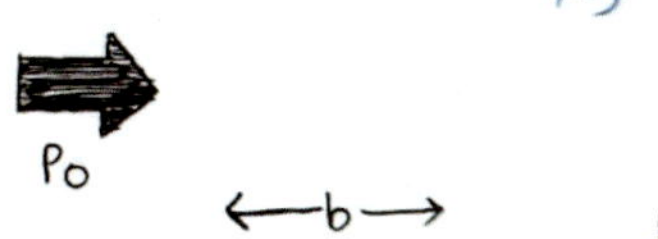

Different chemicals have different energy levels.

- The different ground vs. excited electron states.
- It will have different abilities to absorb light at any given wavelength.

Absorption spectrum is the plot of "absorbance (or e) vs. wavelength" for a compound. The greater the absorbance of a compound at a given wavelength (high e), the easier it will be to detect at low concentrations. By choosing different wavelengths of light (l_A vs. l_B) as different compounds can be measured.

Spectrophotometer

An instrument used to make absorbance or transmittance measurements is known as a spectrophotometer.

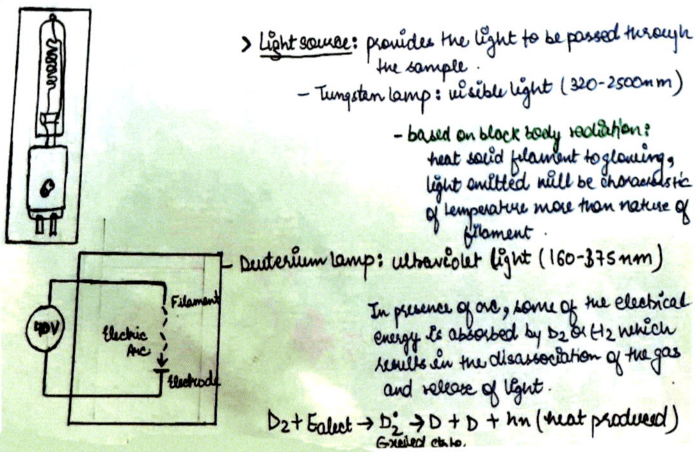

Wavelength selector (monochromator): used to select a given wavelength of light from the light source.

Reflection or Diffraction Grating

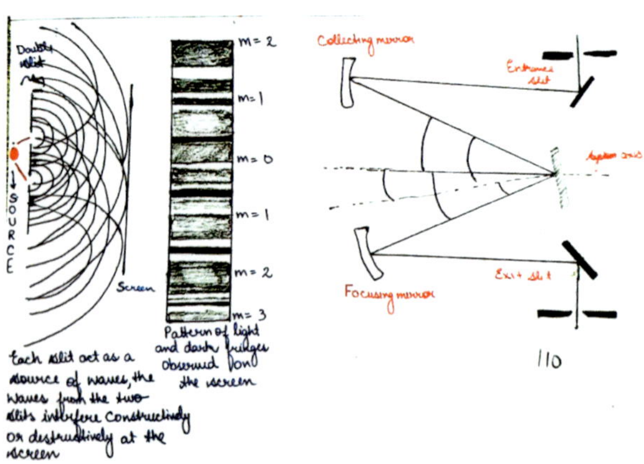

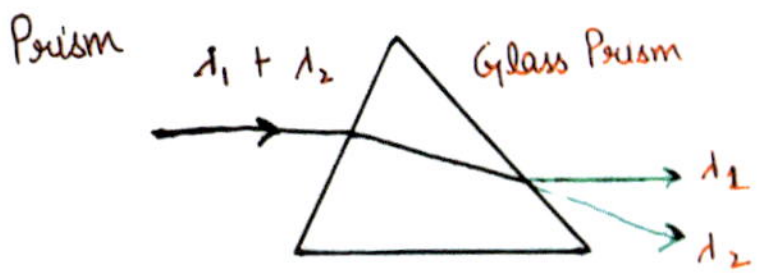

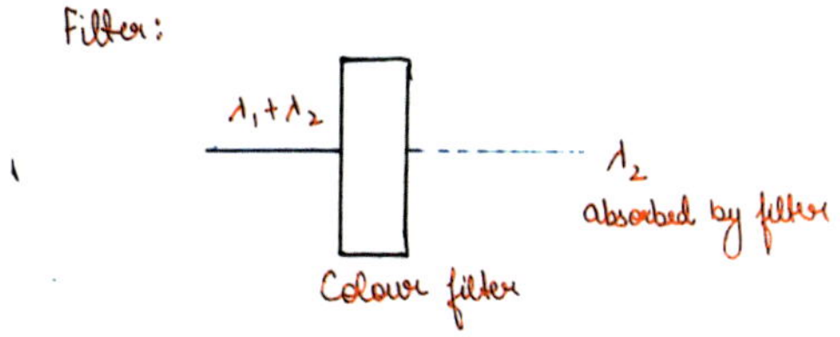

- *Sample Cell:* sample container of fixed length (b).
- It is usually a round or square cuvet.
- It is made from a material that does not absorb light in the wavelength range of interest.
  - Glass – visible region
  - Quartz – ultraviolet
  - NaCl, KBr – Infrared region
- Light Detector: measures the amount of light passing through the sample.
- It usually works by converting the light signal into an electrical signal.

Process

Light hits photoemissive cathode and e^- is emitted. "An emitted e^- is attracted to electrode #1 (dynode 1), which is 90V more positive. It causes several more e^- to be emitted. This e^- is attracted to dynode 2, which is 90V more positive than dynode 1, emitting more e^-. The process continues until e^- is collected at anode after amplification at 9 dynodes. The overall voltage between the anode and cathode is 900V. One photon produces 10^6 – 10^7 electrons". The current is amplified and measured.

Types of Spectrophotometers

Single-Beam Instrument: Sample and blank are alternatively measured in the same sample chamber.

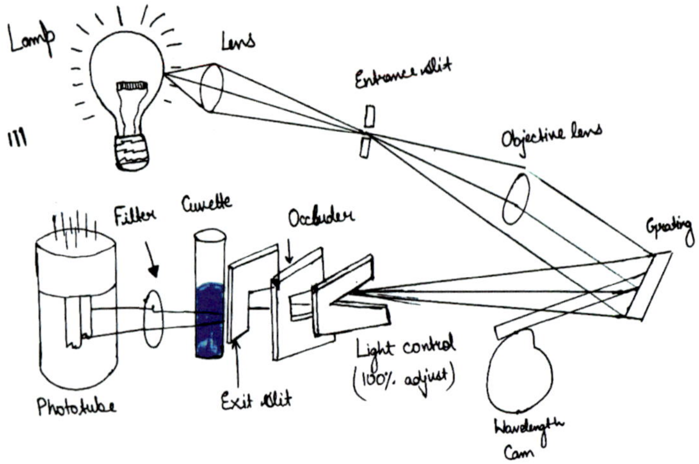

Double-beam Instrument

It continuously compares the sample and blank. It automatically corrects for changes in electronic signal or light intensity of the source.

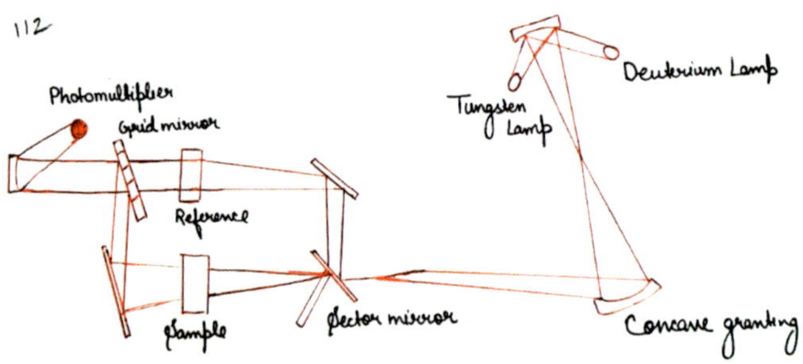

Chemical analysis

1.) Calibration

- To measure the absorbance of a sample, it is necessary to measure P_o and P ratio
  - of light passing through the system with no sample present
  - P_o – the amount P – the intensity of light when the sample is present

- P_o is measured with a blank cuvet
  - Cuvet contains all components in the sample solution except the analyte of interest

P is measured by placing the sample in the cuvet. Then exactly measure an unknown concentration, obtain a calibration curve using a range of known concentrations for the analyte.

2.) Limitations in Beer's Law

The results in a non-linear calibration curve. Solute molecules influence one another because of their proximity at high concentrations.

- Molar absorptivity changes
- Affect on equilibrium, (HA and A^- have difference absorption)

Analyte properties change in different solvents. Errors are found in reproducible positioning of cuvet. It is also problems with dirt & fingerprints. The instrument has electrical noise. Keep A in a range of 0.1 – 1.5 absorbance units (80 -3%T)

Precautions in Quantitative Absorbance Measurements

Choose a wavelength at an absorption maximum. Minimizes deviations from Beer's law, which assumes e- is constant. Pick a peak in the absorption spectrum where the analyte is the only compound absorbing light. Choose a wavelength where the analyte has the largest difference in its absorbance relative to other sample components.

(www.scribd.com., www.slideshare.com., www.coursehero.com., www. biocyclopedia.com., www. molecular-plant-biotechnology.info., www. slideplayer.com., www.documents.mx., www.docplayer.net., www.bmb. psu.edu.,www.archive.org.,www.mafiadoc.com., www.powershow.com., www.bioscience.jbpub.com., www.outhorstream.com., www.docslide. com., www.msluay.com., www.scienceprofonlue.com., www.csus.edu., www.krishikosh.egrouth.ac.in., www.tankonyvtor.hu., www.silo.pub., www.studyblue.com., www.Preservearticles.com., www.biotechuniverse. blogspot.com).

11. CENTRIFUGATION: PRINCIPLES AND APPLICATION, SEDIMENTATION COEFFICIENT, DIFFERENTIAL, DENSITY GRADIENT AND ULTRA-CENTRIFUGATION.

Centrifugation is a process used to "separate or concentrate" materials suspended in a liquid medium. The theoretical basis of this technique is the effect of gravity on particles (including macromolecules) in suspension. The two particles of "different masses will settle in a tube at different rates in response to gravity". It is a process that involves the use of the centrifugal force for the sedimentation of heterogeneous mixtures with a centrifuge used in industry and in laboratory settings. This process is used to separate two immiscible liquids. More dense components of the mixture migrate away from the axis of the centrifuge, while less-dense components of the mixture migrate towards the axis.

Principle

It is a device for "separating particles from a solution according to their size shape, density, viscosity of the medium and rotor speed. The density is higher than that of the solvent sink sediment and the particles that are lighter then it floats to the top". The greater the difference in density, the faster they move. If there is no difference in density i.e. isopycnic conditions, the particles stay steady.

- **Routine centrifuge rotors**
- **Calculation of *g*-force**
- **Differential centrifugation**
- **Density gradient theory**

Centrifuge rotor

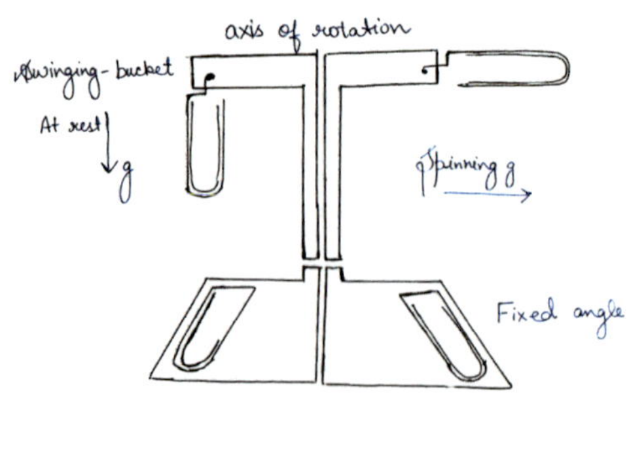

Geometry of rotors

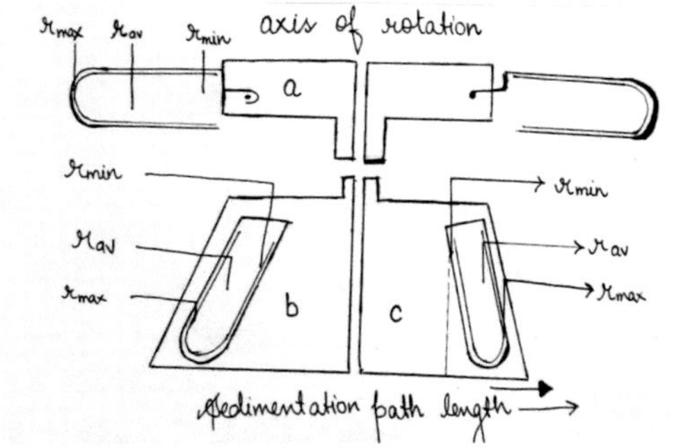

k'-factor of rotors

The *k*'-factor is a measure of the time taken for a particle to sediment through a sucrose gradient. The most efficient rotors which operate at a high RCF and have a low sedimentation path length therefore have the lowest *k*'-factors. The centrifugation times (*t*) and *k*'-factors for two Different.Rotors (1 and 2) are related by:

$$t_1 = \frac{k_1 t_2}{k_2}$$

Calculation of *RCF* and *Q*

$$RCF = 11.18\, x\, r \left(\frac{Q}{1000}\right)^2$$

$$Q = 299\sqrt{\frac{RCF}{r}}$$

RCF = Relative Centrifugal Force (*g*-force)
Q = rpm; *r* = radius in cm

- RCF in swinging-bucket and fixed-angle rotors at 40,000 rpm.
- Beckman SW41 swinging-bucket (13 ml)
- g_{min} = 119,850g; g_{av} = 196,770g;
- g_{max} = 273,690g
- Beckman 70.1Ti fixed-angle rotor (13 ml)
- g_{min} = 72,450g; g_{av} = 109,120g;
- g_{max} = 146,680g

Velocity of sedimentation of a particle

$$v = \frac{d^2(\rho_p - \rho_l)}{18\mu}\ g$$

v = velocity of sedimentation *d = diameter of particle*

ρ_p = density of particle *ρ_l = density of liquid*

g = centrifugal force *μ = viscosity of liquid*

Differential Centrifugation

The density of a liquid is uniform. The density of a liquid is << Density of particles. The viscosity of the liquid is low. Consequence: The rate of particle sedimentation depends mainly on its size and the applied *g*-force.

Size of major cell organelles

• **Nucleus**	**4-12 μm**
• **Plasma membrane sheets**	**3-20 μm**
• **Golgi tubules**	**1-2 μm**
• **Mitochondria**	**0.4-2.5 μm**
• **Lysosomes/peroxisomes**	**0.4-0.8 μm**
• **Microsomal vesicles**	**0.05-0.3μm**

Differential Centrifugation of A Tissue Homogenate

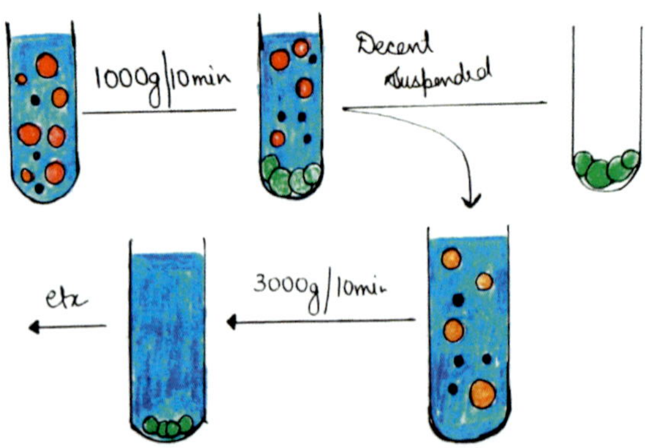

Differential centrifugation of a tissue homogenate

- Homogenate – 1000g for 10 min
- Supernatant from 1 – 3000g for 10 min
- Supernatant from 2 – 15,000g for 15 min
- Supernatant from 3 – 100,000g for 45 min
- Pellet 1 – nuclear
- Pellet 2 – "heavy" mitochondrial
- Pellet 3 – "light" mitochondrial
- Pellet **4 – microsomal**

Differential Centrifugation (III) Expected Content Of Pellets

- 1000g pellet: nuclei, plasma membrane sheets.
- 3000g pellet: large mitochondria, Golgi tubules.
- 15,000g pellet: small mitochondria, lysosomes, peroxisomes.
- 100,000g pellet: microsomes.

Differential Centrifugation

It has poor resolution and recovery because of particle size heterogeneity. The particles starting out at r_{min} have the furthest to travel but initially experience the lowest RCF. Smaller particles close to r_{max} have only a short distance to travel and experience the highest RCF.

The rate of sedimentation can be modulated by particle density. The nuclei have an unusually rapid sedimentation rate because of their size and high density. Golgi tubules do not sediment at 3000g, in spite of their size: they have an unusually low sedimentation rate because of their very low density: $(\rho_p - \rho_l)$ becomes rate limiting.

Density Gradient Centrifugation

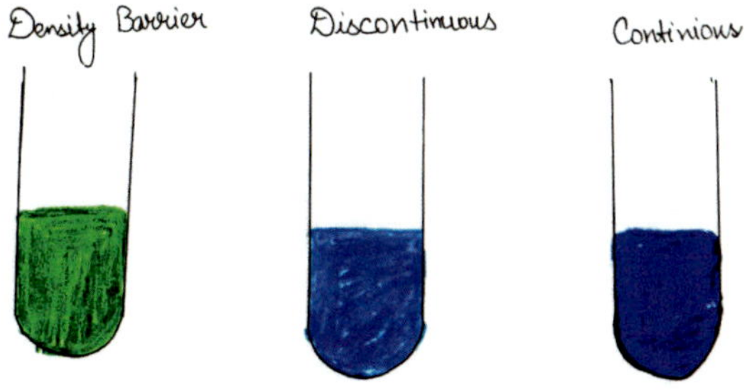

A particle will sediment through a solution if the particle density > solution density. If the particle density < solution density, the particle will float through the solution. When particle density = solution density the particle stops sedimenting or floating. The formats for the separation of particles are according to their density.

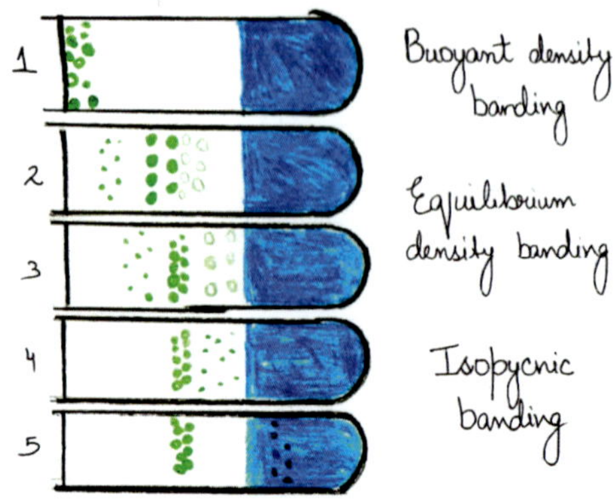

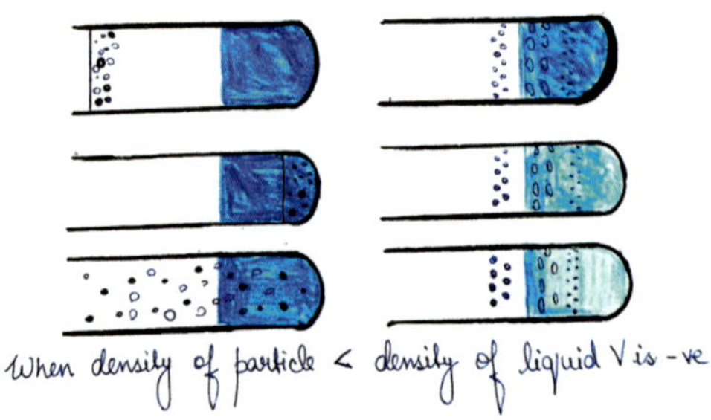

Resolution Of Density Gradients

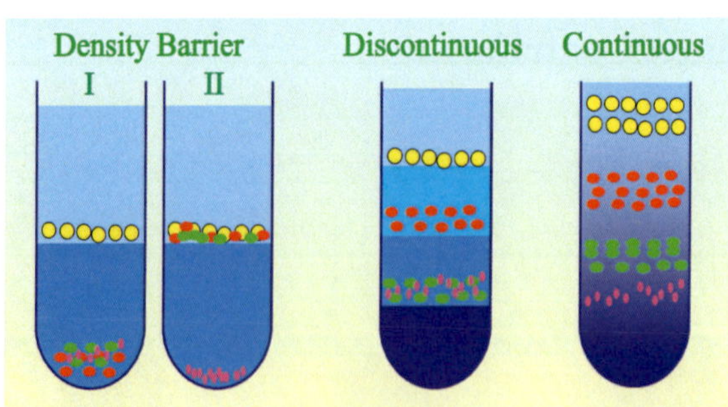

Separation of Particles According to size

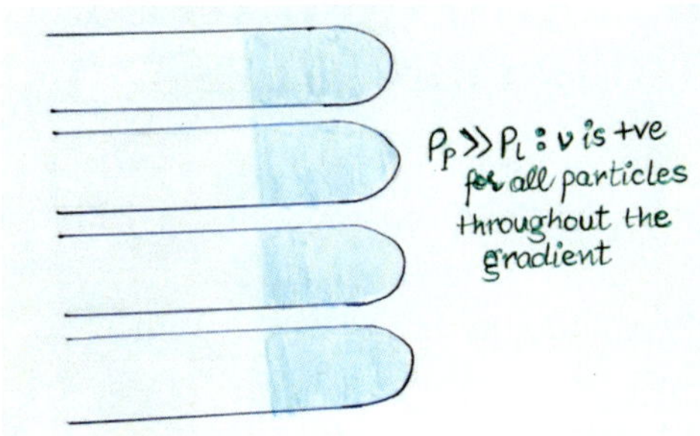

Application of Centrifugation

Centrifugation can be employed to separate a mixture of two different miscible liquids. This "technique can also be used in order to study and analyze macromolecules and their hydrodynamic properties. Mammalian cells can be purified with the help of a special type of centrifuge. Centrifugation is known to have a vital role in the fractionation of many subcellular organelles. Furthermore, centrifugation is also useful in the fractionation of membrane fractions and membranes. Centrifugation also has applications in the fractionation of membrane vesicles. Chalk can be separated from water with the help of a centrifuge. Skimmed milk is a form of milk that has a lower amount of dissolved fats. Skimmed milk can be obtained from regular milk with the help of the process of centrifugation. Here, the centrifuge serves to separate the fat from the milk, leaving the required skimmed milk behind. Cyclonic separation is an important process that has vital applications in the separation of particles from air flows. Another important application of this technique is in the stabilization and clarification of wine. This technique, in combination with other purification techniques, is extremely helpful while separating proteins. Other techniques that are used include salting out techniques such as ammonium sulfate precipitation. Centrifuges are widely used in the field of forensic chemistry. In this field, the technique is employed for the separation of blood components from blood samples. Furthermore, the technique is also employed in certain laboratories for the separation of urine components from urine samples". Differential centrifugation, a distinct type of centrifugation, is known to have applications in the identification of organelles.

12. CHROMATOGRAPHY: ION EXCHANGE, PARTITION, GEL FILTRATION AND AFFINITY CHROMATOGRAPHY THEIR PRINCIPLES AND APPLICATIONS

Mikhail TSWETT invented chromatography in 1901 during his research on plant pigments. He used the technique to separate various plant pigments such as chlorophylls, xanthophylls and carotenoids.

Original chromatography experiment

- **Start:** A glass column is filled with powdered limestone ($CaCO_3$).
- **End:** A series of colored bands was seen to form, corresponding to the different pigments in the original plant extract. These bands were later determined to be chlorophylls, xanthophylls and carotenoids.

Chromatography

Tswett named this new technique chromatography (Greek = chroma "color" and graphein "writing") based on the fact that it separated the components of a solution by color.

Types Of Chromatography

Tswett's technique is based on liquid chromatography. There are now several common chromatographic methods. These include

- Paper Chromatography
- Thin Layer Chromatography (TLC)
- Liquid Chromatography (LC)
- High Pressure Liquid Chromatography (HPLC)
- Ion Chromatography
- Gas Chromatography (GC)

Paper And Thin Layer Chromatography

The solvent moves up paper by capillary action, carrying mixture components at different rates.

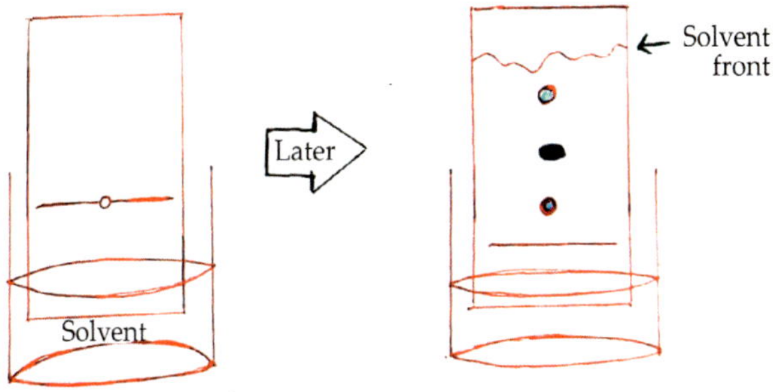

In all "chromatographic separations and the sample is transported in a mobile phase. The mobile phase can be a gas, a liquid, or a supercritical fluid. The mobile phase is then forced through a stationary phase held in a column or on a solid surface. The stationary phase needs to be something that does not react with the mobile phase or the sample. The sample has the opportunity to interact with the stationary phase as it moves past it. Samples that interact greatly then appear to move more slowly". Samples that interact weakly then appear to move more quickly. Because of this difference in rates, the samples can then be separated into their components.

Chromatography

It is based on a "physical equilibrium that results when a solute is transferred between the mobile and a stationary phase. In a chromatography column, flowing gas or liquid continuously replaces saturated mobile phase and results in the movement of A through the column. As material travels through the column, it assumes a Gaussian concentration profile as it distributes between the stationary packing phase and the flowing mobile gas or liquid carrier phase".

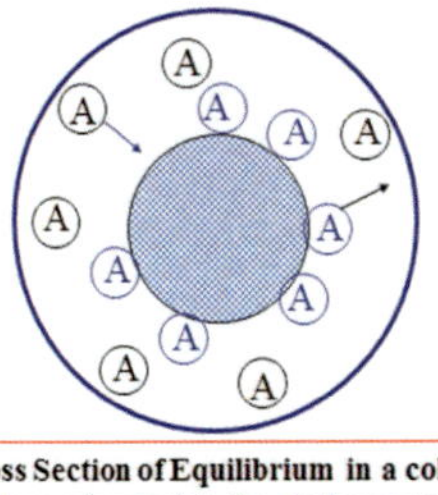

Cross Section of Equilibrium in a column. "A" are adsorbed to the stationary phase. "A" are traveling in the mobile phase.

K = distribution coefficient *or* partition ratio

$$K = \frac{C_S}{C_M}$$

Where C_S is the molar concentration of the solute in the stationary phase and C_M is the molar concentration in the mobile phase.

In a mixture, "each component has a different distribution coefficient, and thus spends a different amount of time absorbed on the solid packing phase vs. being carried along with the flowing gas. More volatile materials are carried through the column more rapidly than less volatile materials, which results in the separation".

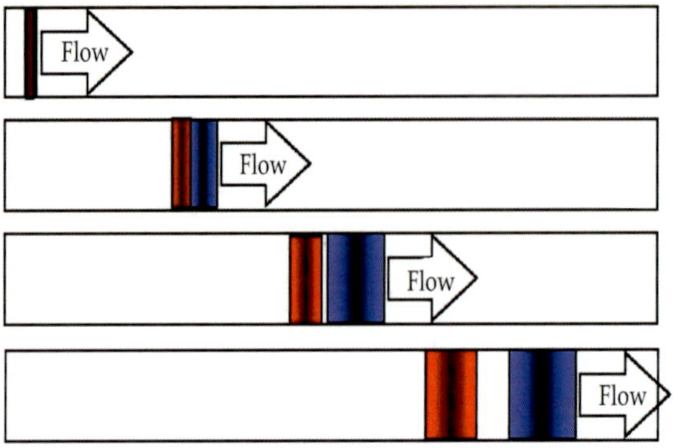

If a detector is used to determine when the components elute from the column, a series of Gaussian peaks are obtained, one for each component in the mixture that was separated by the column.

Note: The first two components were not completely separated.

Peaks in general tend to become shorter and wider with time.

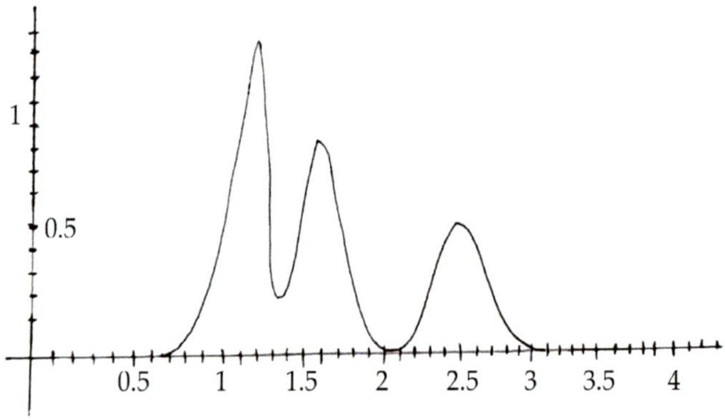

Theoretical plate is a term coined by Martin & Synge. It is based on a study in which they imagined that chromatographic columns were analogous to distillation columns and made up or numerous discrete but connected narrow layers or plates. Movement of the solute down the column then could be treated as a stepwise transfer. Theoretical plates (N) measure how efficiently a column can separate a mixture into its components. This efficiency is based on the retention time of the components and the width of the peaks.

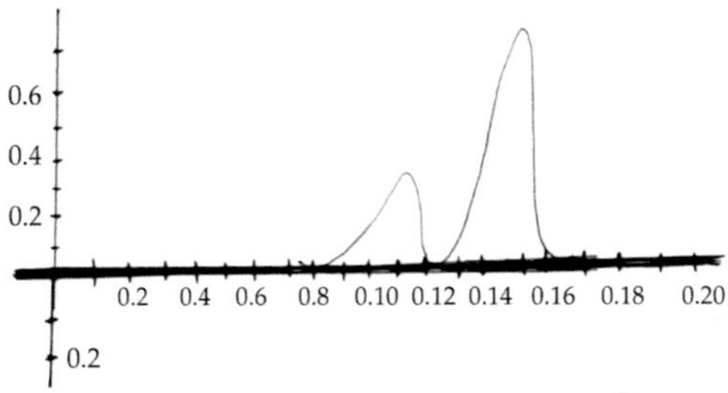

$$N = 16\left(\frac{t_R}{W_b}\right)^2$$

N = Number Of Theoretical Plates (a measure of efficiency) t_R is the retention time and it is measured from the injection peak (or zero) to the intersection of the tangents. W_b is the width of the base of the triangle; it is measured at the intersection of the tangents with the baseline. When the retention time, t_R, is held constant, the column that produces peaks with narrower bases, w_b, will be more efficient have a greater N value. Likewise, a column that produces wider peaks will be less efficient and have a smaller N value. This is because a smaller denominator, w_b, will yield a larger overall number and a larger denominator will yield a smaller number.

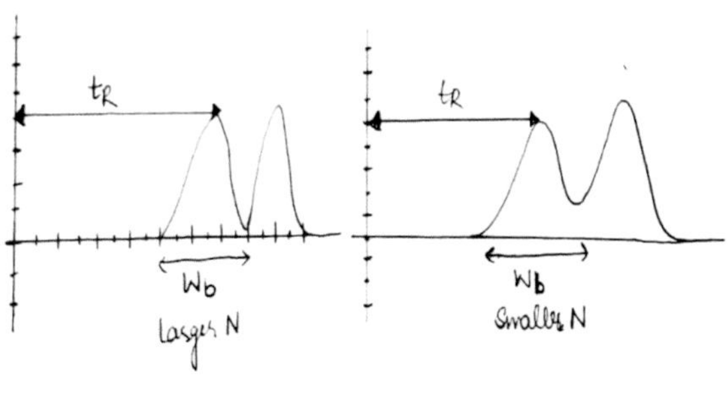

Gas Chromatography

It is good for volatile samples (up to about 250 °C). It require 0.1-1.0 microliter of liquid or 1-10 ml vapor. It can detect <1 ppm with certain detectors. It can be easily automated for injection and data analysis.

Components Of A Gas Chromatography

- **Gas supply:** (usually N_2 or He).
- **Sample Injector:** (syringe / septum).
- **Column:** 1/8" or 1/4" x 6-50' tubing packed with small uniform size, inert support coated with a thin film of nonvolatile liquid
- **Detector:** TC - thermal conductivity and FID - flame ionization detector

13. ELECTROPHORESIS: PRINCIPLE, TYPES (POLYACRYLAMIDE AND AGAROSE GEL ELECTROPHORESIS), APPLICATIONS, ISOELECTRIC FOCUSING

Electrophoresis

Electrophoresis is a method whereby "charged molecules in solution, chiefly proteins and nucleic acids, migrate in response to an electrical field. Their rate of migration through the electrical field depends on the strength of the field, on the net charge, size, and shape of the molecules, and also on the ionic strength, viscosity, and temperature of the medium in which the molecules are moving. As an analytical tool, electrophoresis is simple, rapid and highly sensitive. It can be used analytically to study the properties of a single charged species or mixtures of molecules. It can also be used preparatively as a separating technique. Electrophoresis is usually done with gels formed in tubes, slabs, or on a flat bed. In many electrophoresis units, the gel is mounted between two buffer chambers containing separate electrodes, so that the only electrical connection between the two chambers is through the gel". In most electrophoresis units, the gel is mounted between two buffer chambers containing separate electrodes so that the only electrical connection between the two chambers is through the gel.

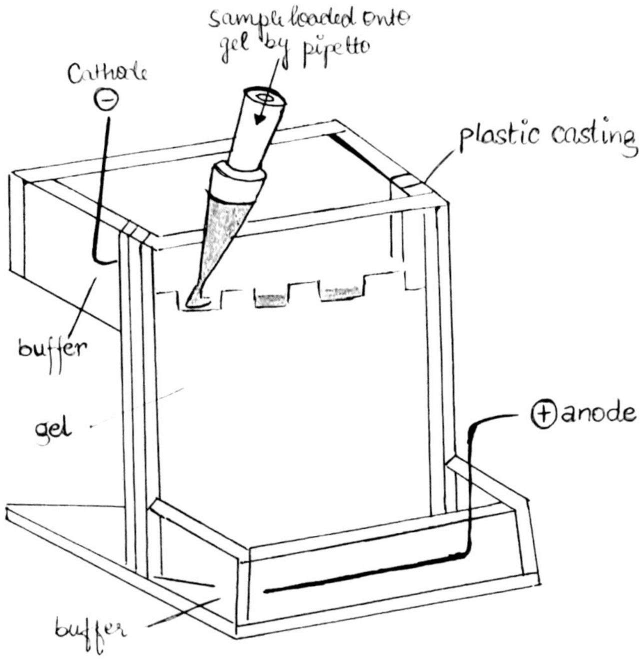

Interrelation of resistance, voltage, current and power and two basic electrical equations are important in electrophoresis.

- The first is Ohm's Law, $I = E/R$
- The second is $P = EI$
- This can also be expressed as $P = I^2R$

In electrophoresis, one electrical parameter, either current, voltage, or power, is always held constant.

Consequences

Under constant current conditions (velocity is directly proportional to current), the velocity of the molecules is maintained, but heat is generated. Under constant voltage conditions, the velocity slows, but no additional heat is generated during the course of the run. Under constant power conditions, the velocity slows but heating is kept constant. The net charge is determined by the pH of the medium. Proteins are amphoteric compounds, that is, they contain both acidic and basic residues. Each protein has its own characteristic charge properties depending on the number and kinds of amino acids carrying amino or carboxyl groups. Nucleic acids, unlike proteins, are not amphoteric. They remain negative at any pH used for electrophoresis.

During Polymerization

- Exothermic reaction
- Gel irregularities
- Pore size

During electrophoresis

- Denaturation of proteins
- Smile effect
- Temperature regulation of buffers
- It inhibits convection and diffusion, which would otherwise impede the separation of molecules. It allows a permanent record of results by staining after a run. It can provide additional separation via molecular sieving.

Agarose And Polyacrylamide

Although "agarose and polyacrylamide differ greatly in their physical and chemical structures, they both make porous gels. A porous gel acts as a sieve by retarding or, in some cases, by completely obstructing the movement of macromolecules while allowing smaller molecules to migrate freely. By preparing a gel with restrictive pore size, the operator can take advantage of molecular size differences among proteins. Because the pores of an agarose gel are large, agarose is used to separate macromolecules such as nucleic acids, large proteins and protein complexes. Polyacrylamide, which makes a small pore gel, is used to separate most proteins and small oligonucleotides. Both are relatively electrically neutral. Agarose is a highly purified uncharged polysaccharide derived from agar. Agarose dissolves when added to a boiling liquid. It remains in a liquid state until the temperature is lowered to about 40° C at which it gels. The pore size may be predetermined by adjusting the concentration of agarose in the gel". Agarose gels are fragile, however they are actually hydrocolloids, and they are held together by the formation of weak hydrogen and hydrophobic bonds.

Polyacrylamide gels are tougher than agarose gels. Acrylamide monomers polymerize into long chains that are covalently linked by a crosslinker. Polyacrylamide is chemically complex, as is the production and use of the gel.

Considerations With PAGE

- Preparing and Pouring Gels
- Determine pore size
  - Adjust the total percentage of acrylamide
  - Vary the amount of crosslinker
- Remove oxygen from the mixture
- Initiate polymerization
  - Chemical method
  - Photochemical method
- Analysis of a gel
  - Staining or autoradiography followed by densitometry.
    - Blotting to a membrane, either by capillarity or by electrophoresis, for nucleic acid hybridization, autoradiography or immunodetection.

SDS Gel Electrophoresis

In SDS gel electrophoresis, "migration is determined not by the intrinsic electrical charge of polypeptides but by molecular weight. Sodium dodecylsulfate (SDS) is an anionic detergent that denatures secondary and non–disulfide–linked tertiary structures by wrapping around the polypeptide backbone. So, SDS confers a net negative charge to the polypeptide in proportion to its length. When treated with SDS and a reducing agent, the polypeptides become rods of negative charges with equal charge densities" or charge per unit length.

Continuous and discontinuous buffer systems

A continuous system has only a "single separating gel and uses the same buffer in the tanks and the gel. In a discontinuous system, a nonrestrictive large pore gel, called a stacking gel, is layered on top of a separating gel. The resolution obtainable in a discontinuous system is much greater than that obtainable in a continuous one. However, the continuous system is a little easier to set up. It can determine of molecular weights of proteins by SDS-PAGE. Run a gel with standard proteins of known molecular weights along with the polypeptide to be characterized. A linear relationship exists between the log_{10} of the molecular weight of a polypeptide and its R_f. Rf = ratio of the distance migrated by the molecule to that migrated by a marker dye-front. The R_f of the polypeptide to be characterized is determined in the same way, and the log_{10} of its molecular weight is read directly from the standard curve".

Blotting

Blotting is used to transfer proteins or nucleic acids from a slab gel to a membrane such as nitrocellulose, nylon, DEAE, or CM paper. The transfer of the sample can be done by capillary or southern blotting for nucleic acids (Southern, 1975) or by electrophoresis for proteins or nucleic acids.

Isoelectric Point

There is a pH at which there is no net charge on a protein and this is the isoelectric point (pI). Above its isoelectric point, a protein has a net negative charge and migrates toward the anode in an electrical field. Below its isoelectric point, the protein is positive and migrates toward the cathode.

Isoelectric Focusing

Isoelectric focusing is a method in which "proteins are separated in a pH gradient according to their isoelectric points. Focusing occurs in two stages; first, the pH gradient is formed. In the second stage, the proteins begin their migrations toward the anode if their net charge is negative or toward the cathode if their net charge is positive. When a protein reaches its isoelectric point (pI) in the pH gradient" it carries a net charge of zero and will stop migrating.

Two-dimensional gel electrophoresis

Two-dimensional gel electrophoresis is widely used to separate complex mixtures of proteins into many more components than is possible in conventional one-dimensional electrophoresis. Each dimension separates proteins according to different properties.

14. MICROSCOPY: PRINCIPLES AND APPLICATIONS OF LIGHT, PHASE CONTRAST, FLUORESCENCE AND SCANNING ELECTRON MICROSCOPY

Microorganisms are very small to be seen with the unaided eye. The study of microorganisms is greatly enhanced by the use of microscopes. The word microscope, first coined by Giovanni in 1625, derived from the Latin word micro (small) and the Greek word scopos (to look at), is an important tool for microbiologists and the technology of making very small things visible to the human eye (Figure 14.1). The simple microscope used by Anton van Leuwenhoek in the 17th century in 1673 had only one lens and was similar to a magnifying glass that could magnify a microbe 300X.

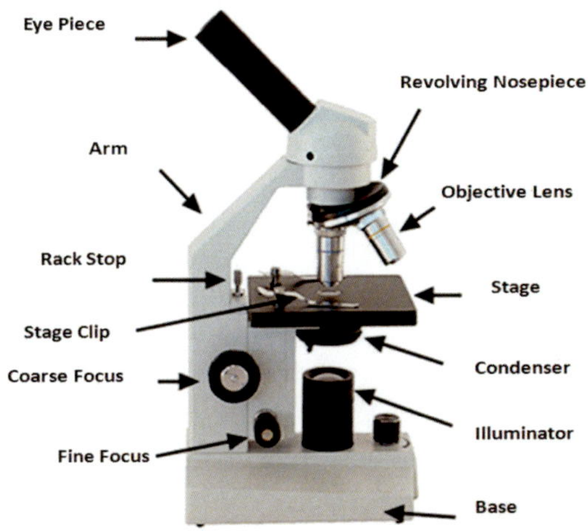

Figure 14.1 Microscope

The simple microscopes enabled us to see bacteria and protozoa. Robert Hooke, a contemporary of van Leeuwenhoek built compound microscopes which have multiple lenses. The microscopes used today are of two categories: light and electron. Light microscopes are sub-divided into bright-field and fluorescent microscopes, while the electron microscope is subdivided into scanning and transmission electron microscopes. The bright-field microscope or compound light microscope that allows light rays to pass directly through to the eye without being deflected by an intervening opaque plate in the condenser is routinely used in all the microbiology laboratory to study the size, shape, arrangement of microbial cells and sporulating structures. Dark-field microscopy is a kind of light microscopy that uses a special dark field condenser that contains an opaque disk that causes the light to reflect off the specimen at an angle rather than pass directly through it. This microscopy is used to examine live microorganisms. Fluorescent microscopy uses ultraviolet light instead of white light to excite molecules within the specimen or fluorescent dye. Fluorescence microscopy must be intrinsically fluorescent or stained with dye such as auramine O, acridine orange, fluorescein and DAPI (diamidino-2-phenylinodole).

Bright-Field Light Microscopy

A microscope is a compound optical system, a compound magnifying glass. The real meaning of the functioning of a microscope is a test object positioned between the single and double focus points of the objective. The light coming from the object and passing through the lens creates a magnified, inverted and real image of the subject on the other side of the objective, behind the double focus point. The eyepiece is at a distance from the objective lens so that the image formed by the objective is generated within the focus of the eyepiece. Thus, looking throughout the eyepieces, one can see a further enlarged, direct but virtual image of this real, inverted and magnified image.

The magnification power of a bright-field microscope can be calculated by multiplying the magnification of the objective and of the eyepiece, respectively. The objective lens system consists of multiple lenses. The first is the front lens facing the object. This determines the magnification and resolution of the microscope. Other elements are responsible for the exclusion of lens errors. The features of an image formed by the objective depending on the optical characteristics of the objective lens. The quality of a lens represents its property of how sharp the object image is drawn. The image defect, i.e. blurring is caused by lens errors (aberrations). The spherical aberration (spherical divergence) is caused when the lens away from the optical axis increasingly breaks the light rays passing through it, therefore a point-like object in the image will be blurred at the edge. The reason for chromatic aberration is that the focal points of different wavelengths of light do not coincide on the optical axis.

- The shorter wavelength rays unite closer, while longer wavelength rays unite farther.
- Thus, a sharp image cannot be obtained using white light, and the image has a colorcolored (rainbow) border.

- A concave flint glass lens with lead content and high refractive index is fitted to the convex lens to correct for chromatic aberration (the achromatic lens has two-color correction, while the apochromatic lens has three-color correction).
- The resolution of the objective lens is the ability of how detailed the image of a subject can be drawn.
- The resolving power is quantified with the minimum distance between two points that are just distinguishable.
- The resolution (d) depends on the illumination wavelength of light used (λ), the half-angular aperture of the objective lens (α) and the refractive index of the material between the front lens and the cover slip (n).

$d = 1.22\ \lambda\ /2\ n\ x\ \sin\alpha$

where $n\ x\ \sin\alpha$ = numerical aperture (NA).

- The greater the resolution the smaller the values of d are.
- This can be achieved by reducing the wavelength of light used, increasing the angular aperture of the objective, or increasing the refractive index of the material between the front lens and the cover slip.

For this purpose, cedar oil (oil immersion) is the most suitable material, because it has almost the same refractive index as that of the glass, so the light passes through a virtually homogeneous medium. The denominator of the above formula and $2\ nX\ \sin\alpha$ is the value of numerical aperture (NA), which may vary between 0.20 and 1.4 (always indicated on the lens). The eyepiece draws a direct image of the test object. The fine structure of an image observed in the microscope depends on the details of the real picture, which in turn is determined by the resolution of the objective. This image, however, is not visible to the naked eye; it can only be visualized in the magnification of the eyepiece.

Microorganisms examined from natural waters by bright-field light microscopy.

Principle

A microscope is a series of finely ground lenses and visible light as a source of illumination are used. The microscope field or area observed is brightly lighted and the objects being studied appear dark because they absorb some of the light. However, microbes absorb light after staining them with a dye greatly increases their light-absorbing ability resulting in greater contrast and color differentiation.

Requirements

- Environmental water sample (e.g. from a lake, a stream, a creek or an aquarium)
- Glass slides
- Cover slips
- Glass dropper dispenser alcohol (for sterilization)
- Bunsen burner
- Light microscope

Procedure

- Dry the surface of a glass slide with alcohol over a Bunsen burner and then label the slide.
- Place one drop of the sample to the slide with a glass dropper.
- Put the edge of a cover slip on the slide so that it touches the edge of the water drop.
- Slowly lower the cover slip to prevent the formation and trapping of air bubbles.
- Put the sample under the microscope, and locate the focal plane.
- Apply first an objective lens of 16X, and then 40X magnification.
- Observe the shape and movement of microbes, in the case of protists and eukaryotic algae, try to identify the different cell organelles (e.g. chloroplasts and contractile vacuoles)
- Make drawings of the observations.

Fluorescence Microscopy

Specimens that absorb light of color and subsequently emit light of another color (fluoresce) can be visualized by using the fluorescence microscope. The basis of fluorescence microscopy is the principle of the removal of incident illumination by selective absorption, where light absorbed by the specimen and reemitted at another wavelength is transmitted. The light source must produce light of appropriate wavelength and ineffective wavelengths that are unable to excite the fluorochrome used are removed by an excitation filter. The phenomenon when a sample (e.g. cell organelle of a microbe) shows fluorescence without prior staining is called auto-fluorescence. Otherwise, bacterial cells can be stained with a fluorescent dye. Using a fluorescent dye, e.g. DAPI (4',6-diamidino-2-phenylindole), cells are stained bright blue because the dye forms a complex with the cell's DNA. Fluorescent dyes are therefore widely applied to visualize and enumerate bacteria in different natural habitats or clinical samples. Microbes are examined with DAPI staining (microbes from water sample) DAPI molecules are able to penetrate cell membranes and bind to the double helix of the DNA.

- Cells can be easily counted, if a known volume of a fixed water sample is filtered through a membrane and, after staining the filter, the surface is investigated by a fluorescence microscope (Figure 14.2).
- A microscope is a series of finely ground lenses and visible light as a source of illumination are used.
- The microscope field or area observed is brightly lighted and the objects being studied appear dark because they absorb some of the light.
- However, microbes do not absorb much light but after staining them with a dye greatly increases their light-absorbing ability resulting in greater contrast and color differentiation.
- The effective magnification of a microscope depends upon its resolving power.

Typical magnification values are:

Scanning 10X × 4X= 40X magnification

Low power 10X × 10X = 100X magnification

High power 10X × 40X = 400X magnification

Oil immersion 10X ×100X = 1000X magnification

Total magnification = Magnifying power of the objective lens* magnifying power of the ocular lens.

Requirements

- Disposable gloves laboratory scales paraformaldehyde (PFA)
- Beaker
- Magnetic stirrer

- Phosphate buffered saline (PBS)
- NaOH solution
- Pasteur pipette
- Membrane filtration apparatus
- Polycarbonate and cellulose nitrate membrane filters (0.22 or 0.45 μm pore size)
- 50 ml Falcon tube
- Plastic Petri dishes
- scalpel
- Pipette with pipette tips
- DAPI solution (1μg/ml)
- 80% ethanol
- Double distilled water
- glass slide
- Cover slip
- Vectashield Mounting Medium (H-1000, Vector Laboratories Ltd) immersion oil (non-fluorescent)
- Fluorescence microscope with a mercury lamp digital camera computer with software

Procedure

- For the preparation of the fixative solution, dissolve 1 gm PFA in 50 ml PBS. (PFA causes irritation when inhaled; therefore the use of a fume hood is recommended.)
- Dissolution can be aided with heating (60°C), permanent stirring and adding some drops of NaOH solution.
- Adjust the pH at 7.0.
- Filter the solution through a 0.22 μm pore size membrane filter. (The prepared 2% PFA solution can be stored in the fridge for one week).
- Filter the water sample (2-50 ml, depending on the type of sample) using a polycarbonate membrane filter (slowly with occasional stirring).

- To help uniform cell distribution, place a 0.45 μm pore size cellulose nitrate membrane filter between the sieve of the filtration unit and the polycarbonate filter.
- Fill up the Falcon tube with fixative (PFA solution) and immerse the filter in it with sterile forceps (PFA solution must cover the entire membrane filter).
- Incubate the filter overnight at 4°C.
- Fill PBS into an empty Petri dish, then transfer the filter into the PBS solution for 1-2 minutes (liquid must cover the entire membrane filter).
- Transfer the filter to another empty Petri dish and let it dry.
- Cut a 0.5 by 0.5 cm piece from the filter with a scalpel or scissors, and pipette 30 μl DAPI solutions onto its surface. From this step onwards, work in a dark place.
- The filter piece can be marked with a soft pencil.
- After 2 minutes, transfer the filter into 80% ethanol for a few seconds.
- Dip the filter paper into double distilled water for a few seconds.
- Dry the filter.
- Place the filter onto the surface of a glass slide, put a drop of Vectashield Mounting Medium onto the filter and then cover with a cover slip.
- Cover with a paper towel and gently press the cover slip to remove any excess mounting medium.
- Examine the slide with an epifluorescence microscope using a 100X objective and immersion oil under UV excitation (the absorption maximum of DAPI is at 358 nm, and the emission maximum is at 461 nm).
- Record images from at least 20 different microscopic fields with a digital camera.
- Count the cells on each picture and determine the mean values.
- Determine the cell count for one ml of the water sample based on the amount of filtered water and the size of the membrane filter field.

Phase-contrast light microscope

Most of the microorganisms are colorless and so are their organelles, which often are too difficult see by dark field microscopy and common bright-field microscopy. The observations are clear by phase-contrast microscopes such as structure of endospores, unstained bacteria and living organisms. The species are often arranged in pairs or chains, rod-shaped and rounded or square ends known as *Bacillus* species and their endospores are mostly oval shape or sometimes cylindrical. *Clostridium* species are repeatedly set in short chains or pairs, mostly with rounded or sometimes pointed ends. The endospores regularly swell up the cell. Consequently, by means of ready slides of *Clostridium* and *Bacillus* the students gain knowledge of a phase-contrast microscope and observe precise bacterial structures such as diverse endospores are present in bacteria. Pond water is generally crowded with protists and bacteria, the internal structure can be seen by using a phase-contrast microscope. Apart from this we can also see the internal structure of paramecium (protist) and slowing down the different microorganisms present in pond water.

Principle

- The colorless and transparent living microorganisms having organelles are generally impossible to see by bright-field microscope due to their absorption, refract and reflect limits.
- Another reason is sufficient amount of light diffract with the nearby environment for the rest of the microbes.
- The microbes are visible when they absorb much more light or to refract and reflect from the environment.
- Thoroughly, we can see microbial parts by this process.
- The observations of living and invisible unstained microbes are by using only phase-contrast microscope.
- The condenser has an annular diaphragm and hollow cone which produces light, glass disk (phase plat) has an objective with a thin film transparent material deposited on it for phase changes produced in the specimen.
- The phase change of images is observed due to the difference of intensity in light.
- The positive phase plate is diffracted due to undiffracted light in dark phase-contrast light microscopy or negative phase plate

same as above given light relative to the directed light producing by phase-contrast light microscopy.

Requirements

- Pond waste water
- Phase-contrast light microscope
- New slides and coverslips
- Pipette
- Microorganisms
- Methyl cellulose "Protoslo, Carolina Biological Supply"
- Tweezers
- Lens paper
- Lens cleaner
- Prepared slides of "*Bacillus* or *Clostridium*"

Procedure

- Collect waste pond water.
- Add one drop of methyl cellulose to slow down the swimming of the microorganisms.
- Prepare the slides of selected microbes given above in the introduction part.
- The slide is placed on the stage of the phase-contrast microscope so that specimen is under the light hole.
- The lenses are set into place and rotate the position (annular diaphragm) of the specimen if it is required.
- It is necessary that the cone of light produced by the annular diaphragm below the condenser be centered.
- Center with the 10X objective and observe the specimen that is under over light.
- Rotate the annular diaphragm into the correct position for the observation with a 40X objective if the image isn't clear.
- Do the same with the oil immersion lens.
- In the report for exercise 4, sketch several of the microorganisms that you have observed.

- If you examine pond water (collection of microbes), use the pictorial guides provided by your teacher to help you in identifying microorganisms present in pond water.

Scanning electron microscope

A scanning electron microscope (SEM) is a very specific type of microscope; it uses scanned electron beams to produce an image of a specimen. A complete labeled picture of the SEM instrument is shown in **figure 14.2**. The scanned electrons when falling on a specimen interact with the atoms of the specimen and starts producing various signals that are captured by the SEM software to produce an image. SEM is used to know the structure of specimens with their composition as well as topography. By using SEM we can achieve a much better resolution (better than 1 nm). Mostly the image formation and detection of the specimen are done with the help of a secondary electron beam of SEM. Different states of the sample can be imaged like in their dry, wet, hot or cool states.

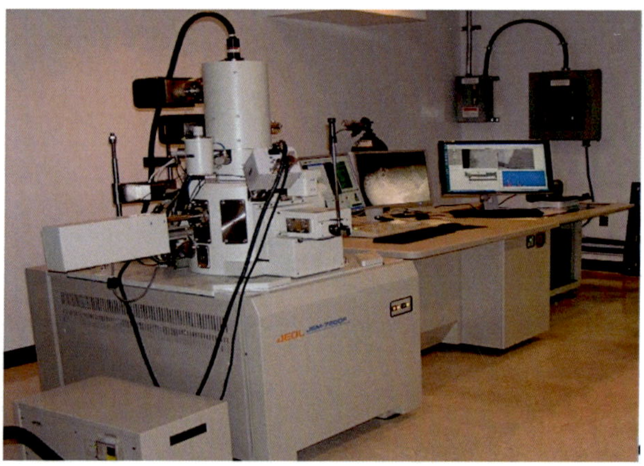

Figure 14.2 A complete labeled picture of SEM instrument.

In 1935, a well-known scientist named M. Knoll first described the concept of a scanning electron microscope, he was working in Germany. Another scientist named **M**. Von Ardenne constructed a scanning transmission electron microscope (STEM) by adding scan coils to a transmission electron microscope (TEM) after 3 years in 1938. ZnO crystals were imaged for the first time by using STEM, the operating voltage was 23 kV and magnification 8000 times with spatial resolution between 50-100 nm. But this instrument had no system to store the image. The first SEM was used to examine the surface of a solid specimen, which was described by Zworykin in 1942 after STEM. A scientist of Cambridge University, England named C. W. Oatley start reinvestigating the SEM with Dennis McMullan and they built their first SEM in 1948. In 1952, SEM had achieved various advancements like a resolution of 50 nm and three-dimensional imaging capacity. +A research student named O.C. Wells in 1953 under Oatley's supervision started building a second SEM, this instrument had the gun at the bottom of the column, a configuration considered better for experimental work. In 1955, Oatley's another research student working on SEM and he improved the secondary electron (SE) detector by using a scintillator to convert electrons to photons, which were then transmitted by a light pipe directly interfaced to the photomultiplier tube.

Principle

- The scanning electron microscope is a kind of specific microscope (**Figure 14.3**).
- In this microscope, the image is produced by using scanned electron beams of the sample.
- The image is produced on both top-down sides.
- There are different ways by which the primary electron beam interacts with the sample.
- The sample of primary electrons generates low energy secondary electrons that fall on the sample and produce an image.
- The primary electrons can also produce images.
- The X-rays produced by relaxed ionized atoms also help in producing the image of the sample.

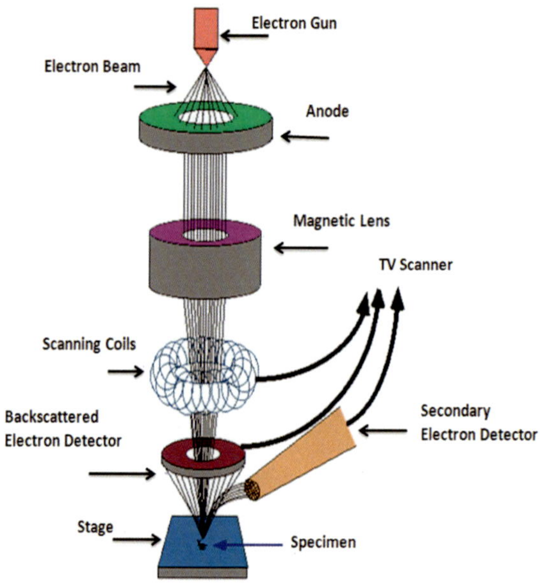

Figure 14.3 A complete schematic diagram of the working principle of SEM

How to work

- Sample Preparation:
- The sample needs to be electrically connected to the sample holder to prevent the electron beam from "charging" the sample and distorting the image: called sample grounding.
- The samples for SEM are sputtered with gold and platinum coating: called sample coating.
- After sample preparation, the SEM software is opened.
- The sample is placed on an SEM sample mount.
- When the sample is properly placed on the SEM stage, close the door of the chamber gently.
- Now click OK in the box on the screen asking "Press OK to pump."
- Click on the "Vacuum" tab to view the vacuum level in the chamber and the gun.
- The voltage and resolution are saved according to the nature of the sample.

- Activate the SCM (sample current monitor) window to check whether it is on or off.
- Remove the sample and collect your results from the system

Usage

- There are multiple usages of SEM, some are mentioned below:

Life sciences

Biomedical engineering: For designing a variety of technologies such as support matrices for cell growth, artificial tissue, and implantable biomedical device. Cell and tissue morphology: For investigating cellular and tissue structure with high resolution (**Figure 14.4**). Typical applications involve observing shape changes of grooves, pores, blebs or microvilli on the cellular in response to the changes in the extracellular environment. Microbiology: High-resolution imaging of the microbial surface using SEM helps to have a better understanding of the morphology of microbial populations, bacteria communication, and biofilm formation. It helps researchers visualize microbial populations with great focal depth and high resolution. Pharmaceuticals: SEM offers very high-resolution imaging and superb analytical performance, which make them perfect and reliable tools for quality control of new preparations.

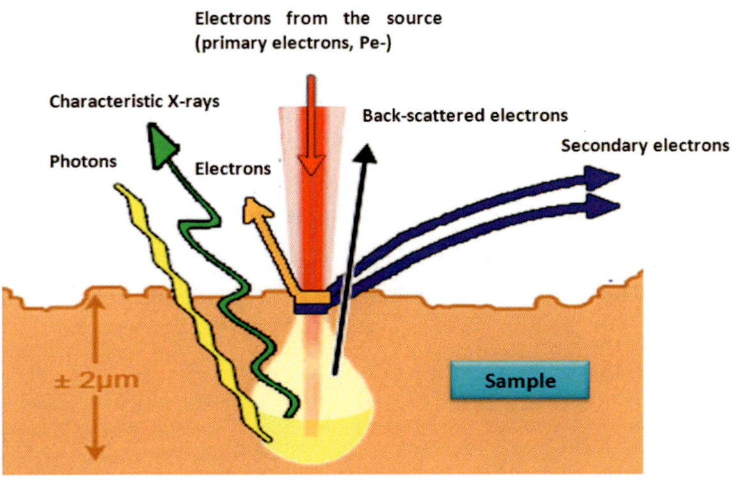

Figure 14.4 An image of electrons scatter during sample inhaled.

Earth sciences

Magnetic and metamorphic petrology and mineralogy: Petrology and mineralogy benefit from the electron microscopy technique because of several reasons. The high magnification factor, as well as the phase-contrast phase-contrast provided by the back, scattered electron imaging play a crucial role in these disciplines. The analytical capabilities of the technique make SEM an ultimate solution in this field. Ore Processing: Scanning electron microscopy helps to characterize the mineral of interest as well as the surrounding gangue. Such a technique is capable of both phase identification and evaluation of textural relationships of individual minerals. Oil and Gas: Scanning electron microscopy is an integral part of reservoir rock characterization workflow. The elementary imaging function of SEM is used when searching for basic rock properties such as porosity or grain shape. Paleontology: Scanning electron microscopy is a key tool in microfossil characterization helping to distinguish features that are otherwise irresolvable by optical microscopy. Sedimentary Petrology: Electron microscopy is helpful in tackling a wide range of petrological problems. Rock characterization mostly relies on optical microscopy, however, there is a growing demand to characterize the rocks with greater detail nowadays. Recycling: Automated mineralogy is currently finding its way to other areas than just ore processing. This technique can be applied e.g. in the field of electronic scrap recycling where it is used to characterize the composition and structure of comminuted e-scrap (www.scribd.com., www.slideshare.com., www.coursehero.com., www. biocyclopedia.com., www. molecular-plant-biotechnology.info., www.slideplayer.com., www. documents.mx., www.docplayer.net., www.bmb.psu.edu.,www.archive. org.,www.mafiadoc.com., www.powershow.com., www.bioscience.jbpub. com., www.outhorstream.com., www.docslide.com., www.msluay.com., www.scienceprofonlue.com., www.csus.edu., www.krishikosh.egrouth. ac.in., www.tankonyvtor.hu., www.silo.pub., www.studyblue.com., www. Preservearticles.com., www.biotechuniverse.blogspot.com).

PART III

CELL BIOLOGY AND GENETICS

15. CELL THEORY, CELL AS BASIC UNIT OF LIFE.

The cell theory states that all organisms are composed of similar units of organization called cells. There are **3** major parts of the cell theory: All organisms are made of cells. All existing cells are produced by other living cells. The cell is the most basic unit of life.

- There are **5** contributors to the cell theory
- Robert Hooke
- Anton van Leeuwenhoek
- Matthias Schleiden
- Theodor Schwann
- Rudolf Virchow

Each of the above contributors produced evidence through investigations and experiments that led to today's cell theory. Before the cell theory, people believed in spontaneous generation, which means living things come from non-living things. Two scientists disproved spontaneous generation. Francisco Redi proved that microorganisms do not spontaneously generate. People thought that maggots came from meat. Redi disproved this in a controlled experiment.

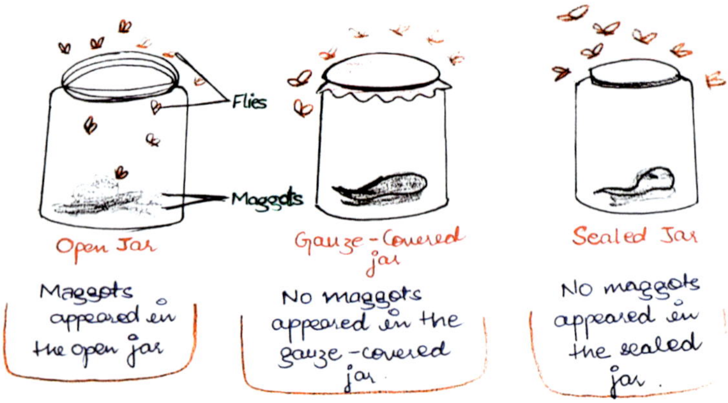

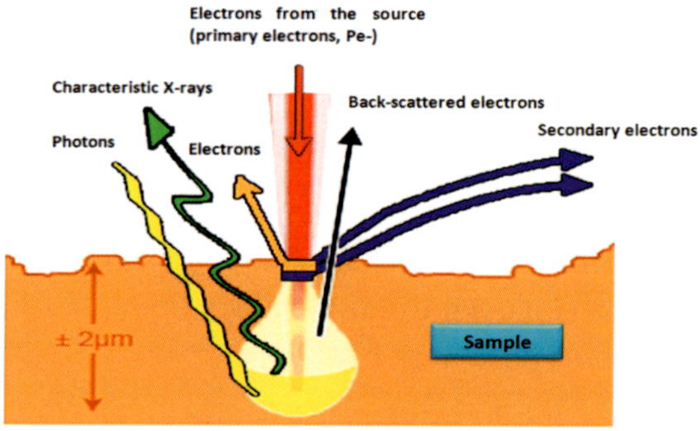

Pasture proved that microorganisms do not spontaneously generate.

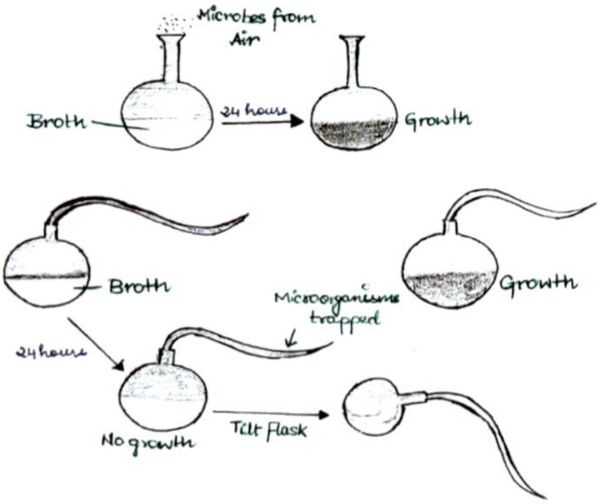

Theodor Schwann and Matthias Schleiden are credited with coming up with the cell theory.

Schwann

- Discovered animal cells.

Schleiden

- Discovered plant cells.
- The cell theory would not exist if it wasn't for the development of the microscope and the work of others.

Janssen

- Invented the first compound microscope.

Robert Hooke

- In 1663 an English scientist, **Robert Hook** improved **Janssen's** invention.
- Used the compound microscope to observe cork.
- Hooke observed that cork is composed of small, hollow compartments.
- The parts prompted Hooke to think of small rooms (cells) in a monastery, so he gave them the same name:
- **Cells**: Investigated cork through experimenting with the compound microscope and came up with the name cells.

Key Note: Hooke discovered the cell in 1665, which started formulating the cell theory.

Anton van Leeuwenhoek

Dutch Tradesman

- Study new approaches for creating lenses to observe cloth.
- Leeuwenhoek's microscope was more powerful than Hooke's compound microscope.
- From investigating and experimenting with his microscope, **Leeuwenhoek** became one of the first scientists to refer to living cells when he observed an abundant number of single-celled organisms, which he called animalcules (plant & animal), swimming in a drop of pond water.

Leeuwenhoek's Investigations/Experiment

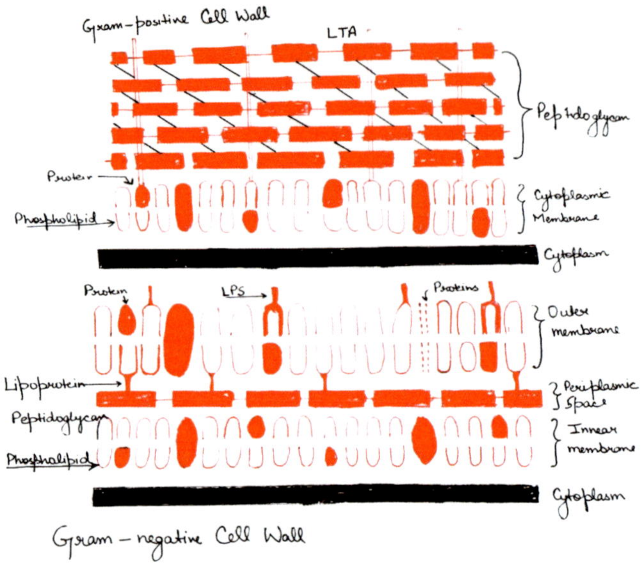

Matthias Schleiden (German Scientist)

- Fascinated with plant cells, **Schleiden** used the compound microscope and studied plant cells.
- From investigating and experimenting with plants, projected plant parts were made of cells.
- Discussed what he observed with his dear friend, German scientist **Theodor Schwann**.

Theodor Schwann

German Scientist

- Studied plant & animal cells, and was intrigued by the similarities between the two.
- From investigating and experimenting with plant & animal cells, Schwann was able to determine that all animals are made of cells.
- Schwann published the 1st statement of the cell theory: all living things are made of cells and cell products.

Rudolf Virchow

- Based on his investigations and experiments, he stated that all cells come from preexisting cells, which is the 2nd part of the cell theory: all existing cells are produced by other living cells.

Key Note: After Virchow's completed his work, the cell theory was finally formulated in 1838.

Two Basic Cell Types

- Prokaryotic Cells Mainly single-celled organisms
- Eukaryotic Cells Mainly multicellular organisms, although some are single-celled (amoeba, algae, yeast).

Prokaryotic Cell

- Prokaryotes have no membrane-bound organelles.

Example – Bacteria cell

- Eukaryotic Cells
- Eukaryotes – contain membrane-bound organelles

Example – Plant and animal cells

16. STRUCTURE AND ORGANIZATION OF PROKARYOTIC AND EUKARYOTIC CELLS. CELL ORGANELLES-STRUCTURE, FUNCTION AND INTEGRATION, MICROBODIES

Prokaryotic cells are single-celled entities that are primitive in structure and function as they lack a membrane-bound nucleus and other organelles. The term "prokaryote" is derived from two Greek words, 'pro' meaning 'before' and 'karyon' meaning 'nucleus'. Prokaryotes are considered to be the first living organisms of the earth as they are the simplest form of life.

- Single-celled.
- Reproduce by binary fission (other copy by dividing).
- No cell nucleus or any other membrane-bound organelles. DNA travels openly around the cell.
- All bacteria are prokaryotes.
- Eu = "true"; karyon = "nucleus"

Most organisms that we can see, such as trees, grass, worms, flies, mice, humans, mushrooms and yeast are eukaryotes. It can either be single-celled or multi-celled. It can reproduce in one of several ways (Ex. meiosis, mitosis). It has a cell nucleus containing its DNA. The nucleus is the most evident distinction between these cell types.

All cells share four Common Components

- A plasma membrane: an outer covering that separates the cell's interior from its surrounding environment.
- Cytoplasm: a jelly-like cytosol within the cell in which other cellular components are found
- DNA: the genetic material of the cell.
- Ribosomes: where protein synthesis occurs.

However, "prokaryotes differ from eukaryotic cells in several ways. A prokaryote is a simple, single-celled (unicellular) organism that lacks an organized nucleus or any other membrane-bound organelle. We will shortly come to see that this is significantly different in eukaryotes. Prokaryotic DNA is found in a central part of the cell: the nucleoid. Most prokaryotes have a peptidoglycan cell wall and many have a polysaccharide capsule. The cell wall acts as an extra layer of protection, helps the cell maintain its shape, and prevents dehydration. The capsule enables the cell to attach to surfaces in its environment. Some prokaryotes have flagella, pili, or fimbriae. Flagella are used for locomotion. Pili are used to exchange genetic material during a type of reproduction called conjugation". Fimbriae are used by bacteria to attach to a host cell.

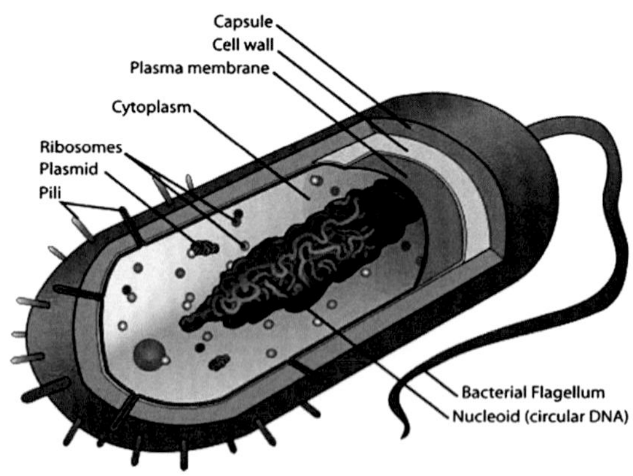

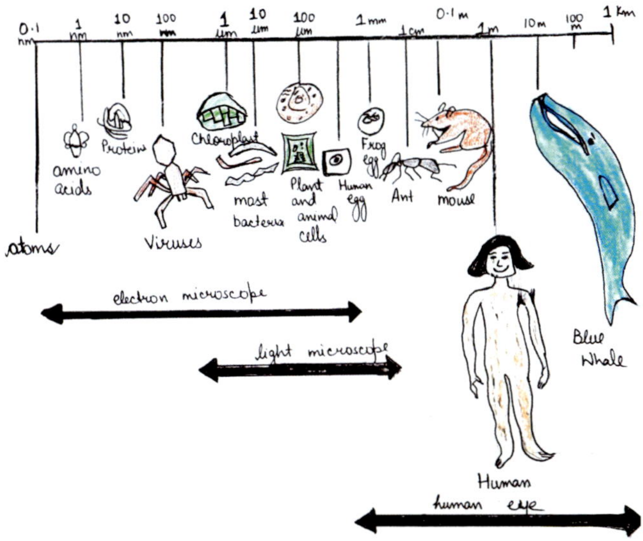

Prokaryote Genetics

Nucleoid

The region of cytoplasm where prokaryote's genome (DNA) is located and it is usually a singular, circular chromosome.

Plasmid

The small extra piece of chromosome/genetic material is called a plasmid. It has 5 - 100 genes. It can provide genetic information to promote:

- Antibiotic resistance
- Virulence factors

The molecules produced by pathogens specifically influence the host's function to allow the pathogen to thrive. It promotes conjugation (transfer of genetic material between bacteria through cell-to-cell contact)

Cytoplasm

It is also known as protoplasm. Gel-like matrix of water, enzymes, nutrients, wastes, and gases and contains cell structures. The site of growth, metabolism, and replication.

Granules

Bacteria's way of storing nutrients and staining of some granules aids in identification.

Cytoskeleton

The cellular "scaffolding" or "skeleton" is present within the cytoplasm. The major advance in prokaryotic cell biology in the last decade has been the discovery of the prokaryotic cytoskeleton.

Ribosomes

It is found within the cytoplasm or attached to the plasma membrane. It is made of protein & rRNA. It is composed of two subunits. The cell may contain thousands of ribosomes.

Prokaryotic plasma membrane

It separates the cell from its environment. Phospholipid molecules are oriented so that hydrophilic water-loving heads are directed outward and hydrophobic water-hating tails are directed inward. Proteins are embedded in two layers of lipids (lipid bilayer). The membrane is semi-permeable.

Prokaryotes: plasma membrane as a barrier

Osmosis

The diffusion of water across a semi-permeable membrane is called osmosis. The environment surrounding cells may contain amounts of dissolved substances (solutes).

Tonicity and Osmosis

Isotonic: equal concentration of a solute inside and outside of the cell.

Hypertonic: a higher concentration of solute.

Hypotonic: a lower concentration of solute.

Water will always move towards a hypertonic environment.

Plasma membrane as a barrier

Active transport

The molecules move across the plasma membrane. It is a pump moving water uphill. Types of active transport are classified by the type of energy used to drive molecules across membranes. ATP-driven active transport. Energy from adenosine triphosphate (ATP) drives substances across the plasma membrane with the aid of carrier molecules.

Prokaryotes: Cell wall

All bacteria are very similar inside the peptidoglycan wall, the bacterial world is divided into two major classes gram-positive and gram-negative.

Gram-positive

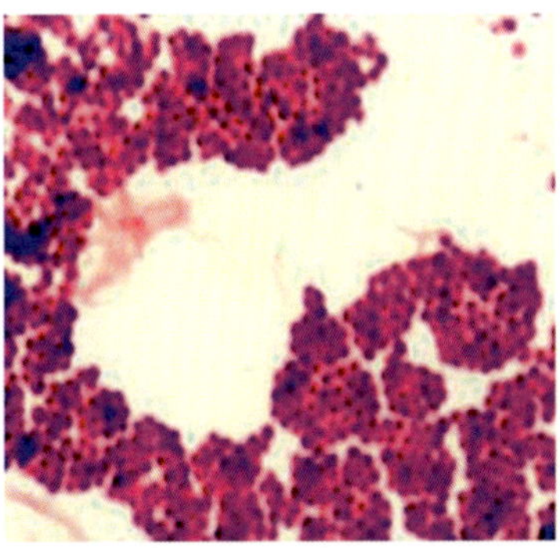

Gram-negative

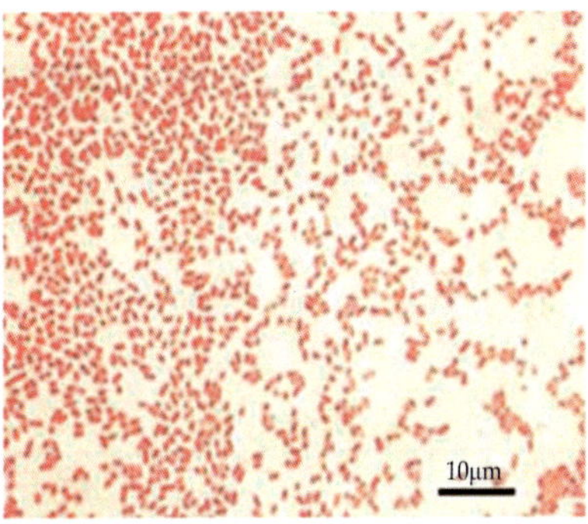

Bacterial Cell Wall

Peptidoglycan is a huge polymer of interlocking chains of alternating monomers. It provides a rigid support while freely permeable to solutes. The backbone of peptidoglycan molecule is composed of two amino sugar derivatives of glucose. The "glycan" part of peptidoglycan.

- N-acetylglucosamine (NAG)
- N-acetlymuramic acid (NAM)

NAG / NAM strands are connected by interlocking peptide bridges. The "peptid" part of peptidoglycan is shown in **figure 16.1**.

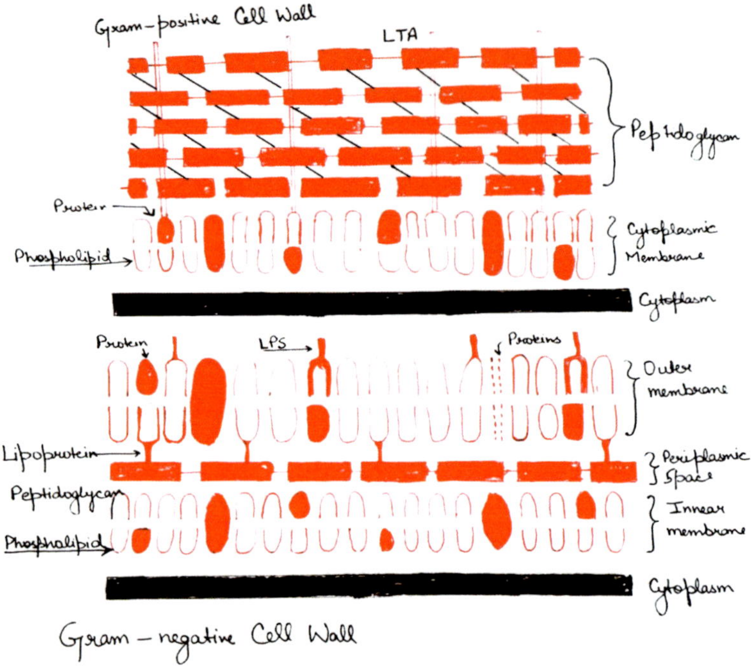

Figure 16.1 Prokaryotes - Cell Wall

Prokaryotes – glycocalyx

Some bacteria have an additional layer outside of the cell wall called the glycocalyx. This additional layer can come in one of two forms:

1. Slime Layer

Glycoproteins are loosely associated with the cell wall. The Slime layer causes bacteria to adhere to solid surfaces and helps prevent the cell from drying out. The slime layer of Gram+ *Streptococcus mutans* allows it to accumulate on tooth enamel (yuck mouth and one of the causes of cavities). Other bacteria in the mouth become trapped in the slime and form a biofilm & eventually a buildup of plaque.

2. Capsule

Polysaccharides are firmly attached to the cell wall. The capsules adhere to solid surfaces and to nutrients in the environment. The adhesive power of capsules is a major factor in the initiation of some bacterial diseases. The capsule also protects bacteria from being phagocytized by cells of the host's immune system.

Prokaryotes:Eendospores

It is a dormant, tough, non-reproductive structure produced by a small number of bacteria. It is resistant to radiation, desiccation, lysozyme, temperature, starvation, and chemical disinfectants. Endospores are commonly found in soil and water, where they may survive for very long periods of time.

Prokaryotes – SurfaceAappendages

Some prokaryotes have distinct appendages that allow them to move about or adhere to solid surfaces. They consist of delicate strands of proteins. **Flagella:** It is long, thin extensions that allow some bacteria to move about freely in aqueous environments. **Axial filament (endoflagella):** Wind around bacteria, causing movement in waves. **Fimbriae:** Most Gram-negative bacteria have these short, fine appendages surrounding the cell. Gram-positive bacteria don't have. It has no role in motility. It helps bacteria adhere to solid surfaces. They have a major factor in virulence. **Pili:** Tubes are longer than fimbriae, usually shorter than flagella. Use for movement, like grappling hooks, and also use conjugation pili to transfer plasmids.

Prokaryotes – Cell Shapes

Most bacteria are classifies according to shape:

1. Bacillus *(bacilli)* = rod-shaped
2. Coccus *(cocci ... sounds like cox-eye)* = spherical

- Spiral shaped
- Spirillum *(spirilla)* = spiral with rigid cell wall, flagella.
- Spirochete *(spirochetes)* = spiral with flexible cell wall, axial filament.

There are many more shapes beyond these basic ones. A few examples

- Filamentous Coccobacilli = elongated coccal form
- = bacilli that occur in long threads
- Vibrios = short, slightly curved rods
  - Fusiform = bacilli with tapered ends

Prokaryotes: Arrangements Of Cells

Bacteria sometimes occur in groups, rather than singly. Bacilli divide along a single axis, seen in pairs or chains. Cocci divide on one or more planes, producing cells in:

- Pairs (diplococci)
- Chains (streptococci)
- Packets (sarcinae)
- Clusters (staphylococci).

Size, shape and arrangement of cells often are the first clues in the identification of a bacterium. Many "look-alikes", so shape and arrangement are not enough for identifying the genus and species.

Eukaryotic Cells

Like prokaryotes, and most living things, the genome is made of DNA. It may include several to many linear chromosomes within a membrane-bound nucleus.

Cytoplasm

The cytoplasm is present inside the plasma membrane; the nucleus is surrounded by cytoplasm. A water-like substance is that fills inside the cells. The cytosol is made up of water, salts, organic molecules and many enzymes that catalyze reactions.

Membrane-bound organelles

Eukaryotic cells have many organelles. Prokaryotes only have ribosomes, which are not bound by a membrane. Membrane-bound eukaryotic organelles organize functions within the cell. The system is present of internal membranes within eukaryotic cells that divide the cell into compartments, or organelles. Transport system, for moving molecules through interior of cell, as well as interactive surfaces for lipid and protein synthesis. The membranes of the endomembrane system are made of a lipid bilayer, with proteins.

Endomembrane system organelles: Nucleus

It is a control center and separates the genetic material (DNA) from the rest of the cell. DNA, the genetic material, is a blueprint, or code for making proteins.

The plasma membrane/cell membrane

the flexible boundary of a cell

separates a cell from its surroundings

It allows nutrients to enter the cell and waste to be removed

- This is referred to as selective permeability (Selective=Chooses, Permeability=filter through).
- Keeping a healthy balance of nutrients and water within the cell is called homeostasis

Overview of Organelles

Nucleus

Largest organelle in the cell and it is the most inner compartment of the cell; contains chromatin (DNA); genetic information on strands called chromosomes "control center" for cell metabolism and reproduction.

Chromatin

The directions on how to make proteins.

Nucleolus

It is found inside the nucleus; ribosomes are made here.

Ribosomes

It makes proteins (made up of RNA and protein); thought of as "factories"

Cytoplasm

It is a clear gel-like fluid inside the cell, which suspends all organelles.

Endoplasmic reticulum

An extensive network of membranes

- **Rough ER:** with ribosomes.
- **Smooth ER:** with no visible ribosomes.
- **Golgi apparatus**: Sorts proteins made by the ribosomes and sends them to needed places in the cell.
- **Lysosomes-** Organelles that are filled with digestive enzymes to remove waste and invading bacteria.

Mitochondria

It is referred to as the "powerhouse" of the cell to release energy for the cell. It converts the energy stored in glucose into ATP for the cell.

Vacuoles are fluid which filled organelles enclosed by a membrane. It is store materials such as food, sugar, water, and waste products.

Eukaryotic plant cell

Plant cells are also eukaryotic cells, but plant cells contain some organelles that are not found in animal cells.

Plant cell organelles

Cell wall

- It is a rigid wall outside the plasma membrane.
- It provides the cell with extra support.

Chloroplasts

It captures light and energy; and converts it into chemical energy.

Chlorophyll

It is the green pigment found inside the chloroplast.

Plastids

- The organelles that store things such as food in the plant cell.

17. CELL DIVISION- MITOSIS AND MEIOSIS, CELL CYCLE, ITS REGULATION AND CANCER, CHARACTERISTIC OF CANCER CELLS.

Mitosis

The division of somatic cells (non-reproductive cells) in eukaryotic organisms is called mitosis. A single cell divides into two identical daughter cells. Daughter cells have same numbers of chromosomes as parent cells.

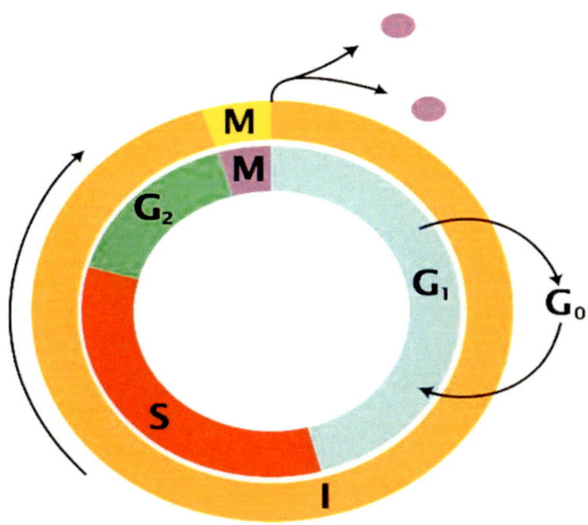

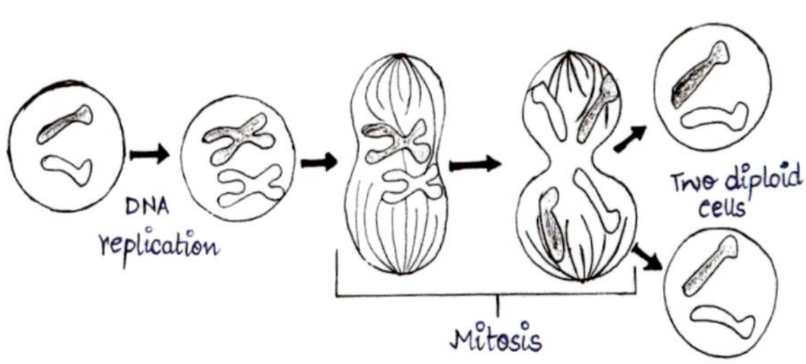

The DNA molecules are in extended form. The uncondensed form is called **chromatin**. The cell can only replicate and transcribe DNA when in an extended state. DNA molecules condense to form **chromosomes** prior to division. Each chromosome is a single molecule of DNA. Easier to sort and organize the replicated DNA into the daughter cells.

There are 4 sub-phases

- 1st – Prophase
- 2nd – Metaphase
- 3rd – Anaphase
- 4th – Telophase
- Cytokinesis

Prophase

The chromosomes condense. Spindle fibers form (Spindle fibers are specialized microtubules radiating out from centrioles). Chromosomes are captured by the spindle.

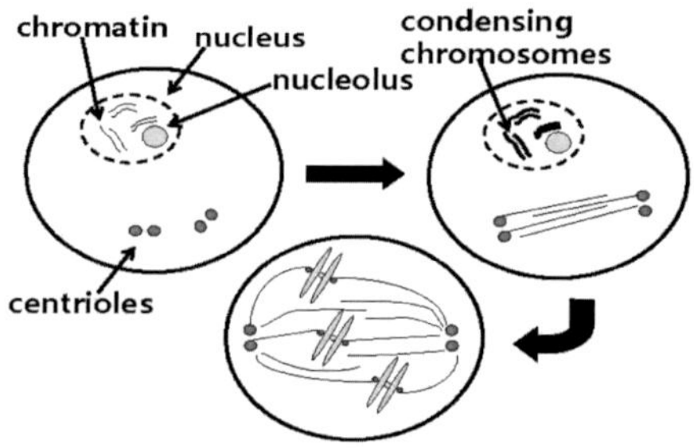

Prophase

Metaphase

Chromosomes align along the equator of the cell, with one kinetochore facing each pole.

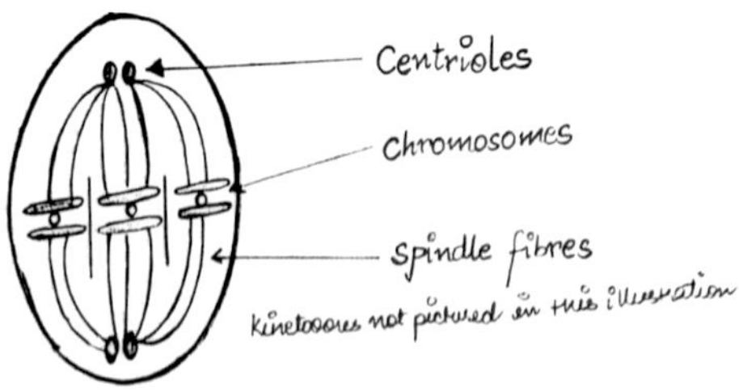

Metaphase

Anaphase

Spindle fibers attached to kinetochores shorten and pull the chromatids towards the poles. The free spindle fibers lengthen and push the poles of the cell apart.

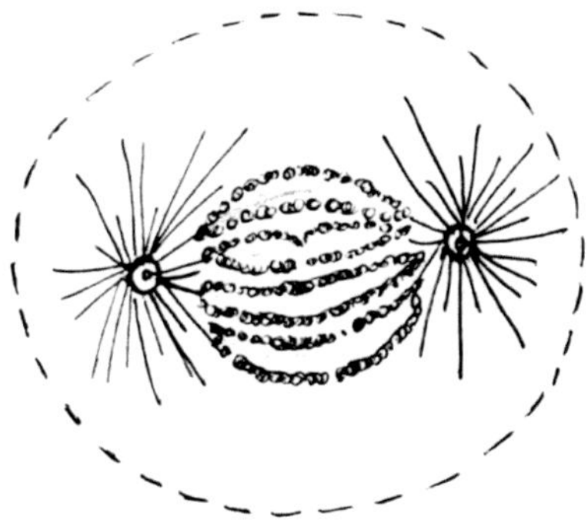

Telophase

Spindle fibers disintegrate and nuclear envelopes form around both groups of chromosomes. Chromosomes revert to their extended state. Cytokinesis occurs, enclosing each daughter nucleus into a separate cell

Cytokinesis

It undergoes cytokinesis by forming a cell plate between the two daughter nuclei in plant cells. They undergo cytokinesis through the formation of a cleavage furrow. A ring of microtubules contract, pinching the cell in half in animal cells.

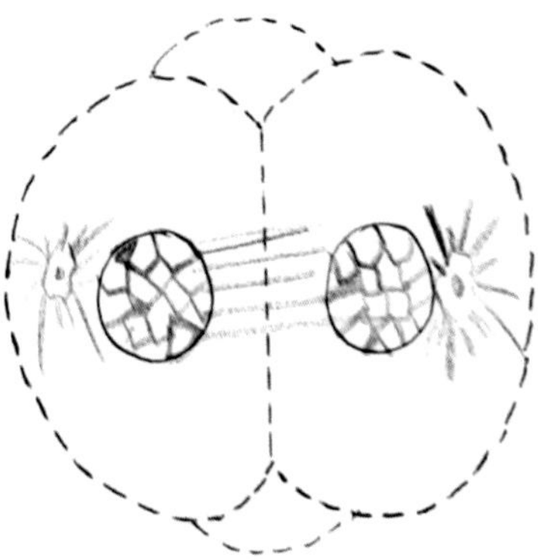

Cytokines

Stages of Mitosis

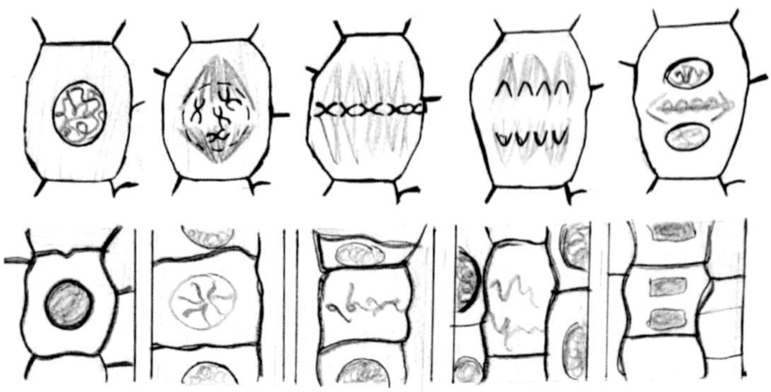

Stage of Mitosis

The cell division of gametes is called meiosis. A single germ cell divides into four unique daughter cells. Daughter cells have half of the chromosomes as a parent cell, so they are considered haploid.

Genetics Terminology: Ploidy. It refers to the number of sets of chromosomes in cells.

Haploid

- One copy of each chromosome.
- Designated as "n", the number of chromosomes in one "set" gametes.

Diploid

Two sets of chromosomes and it is designated as "2n".

Somatic cells

Diploid organisms receive one of each type of chromosome from the female parent (maternal chromosomes) and one of each type of chromosome from the male parent (paternal chromosomes).

Genetics: homologs

The chromosomes exist in homologous pairs in diploid (2n) cells.

Exception: Sex Chromosomes (X, Y).

Other chromosomes, known as autosomes, they have homologs.

Asexual reproduction

The single-celled organisms reproduce by splitting and budding. Some multicellular organisms can reproduce asexually and produce **clones** (offspring genetically identical to parent).

Sexual reproduction

The fusion of two **gametes** to produce a single zygote is called sexual reproduction. Presents greater genetic variation, allows genetic recombination. With the exception of self-fertilizing organisms, the zygote has gametes from two different parents.

Sexual reproduction in humans

It is donated by each parent *(*total = 46 or 23 pairs*)*.

- **Gametes** (sperm/ova): Contain 22 autosomes and 1 sex chromosome.
- Are haploid (haploid number
- "n" = 23 in humans).
- Fertilization results in the diploid zygote.
- Diploid cell; 2n = 46. (n = 23 in humans)

Meiosis

In meiosis, sex cell (Gamete) is formation. In meiosis, there are 2 divisions of the nucleus: meiosis I and meiosis II.

Meiosis & Sexual Reproduction Life Cycle

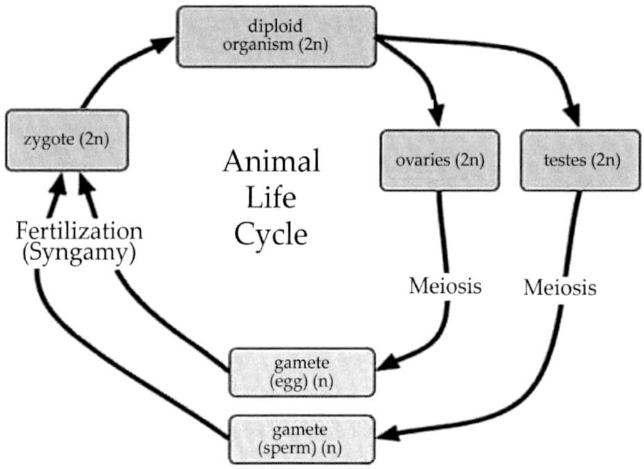

Genetic variation in diploid organisms

The fusion of the sperm and egg results in a unique offspring. So, the young are a product of two individuals with different genetic makeup. Meiosis "shuffles" the genes so that the individual's gametes are genetically different from one another.

Genetic shuffling of Meiosis I

In addition to a new combination of chromosomes resulting from **fertilization**, there are also events in Meiosis I that shuffle the genes.

1. **Crossing over** in Prophase I.
2. **Independent assortment** in Metaphase I.

Crossing over

Homologs break at identical locations and then rejoin opposite partners. This creates new combinations of the alleles on each chromosome. It occurs randomly several times on every chromosome. So, the results in mixing of the genes you inherited from your parents are called crossing over.

Spermatogenesis

Males produce sperm throughout life, after the onset of puberty, about 1,500 sperm per second.

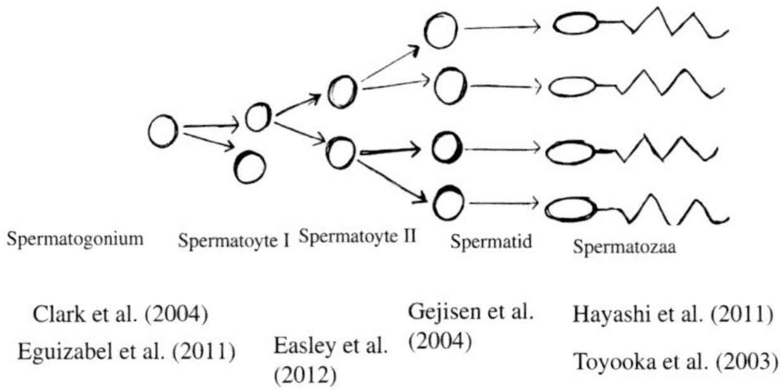

Structure of spermatogenesis

Oogenesis

Oogenesis in females is probably complete either before or shortly after birth. During oogenesis, three polar bodies develop as the mature ovum is generated. Polar bodies contain little cytoplasm and eventually degenerate.

18. MAJOR SIGNALING PATHWAYS OF EUKARYOTIC CELLS

Cell signaling

The general principles of cell signaling are the molecule binds to a receptor molecule on the target cell.

- Endocrine
- Paracrine
- Neuronal
- Contact dependent

Receptor

Each cell responds to a limited set of signals (origin of complexity). It must have receptor molecules for the signaling molecule (thousands of receptors). The different receptors are present for the same signal on different cells.

Signaling distances

There are four types of Signaling.

Intracellular

Signaling pathways

One receptor molecule can interact with many intracellular relay systems so the same signal & same receptor have different effects in different cells. The same relay system act on many different intracellular targets. Receptors sense diverse stimuli but initiate a limited repertoire of cellular signals.

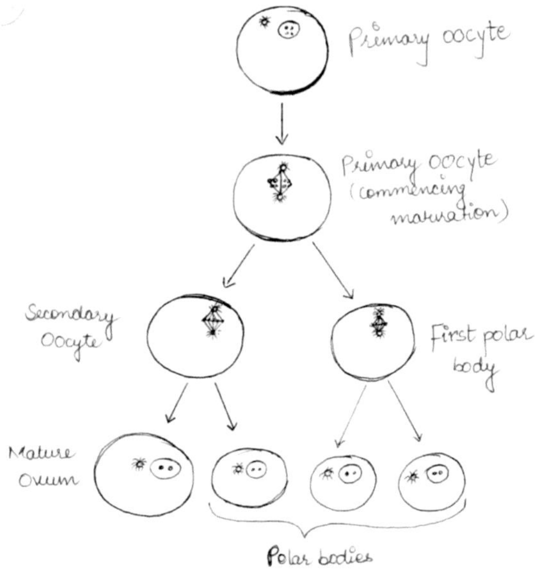

Oogenesis

Receptors

A receptor is a ligand-binding domain and also called effector domain. Receptor modularity allows a wide variety of signals to use a limited number of regulatory mechanisms. Cells may express different receptors for the same ligand. The same ligand may have different effects on the cell depending on the effector domain of its receptor.

Receptors are catalysts and amplifiers

The receptors act by increasing the rates of key regulatory reactions and act as molecular amplifiers.

Ligand binding changes receptor conformation

It can exist in active or inactive conformations and also ligand binding drives the receptor toward the active conformation.

Signals are sorted and integrated in signaling pathways and networks

Signaling pathways usually have multiple steps and can diverge and/or converge. The divergence allows multiple responses to a single signal. The convergence allows signal integration and coordination.

Cellular signaling pathways can be thought of as biochemical logic circuits

The signaling networks are composed of groups of biochemical reactions. The reactions function as mathematical logic functions to integrate information. The combinations of such logic functions combine as signaling networks to process information at more complex levels.

Scaffolds increase the signaling efficiency and enhance the spatial organization of signaling

Scaffolds

It organizes groups of signaling proteins. It may create pathway specificity by sequestering components that have multiple partners. The scaffolds increase the local concentration of signaling proteins. The scaffolds localize signaling pathways to the sites of action.

Independent, modular domains specify protein-protein interactions

The protein interactions may be mediated by small, conserved domains. The modular interaction domains are essential for signal transmission. The adaptors are consist exclusively of binding domains or motifs.

Cellular signaling is remarkably adaptive

The sensitivity of signaling pathways is regulated to allow responses to change over a wide range of signal strengths. The mechanisms are executing this function in all signaling pathways. Most of the pathways are contain multiple adaptive feedback loops to cope with signals of various strengths and durations.

Signaling proteins are frequently expressed as multiple species

The different species (isoforms) of similar signaling proteins expand the regulatory mechanisms possible in signaling pathways. It may differ in:

Function

- Susceptibility to regulation
- Expression
- Cells may express one or several isoforms to fulfill their signaling needs.

Activating and deactivating reactions are separate and independently controlled

The activating and deactivating reactions are usually executed by different regulatory proteins. It is separating activation and inactivation allows for fine-tuned regulation of amplitude and timing.

Cellular signaling uses both allostery and covalent modification

It refers to the ability of a molecule to alter the conformation of a target protein when it binds noncovalently to that protein. The modification of a protein's chemical structure is also frequently used to regulate its activity.

Secondary messengers provide readily diffusible pathways for information transfer

The secondary messengers can propagate signals between proteins that are at a distance. The cAMP and Ca^{2+} are widely used secondary messengers.

Ca^{2+} signaling serves diverse purposes in all eukaryotic cells

Ca^{2+} serves as a secondary messenger and regulatory molecule in essentially all cells. Ca^{2+} acts directly on many target proteins. It also regulates the activity of a regulatory protein calmodulin. The cytosolic concentration of Ca^{2+} is controlled by organellar sequestration and release.

Lipids and lipid-derived compounds are signaling molecules

The multiple lipid-derived second messengers are produced in membranes. Phospholipase Cs release soluble and lipid second messengers in response to diverse inputs. The channels and transporters are modulated by different lipids in addition to inputs from other sources. PI 3-kinase synthesizes PIP_3 to modulate cell shape and motility. PLD and PLA_2 create other lipid second messengers.

PI 3-kinase regulates both cell shape and the activation of essential growth and metabolic functions

The phosphorylation of some lipid second messengers changes their activity. PIP_3 is recognized by proteins with a pleckstrin homology domain.

Signaling through ion channel receptors is very fast

Ion channels are allowing the passage of ions through a pore. It is rapid (microsecond) changes in membrane potential. The channels are selective for particular ions or for cations or anions. The channels regulate intracellular concentrations of regulatory ions, such as Ca^{2+}.

Nuclear receptors regulate transcription

The nuclear receptors are modulating transcription by binding to distinct short sequences in chromosomal DNA known as response elements. The receptor binding to other receptors, inhibitors, or coactivators leads to complex transcriptional control circuits. The signaling through nuclear receptors is relatively slow, consistent with their roles in adaptive responses.

G protein signaling modules are widely used and highly adaptable

The basic module is a receptor. A G protein is an effector protein. The cells are express several varieties of each class of proteins. Effectors are heterogeneous and initiate diverse cellular functions.

Heterotrimeric G proteins regulate a wide variety of effectors

The heterotrimeric G proteins transmit signals by regulating the activities of multiple intracellular signaling proteins known as effectors. Effectors are structurally and functionally diverse. A common G-protein binding domain has not been identified among effector proteins. The effector proteins integrate signals from multiple G protein pathways.

Heterotrimeric G proteins are controlled by a regulatory GTPase cycle

The heterotrimeric G proteins are activated when the Gα subunit binds GTP. GTP hydrolysis to GDP inactivates the G protein. GTP hydrolysis is slow but is accelerated by proteins called GAPs. The receptors are promoting activation by allowing GDP dissociation and GTP association. The spontaneous exchange is very slow. The RGS proteins and phospholipase C-βs are GAPs for G proteins.

Small, monomeric GTP-binding proteins are multiuse switches

Small GTP-binding proteins are active when bound to GTP, inactive when bound to GDP. GDP/GTP exchange catalysts known as GEFs (guanine nucleotide exchange factors) promote activation. GAPs accelerate hydrolysis and deactivation. GDP dissociation inhibitors (GDIs) are slow down the spontaneous nucleotide exchange.

Protein phosphorylation/ dephosphorylation is a major regulatory mechanism in the cell

The protein kinases are a large protein family. Protein kinases phosphorylate (Ser and Thr or Tyr or all three). The protein kinases may recognize the primary sequence surrounding the phosphorylation site. Protein kinases may preferentially recognize phosphorylation sites within folded domains.

Phosphoprotein phosphatases reverse the actions of kinases and are independently regulated

Phosphoprotein phosphatases reverse the actions of protein kinases. Phosphoprotein phosphatases may dephosphorylate.

- Phosphoserine/threonine
- Phosphotyrosine or all three
- Phosphoprotein phosphatase specificity is often achieved through the formation of specific protein complexes.

Covalent modification by ubiquitin and ubiquitin-like proteins is another way of regulating protein function

Ubiquitin and related small proteins may be covalently attached to other proteins as a targeting signal. Ubiquitin is recognized by diverse ubiquitin-binding proteins. Ubiquitination can cooperate with other covalent modifications. Ubiquitination regulates signaling in addition to its role in protein degradation. Diverse signaling mechanisms are regulated by protein tyrosine kinases.

Src family protein kinases cooperate with receptor protein tyrosine kinases

The Src is activated by the release of intrasteric inhibition. Activation of Src involves the release of modular binding domains for activation-dependent interactions. Src often associates with receptors, including receptor tyrosine kinases.

MAPKs are central to many signaling pathways

MAPKs are activated by Tyr and Thr phosphorylation. The requirement for two phosphorylations creates a signaling threshold. The ERK1/2 MAPK pathway is usually regulated through Ras.

Cyclin-dependent protein kinases control the cell cycle

The cell cycle is regulated by cyclin-dependent protein kinases (CDKs) and activation of CDKs involves

- Protein binding
- Dephosphorylation
- Phosphorylation

Diverse receptors recruit protein tyrosine kinases to the plasma membrane

Protein tyrosine kinases are bind to use combinations of effectors similar to those used by receptor tyrosine kinases. These receptors often bind directly to transcription factors.

19. MENDEL'S LAW OF INHERITANCE, GENE INTERACTIONS. SEX DETERMINATION, LINKAGE, CROSSING OVER, RECOMBINATION AND GENE MAPPING

Gregor Mendel

He was born in 1822 in Moravia (now part of the Czech Republic). The son of a tenant farmer; joined a monastery to get an education. He was deeply interested in science, particularly heredity. At the monastery in Brno, Moravia, Mendel received the support of Abbot Napp. From "1851-1855, studied at the University of Vienna, but did not receive a degree. Heredity appeared random and unpredictable. Many traits seemed to blend in the offspring, suggesting a liquid factor controlled heredity. Yet some traits, such as red hair, did not blend away.

With Abbot Napp's encouragement, Mendel studied heredity in peas, carefully choosing traits that did not appear to blend. Collected data from 1856 - 1865. Mendel's creative contribution: he was the first to follow single traits from generation to generation instead of trying to document and follow every trait in the plants. Mendel presented his findings to the Association of Natural Research in Brno in 1865. Few people recognized the significance of Mendel's research. His quantitative methods were uncommon at the time, and the "blending" theory was widely accepted. In 1868, Mendel became abbot of his monastery. His religious work left little time for research, which he set aside, though he was always convinced he had made a valuable contribution to science. Mendel died in 1884. Sixteen years later, in 1900" his work was rediscovered by Hugo de Vries and others looking for clues into the puzzle of heredity. Though criticized in some details, the main body of Mendel's work still stands.

Gametes: reproductive cells produced by sexually reproducing organisms.

Two types:

Male gametes = sperm

In plants: contained in pollen

Female gametes = eggs

In plants: contained in ovules

Ovules contained in carpels

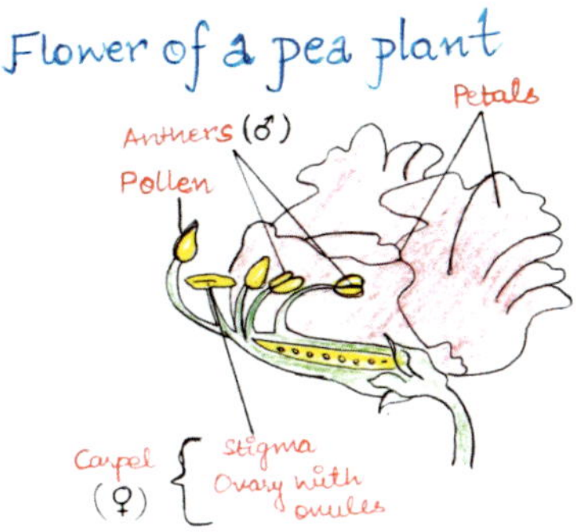

Fertilization

Fertilization: The Fusion of egg and sperm is called fertilization.

Self-fertilized: The fusion of sperm and egg from the same plant is called self-fertilization.

Cross fertilized: Fusion of egg and sperm from two different plants.

Produced hybrids

F_1: First generation

F_2: Second generation

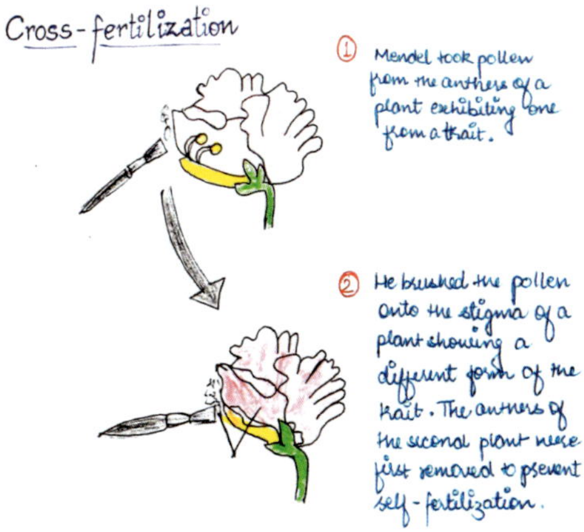

Mendel's Laws

A scientific law is an evidence-based **description** of a natural phenomenon in a given set of circumstances. Mendel's three Laws of Heredity describe what Mendel observed in patterns of inherited traits.

Laws of heredity

- Law of Dominance
- Law of Segregation
- Law of Independent Assortment

Law of dominance

Traits are controlled by two factors that can be called "dominant" or "recessive." A "dominant" trait shows if the offspring inherits at least one dominant factor from one parent. A "recessive" trait shows only if the offspring inherits two recessive factors, one from each parent. "In this cross between two purple-flowered pea plants, one-quarter of the offspring have white flowers. Based just on this information, which are dominant white or purple flowers. Hint: "Dominance" is not based on the numbers of individuals with the trait. It is based on the number of copies of the allele that must be inherited to show the trait.

Mendel's Interpretations

Genes are the hereditary information that determines a single trait. Alleles are alternate forms of a gene. When an organism inherits two identical alleles for a trait, it is said to be homozygous for the trait". When an organism inherits different alleles for one trait, the organism is called heterozygous for the trait.

The offspring of a purple-flowered pea plant and a white-flowered pea plant all have purple flowers. The purple trait is dominant.

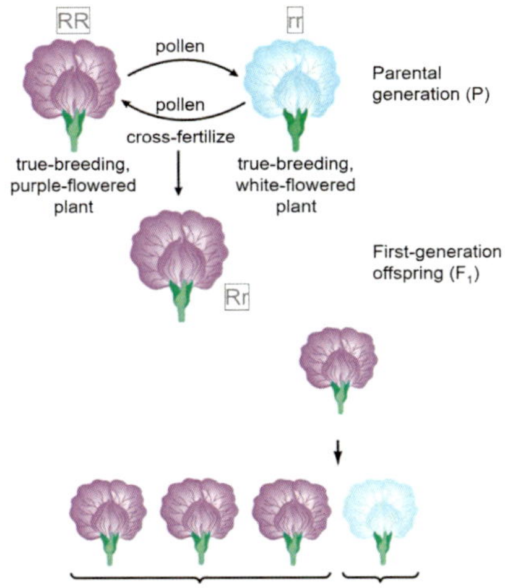

The offspring of the F1 generation (the hybrids) may be purple-flowered if they inherit at least one factor for purple flowers, or maybe white flowered if they inherit the white factor from both parents.

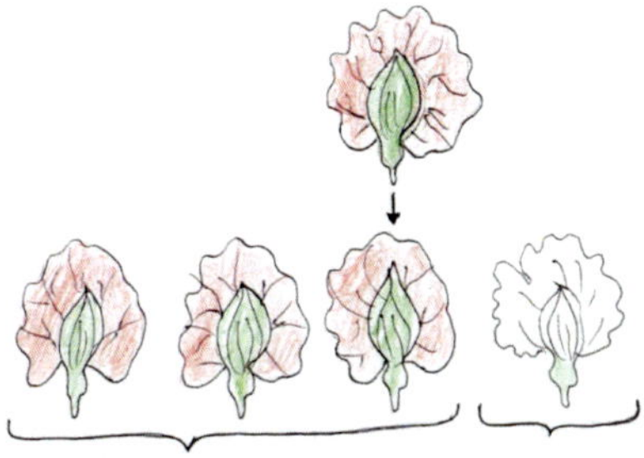

The "purple-flowered trait is dominant because each individual that inherits at least one copy of the purple allele (R) shows the purple phenotype. The white-flowered trait is recessive because an individual must inherit two copies of the white allele (r) to show the white phenotype".

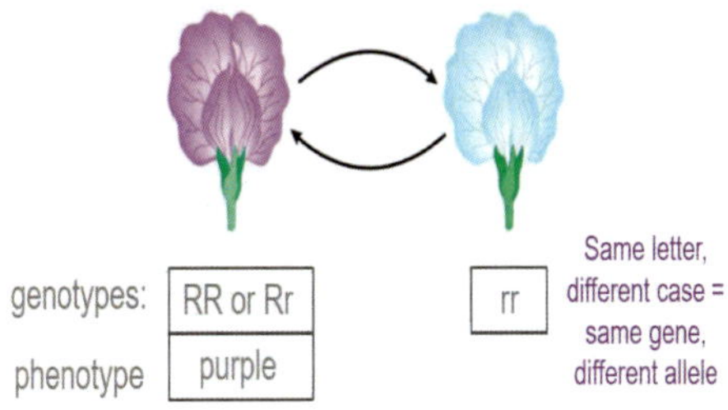

Law of segregation

Each individual has a pair of factors controlling each trait, one inherited from each biological parent. These two factors separate are during the formation of gametes (sex cells). Only one ends up in each sex cell. The homozygous parents in the P generation can pass one kind of allele to their offspring.

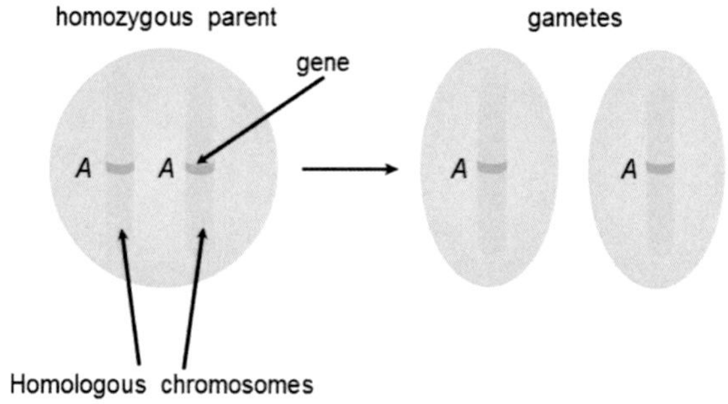

The heterozygous parents of the F1 generation have two alleles for the gene in question and can pass one or the other, but not both, to their offspring.

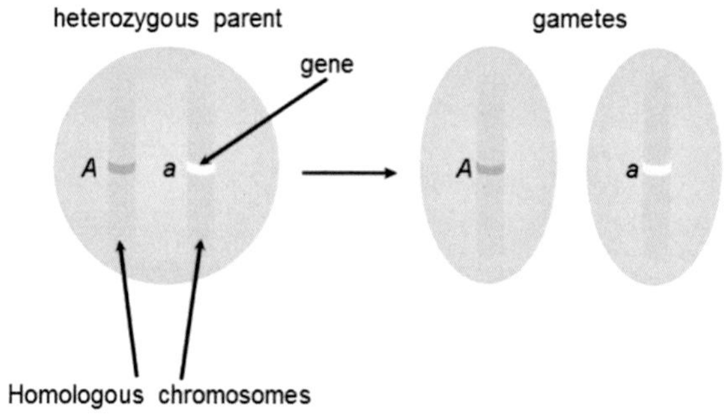

The genotypes can be represented with letters, which symbolize the alleles: capital for dominant alleles, a small case for recessive. When the gametes join to produce the F1 generation, all offspring of homozygous dominant and homozygous recessive parents are heterozygous.

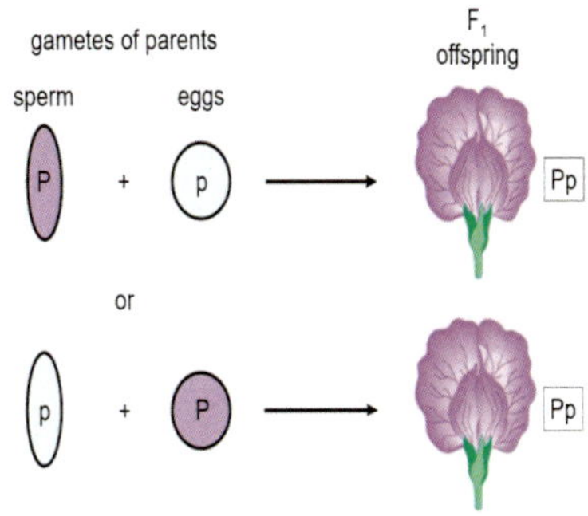

The heterozygous F1 individuals can transfer either a dominant OR a recessive allele in each of their gametes.

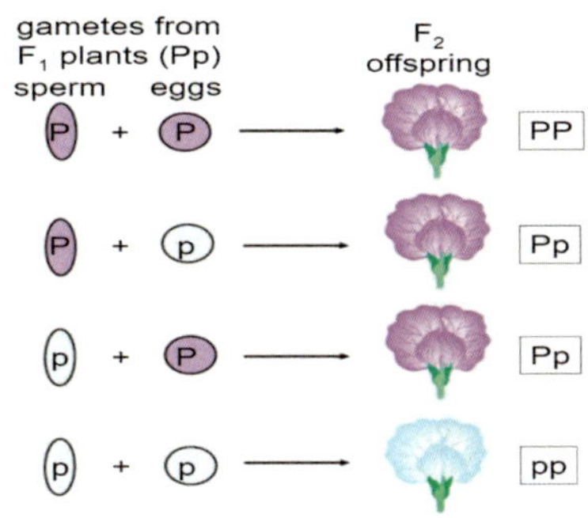

A Punnet's square is one way to predict the outcome of a cross by showing all the possible combinations of all the possible gametes.

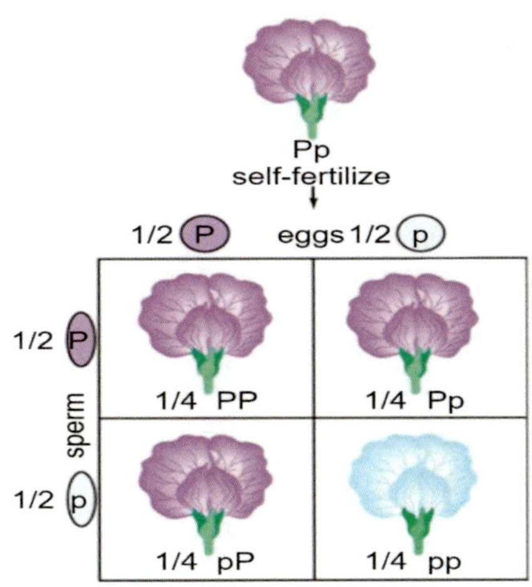

Law Of Independent Assortment

They segregate independently of one another when genetic factors segregate in the gametes. "A dominant allele for one trait does not guarantee the inheritance of a dominant allele for a different trait. All organisms have multiple inheritable traits controlled by genes. Each trait is inherited independently of the others. A pea plant may, for example, have yellow seeds "dominant" but white flowers "recessive".

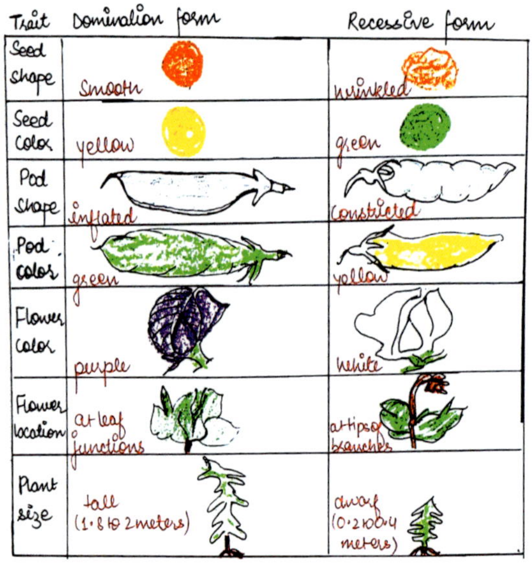

Laws: "proven" forever

Mendel's Laws were good descriptions of what he observed in the peas and other plants he worked with. New knowledge accumulated since Mendel's time has refined his ideas. "While his laws still hold true in some instances, there are many exceptions that we will explore in the next presentations. Genes may have multiple alleles, such as dominant and recessive alleles. Chromosomes, which carry genes, separate from one another during gamete formation. Chromosomes sort independently of one another during gamete formation, but each gamete gets one of each kind of chromosome".

Mutations

The change in the DNA sequence is resulting in the appearance of a new allele. It can occur at any time, but not always beneficial. It can be harmful and result in death. It can result in the organism having traits that make them better suited to the environment.

Incomplete Dominance

A condition where all three genotypes are expressed and phenotypic level contradicts Mendel's conclusions. The genotypic level is consistent with Mendel's laws.

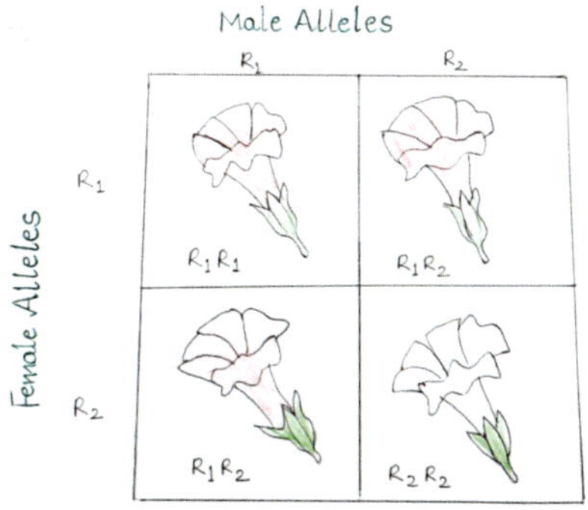

Lethality

A condition in which the inheritances of a lethal combination of alleles results in the death of the organism example is Huntington's disease. Lucien Cuenot studied the inheritance of coat color in mice.

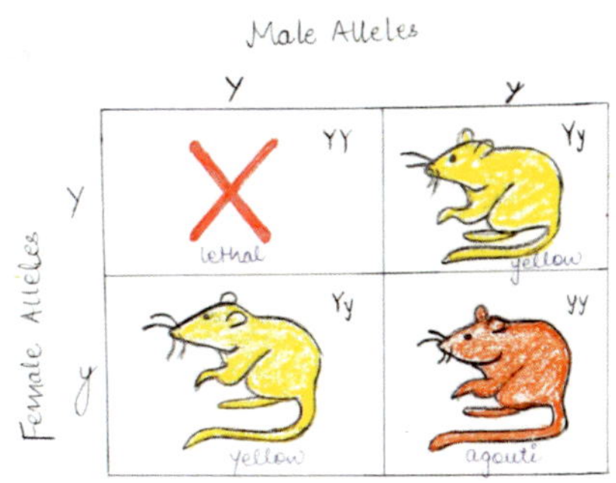

Pleiotropy

A single gene affects two or more traits **example:** One gene affects whether the seed coat is round/wrinkled and Cat fur.

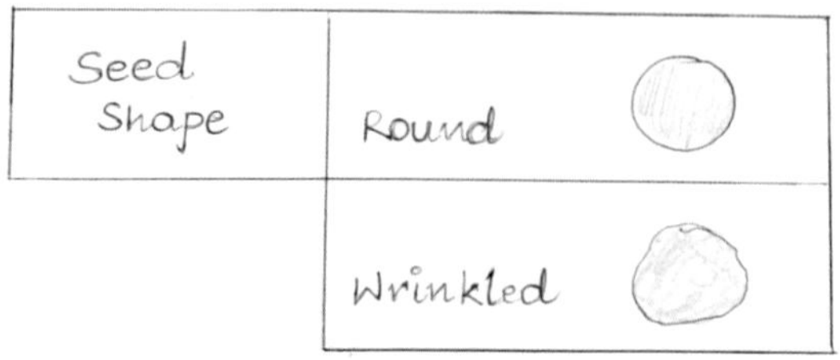

Polygenic.Inheritance

A trait affected by many genes examples: Height, weight, skin color.

Monogenic Inheritance

Traits determined by a single gene with two alleles **examples:** Flower color in four-o'clock plants.

20. CHROMOSOME STRUCTURE AND BEHAVIOR THROUGH THE CELL CYCLE, KARYOTYPE. CHROMATIN ORGANIZATION. POLYTENE AND LAMPBRUSH CHROMOSOME, BANDING PATTERNS IN HUMAN CHROMOSOME, STRUCTURAL AND NUMERICAL CHANGES IN CHROMOSOMES, HEREDITARY DEFECTS

The cellular structure holds the genetic information.

Chromosomes

It contains the genetic material: DNA and RNA

Chromatin

The chromosomal material in its decondensed and threadlike state is called chromatin.

Mitosis

It is a form of asexual reproduction. It occurs when an organism grows or replaces damaged cells. Prior to mitosis, cells undergo replication. The process in which chromatin is copied. It produces diploid cells.

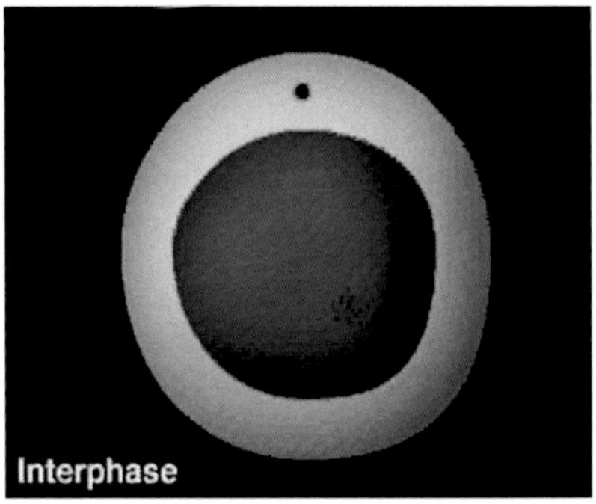

Prophase

It is the start of mitosis. The chromatin condenses into rod-like chromosomes. Each chromosome consists of sister chromatids, connected at the centromere. Nuclear membrane disappears

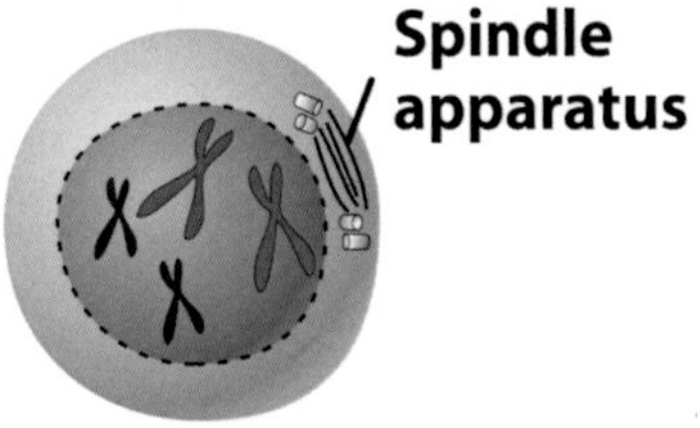

Metaphase

Chromosomes align themselves in a flat plane at the cell equator.

Anaphase

Centromeres are split in anaphase and sister chromatids-now chromosomes-are pulled to opposite poles of the cell.

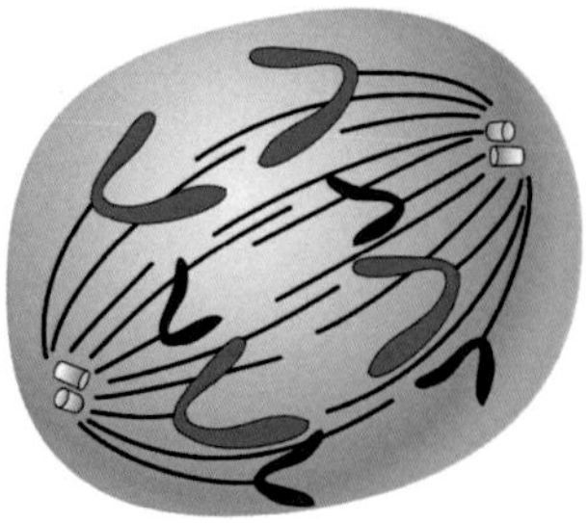

Telophase

Chromosomes unravel, returning the chromatin to its non-dividing threadlike state and the nuclear membrane assembles.

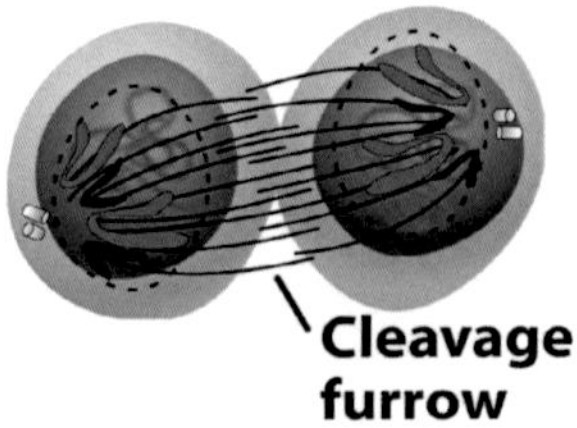

Cytokinesis

The division of the cytoplasm is in cytokinesis. It begins during anaphase and telophase. It differs in animals and plant cells. Plant cells form a cell plate. Membranous vesicles congregate at the center of the cell. The vesicles are containing cell wall material. Animal cells form a cleavage furrow. It forms around the periphery of the dividing cell. The furrow becomes deeper and deeper until the membrane pinches off forming two cells.

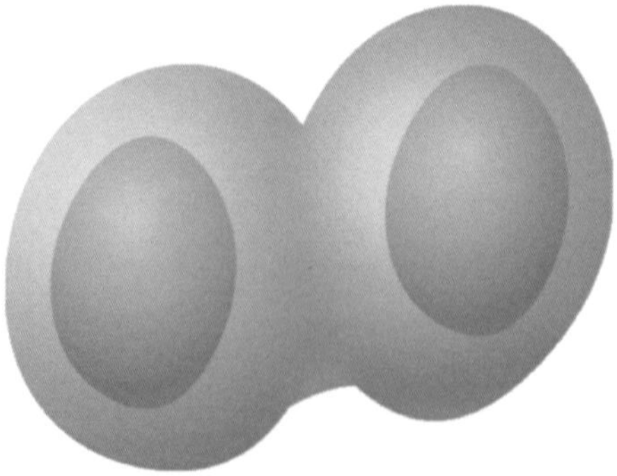

Chromosomes Pairs

Homologous Pairs: chromosomes that are closely matched in size and shape and determine the same traits. Sex chromosomes are determining the gender of the organism.

Chromosomal theory of inheritance

The two members of each pair of a homologous pair of chromosomes carry alleles for the same genes and, therefore, affect the same traits. It was proposed in 1903.

Mapping genes

The location of a gene on a chromosome is currently trying to map all human genes to the appropriate chromosome **example:** gene responsible for sickle cell anemia is located on chromosome 11.

Sexual eproduction

It is the type of reproduction in which genetic information from females combines with the male. It requires fertilization. The fusion of egg and sperm is called gametes. These cells need to have half the amount of genetic information.

Meiosis

It produces haploid cells (gametes-eggs and sperm). It is posses only one member of each pair of homologous chromosomes. Chromosomes replicate before meiosis. It requires two rounds of division.

Prophase I

The chromatin condenses into compact chromosomes. The nuclear envelope disappears. The synapsis occurs. The homologous pairs of chromosomes are closely aligned allowing the exchange of chromosome segments.

Metaphase I

The aligned pairs of replicated chromosomes move to the equator of the dividing cell.

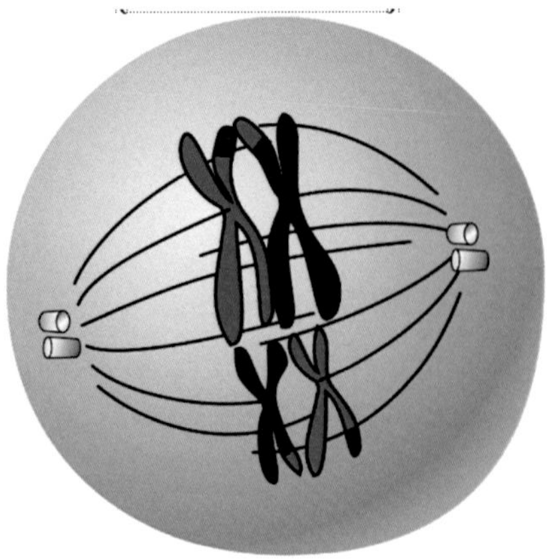

Anaphase I

The members of homologous pairs of chromosomes separate from each other. They move to the opposite poles of the cell. The chromosomes experience independent assortment.

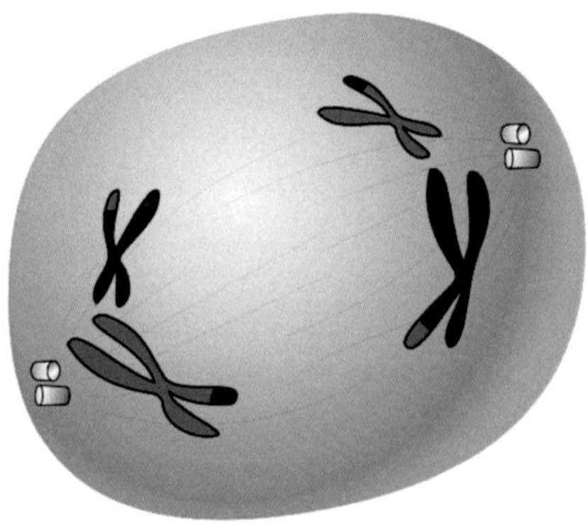

Telophase I

The chromosomes cluster at the opposite poles of the cell and begin to decondense in telophase I. The nuclear envelope may reform and cytokinesis occurs.

Interkinesis: phase between meiosis I and II. The daughter cells are now haploid.

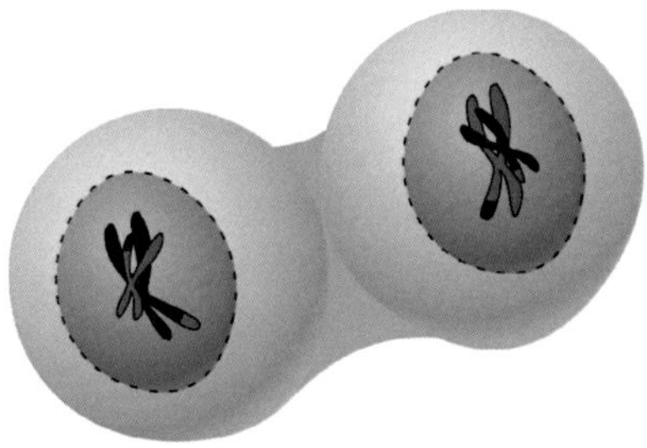

Prophase II The partially unraveled chromosomes condense again.

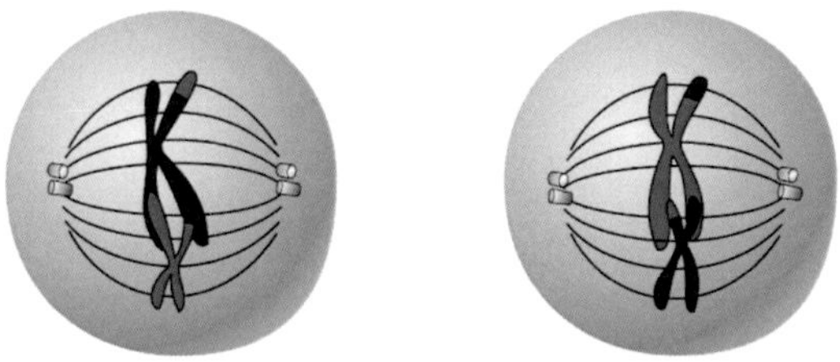

Metaphase II

The chromosomes are move to the cell equator. It has no longer homologous pairs, so chromosomes line up singly at the middle of the cell.

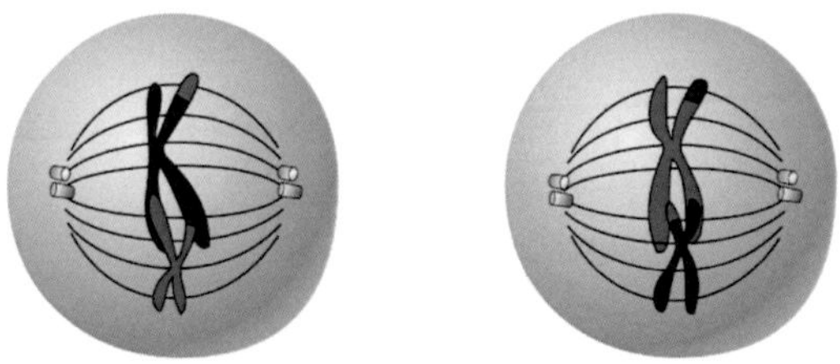

Anaphase II

The centromeres are dividing and chromosomes are separated. It moves to the opposite poles of the cell.

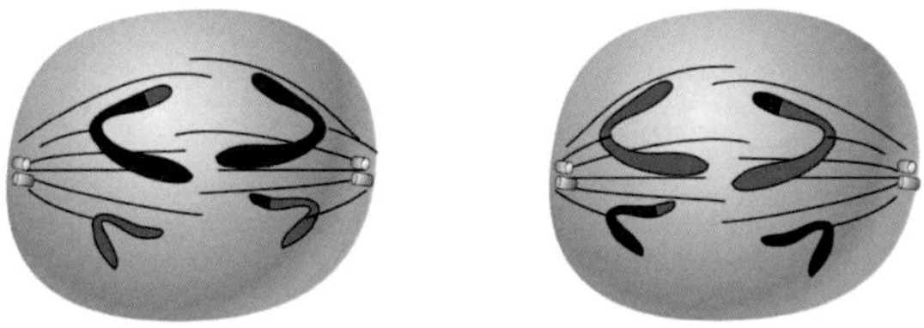

Telophase II

The clustered chromosomes at the cell pole begin to decondense and the nuclear membrane develops. Cytokinesis occurs and produces four non-identical haploid cells.

Cell Cycle

It is repetitive sequence of events that characterizes the life of a cell and consists of two main phases are:

Interphase

- Cells are not dividing.
- 90% of the cell cycle is this phase

M phase

- It includes mitosis and cytokinesis

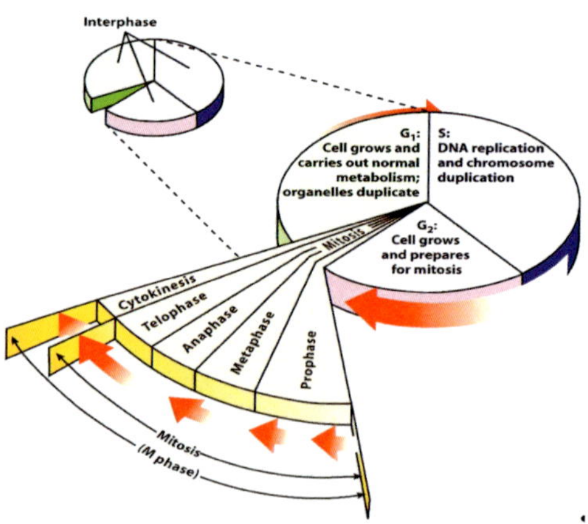

Interphase has three subphases

- **G_1 (first gap):** Cell makes copies of organelles and grows larger
- **S (synthesis):** Genetic material is copied

- **End of this phase, cells chromosomes are doubled**
- **Copies are attached:** thus the total number of chromosomes remains the same
- **G_2 (second gap):** Cell prepares upcoming M phase

Regulation of cell cycle

The cell cycle must be regulated or can result in cancer and uncontrolled cell growth.

Metastasize: A cell breaks free from the original cancerous mass and resides in a new area in the body

- Two checkpoints
- First between G_1 and S
- Second between G_2 and M
- To pass checkpoints, the cell must possess appropriate amounts of protein in the cytoplasm.
- These proteins activate other proteins necessary for the production of genetic material and mitosis.

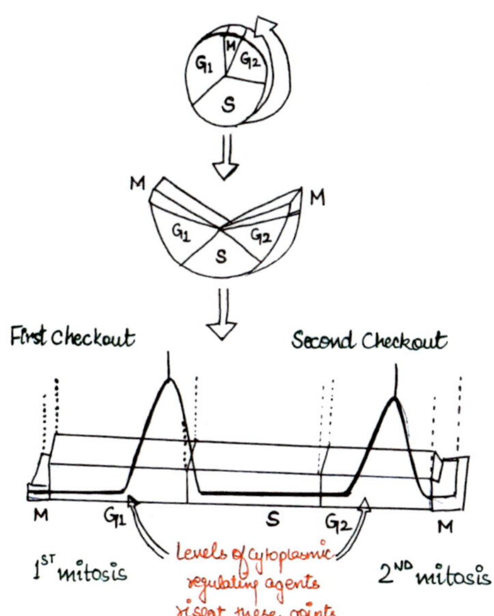

To pass checkpoints, a cell must possess appropriate amounts of regulating proteins in the cytoplasm when the regulator concentration is high, the cell cycle progresses. The cell cycle is suspended at that stage when genetic factors segregate in the gametes. External and internal regulatory agents also influence passage through checkpoints. Chromosomes contain genes. Chromosomes follow the law of independent assortment, not genes. If two genes are on the same chromosome, the two genes are inherited together or are said to be linked or in linkage groups. example: sex linkage

Sex Chromosomes

It contains other genes aside from those to determine gender **example:** eye color and gender in fruit flies.

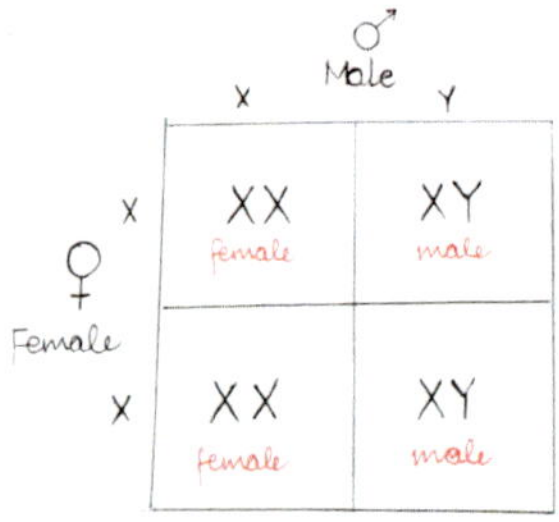

Crossing over

The chromosomes can exchange segments during meiosis. The exchange of the genetic material between chromatids of homologous chromosomes and occurs at the chiasmata. It is an important mechanism for creating new combinations of genes and disrupts linkage groups.

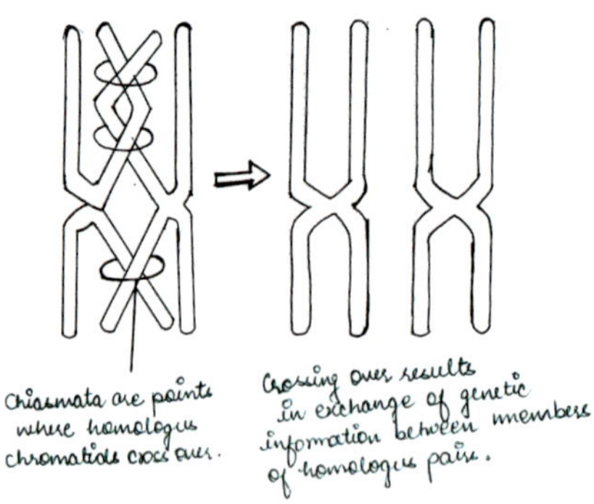

DNA is the genetic material

The first genetic material is generally believed to have been RNA, initially manifested by self-replicating RNA molecules floating on bodies of water. This hypothetical period in the evolution of cellular life is known as the

RNA world. Frederich Meisner has studied the fundamental constituents of life in the 1860s. It was discovered an unknown substance containing carbon, nitrogen, oxygen, and phosphorus. It came from the nucleus of the cell and called it nuclein. His students renamed the substance nucleic acid after finding it was acidic in 1881. He discovered that nucleic acids were contained in chromatin.

Fred Griffiths, medical officer for the British Ministry of Health in 1928. It was studied in *Pneumoncoccus* bacteria. Two kinds: smooth and rough and experimented by injecting the two types in mice.

- Smooth: Virulent form that appears smooth and shiny when grown on an agar plate
- Rough: the harmless form that appears rough when grown on an agar plate.

Griffith performed transformation experiments with two different strains of the bacterium Diplococcus pneumonia (virulent strains), which cause pneumonia in certain vertebrates (such as human and mice), and avirulent strains which does not cause pneumonia. The difference in virulence is related to the polysaccharide capsule of the bacterium. Virulent strains have this capsule, whereas avirulent strains do not. The non-capsulated bacteria are readily engulfed and destroyed by phagocytic cells. The virulent strain bacteria is posse's polysaccharide capsule is not easily engulfed. So they are able to multiply and cause pneumonia (Further readings are suggested in Pathfinder Life sciences book-Parnav Kumar and Usha Mina).

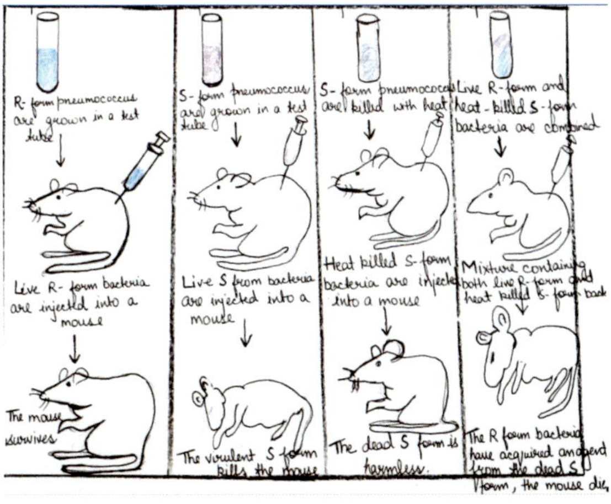

Griffiths identified the material as the transforming principle. Avery, MacLeod and McCartney studied the transforming principle for 20 years.

They determined that the transforming agent was DNA. Alfred Hershey and Martha Chase studied viruses that infect bacteria. Viruses are called bacteriophages. Viruses are made of Protein coat and nucleic acid. Viruses mix their genes with host genes to hijack cell machinery and use it to produces new viruses. Usually kills a host cell. Hershey and Chase labeled protein and DNA differently with isotopes. Isotopes are variants of elements that share the same chemical properties but differ in the number of neutrons. It labeled DNA with ^{32}P and protein with ^{35}S.

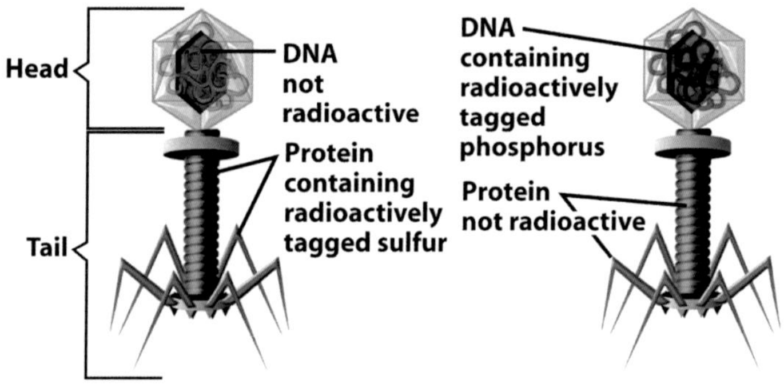

Virus structure and the experiment

Eukaryotic chromosome structure

Chromatin: the complex of DNA and chromosomal proteins

- It is twice as much protein as DNA

Two major types of proteins

- **Histones:** Abundant, basic proteins with a positive charge that bind to DNA 5 main types: H1, H2A, H2B, H3, H4, equal in mass to DNA, evolutionarily conserved
- **Non-histones:** All the other proteins associated with DNA differ markedly in type and structure amounts vary widely, >> 100% DNA mass and **<< 50% DNA mass**

Histones (Components of the nucleosome)

The major protein component of chromatin

↳ small, very basic (lys, arg) proteins (to neutralize the highly charged DNA)

↳ assembly of histones with DNA requires, a molecular chaperone, nucleoplasmin

Five Types of Histones:

H1 Clamp for DNA wrapped around nucleosome

"Core" Histones: H2A, H2B, H3, H4 } Found in 2 copies of each in the nucleosome. The DNA is then wrapped around the protein core made of these proteins – together they form an octamer.

H4: Is very highly conserved between species (98% identical between cows and peas) ~ 1% change in 600 million years

H3: Is also very conserved (97% identical)

Nucleosomal packaging of DNA

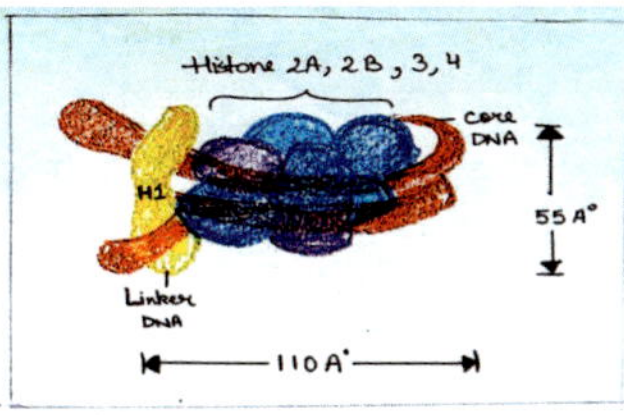

In the presence of Histone H1 175-200 bp DNA is associated with the nucleosome – but, only 1-16 bp is wrapped around the ~~octamer~~ (~~ie~~ if H1 ~~is removed~~ ~ 146 bp is observed in nuclease digits)

Histone H1 – helps 'champ' the DNA onto the nucleosome and participates in higher-order chromatin folding.

Histone modifications

Histones are also modified in order to regulate

1. chromosome structure; compaction
2. gene activity

The main types of modifications are

a) Phosphorylation of serines

b) Methylation of lysines

c) Acetylation of lysines: neutralizes + charges

$$--NH_3^+ \rightarrow --NH-\underset{\underset{O}{\|}}{C}-CH_3$$

11nm fiber is coiled into 30nm solenoid - requires histone H1

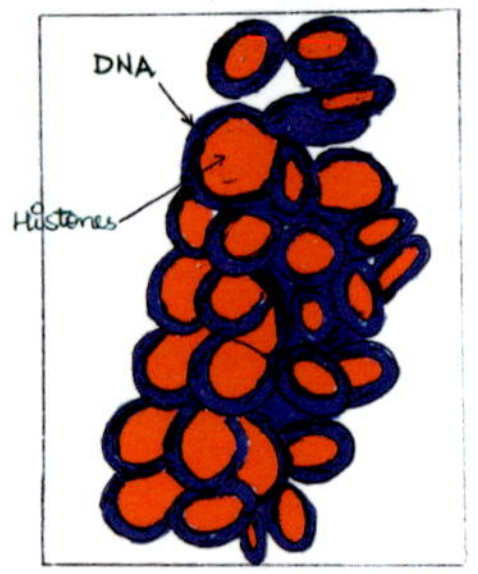

Schematic of formation of 30nm chromatin fiber

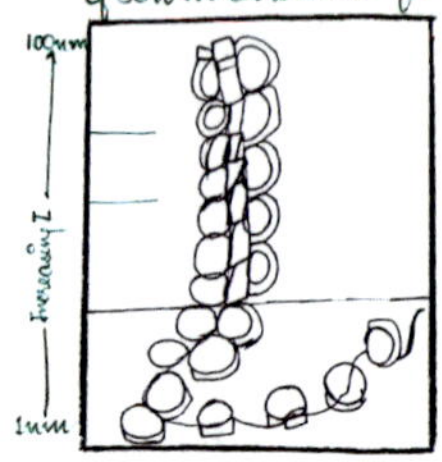

Packing of DNA into chromosomes :

1. Level 1 Winding of DNA around histones to create a nucleosome structure.

2. Level 2 Nucleosomes connected by strands of linker DNA like beads on a string.

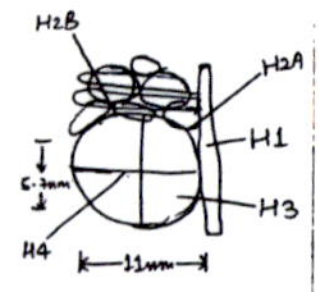

3. Level 3 Packaging of nucleosomes into 30-nm chromatin fibre.

4. Level 4 Formation of looped domains.

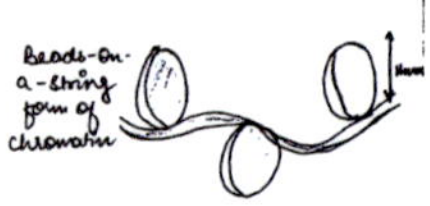

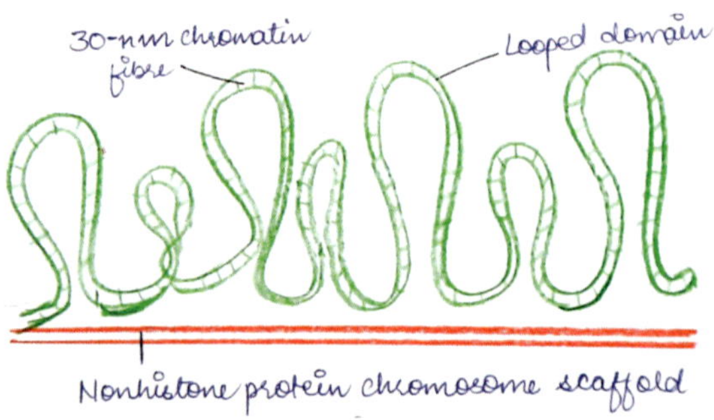

Centromeric DNA (CEN): Specialized sequences that function with microtubules.

Telomeric DNA: At the extreme ends of the chromosome, maintain stability, and consist of tandem repeats. Demonstrated to play a role in DNA replication and the stability of DNA.

Unique-sequence DNA: Often referred to as single-copy and usually code for genes.

Repetitive-sequence DNA: Maybe interspersed or clustered and vary in size.

- **SINEs:** Short interspersed repeated sequences (100-500 bp).
- **LINEs:** Long interspersed repeated sequences (>5,000 bp).
- **Microsatellites:** short tandem repeats (e.g., TTATTATTATTA).

Chromatin is divided into euchromatin and heterochromatin

It can be seen only during mitosis in individual chromosomes. The general mass of chromatin is in the form of euchromatin during interphase. Euchromatin is less tightly packed than mitotic chromosomes. The regions of heterochromatin remain densely packed throughout the interphase.

Chromosomes have banding patterns

Certain staining techniques cause the chromosomes to have the appearance of a series of striations called G-bands. The bands are lower in G • C content than the interbands. Genes are concentrated in the G • C-rich interbands.

The DNA of interphase chromatin is negatively supercoiled into independent domains of ~85 kb. Metaphase chromosomes have a protein scaffold to which the loops of supercoiled DNA are attached.

Specific sequences attach DNA to an interphase matrix

DNA is attached to the nuclear matrix at specific sequences called MARs or SARs. The MARs are A • T-rich but do not have any specific consensus sequence. The centromere is essential for segregation. A eukaryotic chromosome is held on the mitotic spindle by the attachment of microtubules to the kinetochore that forms in its centromeric region. Centromeres often have heterochromatin that is rich in satellite DNA sequences.

Centromeres have short DNA sequences in *S. cerevisiae*

CEN elements are identified in *S. cerevisiae* by the ability to allow a plasmid to segregate accurately at mitosis. CEN elements consist of short conserved sequences CDE-I and CDE-III that flank the A • T-rich region CDE-II.

The centromere binds a protein complex

A specialized protein complex that is an alternative to the usual chromatin structure is formed at CDE-II. The CBF3 protein complex that binds to "CDE-

III is essential for centromeric function. The proteins that connect these two complexes may provide the connection to microtubules. Centromeres may contain repetitious DNA. Centromeres in higher eukaryotic chromosomes contain large amounts of repetitious DNA. The function of the repetitious DNA is not known.

Telomeres are replicated by a special mechanism

The telomere is required for the stability of the chromosome end. A telomere consists of a simple repeat where a C+A-rich strand has the sequence $C_{>1}(A/T)_{1\text{-}4}$.

Telomeres seal the chromosome ends

The protein TRF2 catalyzes a reaction in which 3 repeating unit of the G+T-rich strand forms a loop by displacing its homolog in an upstream region of the telomere.

Lampbrush chromosomes are extended

Sites of gene expression on lampbrush chromosomes show loops that are extended from the chromosomal axis.

Polytene chromosomes form bands

Polytene chromosomes of Dipterans have a series of bands that can be used as a cytological map.

Polytene chromosomes expand at sites of gene expression

Bands that are sites of gene expression on polytene chromosomes expand to give "puffs." The nucleosome is the subunit of all chromatin. Micrococcal nuclease releases individual nucleosomes from chromatin as 11S particles. A nucleosome contains ~200 bp of DNA. Two copies of each core histone (H2A, H2B, H3, and H4). One copy of H1. DNA is wrapped around the outside surface of the protein octamer.

DNA is coiled in arrays of nucleosomes

Greater than 95% of the DNA is recovered in nucleosomes or multimers when micrococcal nuclease cleaves the DNA of chromatin. The length of DNA per nucleosome varies for individual tissues in a range from 154-260 bp.

Nucleosomes have a common structure

Nucleosomal DNA is divided into the core DNA and linker DNA depending on its susceptibility to micrococcal nuclease. The core DNA is the length of 146 bp that is found on the core particles produced by prolonged digestion with micrococcal nuclease. Linker DNA is the region of 8-114 bp that is susceptible to early cleavage by the enzyme. Changes in the length of linker DNA account for the variation in the total length of nucleosomal DNA. H1 is associated with linker DNA and may lie at the point where DNA enters and leaves the nucleosome.

Polytene chromosome and Lambrush chromosome

Giant chromosomes were first observed by **E. G. Balbiani** in 1881 in salivary gland cells of Dipteran species. "**Polytene chromosomes** are giant chromosomes found in the salivary glands, malpighian tubules, gut lining and adipose tissue cells of some Dipteran larvae. These chromosomes may reach a size up to 200 times (or more) the size of corresponding chromosomes at meiosis or in nuclei of normal mitotic cells. These giant chromosomes are formed by repeated cycles of endo-reduplication of single chromatids. This means that the chromatids replicate without cell division, as a result, the number of chromatids keeps increasing. This process is called Polyteny and therefore, these chromosomes are called polytene chromosomes. Polytene chromosomes show another characteristic feature i.e., somatic pairing. As a result of this, chromosomes always appear to be half of that in normal somatic cells.

Lampbrush Chromosomes were first seen in salamander oocytes by Fleming in 1882. The name lampbrush was given by Ruckert in 1892. The nomenclature comes from its similarity in appearance to the brushes used to clean lamp chimneys in old days. These chromosomes are transitory structures that exist during the extended Diplotene in oocytes of most invertebrates and vertebrates except mammals. The loops in these chromosomes are sites of RNA transcriptions". The condensed chromatin in chromomeres is generally not transcribed. The position of each loop is fixed. A given loop always contains the same DNA sequence. They have been widely used in studies of chromosome organization, gene expression, molecular morphology of RNA transcription, etc.

Chromosome banding

Chromosomes are stained with various dyes enabling the chromosome segments to be identified. Most methods can distinguish 550 bands/ haploid set. High-resolution methods can distinguish up to 850 bands/ haploid set that can allow the identification of small interstitial deletions.

G-Banding

Dye gives chromosomes a striped appearance because it stains the regions of DNA that are rich in adenine (A) and thymine (T) base pairs. Regions that stain as dark G bands replicate late in the S phase of the cell cycle and contain more condensed chromatin. While light G bands generally replicate early in the S phase and have less condensed chromatin.

Numerical changes

Mutations and polymorphisms

Mutation: a permanent transmissible change in the genetic material, usually in a single gene. Polymorphism: two or more genetically determined, proportionally represented phenotypes in the same population.

Types of mutations

- Genomic: abnormal chromosome number (monosomy, polysomy, aneuploidy).
- Chromosomal: abnormal chromosome structure.
- Gene: DNA sequence changes in specific genes.

Chromosome morphology

Telomere: chromosome ends. Centromere: site of spindle attachment. Constriction of the metaphase chromosome at the centromere defines two arms. Nucleosome: DNA double helix wrapped around histone proteins.

Chromosome morphology changes during the cell division cycle

DNA double helix: 2nm diameter Interphase (G1, S, G2). Chromatin "beads on a string:" 11 nm. Chromatin in nucleosomes: 30 nm in Metaphase (Mitosis). Extended metaphase chromosomes: 300 nm. Condensed metaphase chromosomes: 700 nm

Linkage and crossing over

Each species of the organism must contain hundreds to thousands of genes. "Yet most species have at most a few dozen chromosomes. Therefore, each chromosome is likely to carry many hundreds or even thousands of different genes. The transmission of such genes will violate Mendel's law of independent assortment. In eukaryotic species, each linear chromosome contains a long piece of DNA. A typical chromosome contains many hundred or even a few thousand different genes. The term linkage has two related meanings

1. Two or more genes can be located on the same chromosome
2. Genes that are close together tend to be transmitted as a unit

Chromosomes are called linkage groups. They contain a group of genes that are linked together. The number of linkage groups is the number of types of chromosomes of the species.

For example, in humans

- 22 autosomal linkage groups
- An X chromosome linkage group
- A Y chromosome linkage group

Genes that are far apart on the same chromosome may independently assort from each other. This is due to crossing-over.

Crossing over may produce recombinant phenotypes

In diploid eukaryotic species, "linkage can be altered during meiosis as a result of crossing over. It occurs during prophase I of meiosis at the bivalent stage. Non-sister chromatids of homologous chromosomes exchange DNA segments. **Bateson and Punnett** discovered two traits that did not assort independently. In 1905, **William Bateson and Reginald Punnett** conducted a cross in sweet pea involving two different traits. Flower color and pollen shape. This is a dihybrid cross that is expected to yield a 9:3:3:1 phenotypic ratio in the F_2 generation. However, **Bateson and Punnett** obtained surprising results. They suggested that the transmission of the two traits from the parents was somehow coupled. The two traits are not easily assorted in an independent manner". However, they did not realize that the coupling is due to the linkage of the two genes on the same chromosome.

21. EXTRA-CHROMOSOMAL INHERITANCE, SEX-LINKED INHERITANCE IN HUMANS AND MUTATION

Extrachromosomal inheritance: Phenotype of maternal parent is solely responsible for phenotype of offspring - example of maternal inheritance of chloroplasts.

Mitochondrial heredity

About 30,000 genes on 23 human chromosomes (3.3 billion base pairs/ haploid cell). Mitochondrial Genome contains 37 genes.

- 13 code for some of the proteins involved in oxidative respiration
- 22 tRNA genes
- 2 rRNA genes
- some proteins involved in mt protein synthesis
- 16,569 base pairs, circular

Human mitochondrial diseases

It maternally inherited: only offspring of affected mothers are affected. Show deficiency in mitochondrial function. It is caused by a mutation in a mitochondrial gene.

Example: myoclonic epilepsy and ragged red fiber disease (MERRF)

- Deafness, dementia, seizures
- Point Mutation in a mitochondrial tRNA

Human mitochondrial diseases

- Leber's Hereditary Optic Neuropathy (LHON)
- Sudden bilateral blindness.
- Point mutation in small subunit of NADH dehydrogenase.
- Kearns-Sayre Syndrome (KSS).
- Symptoms in eyes, muscles, heart, brain.
- Deletion mutation in mtDNA.

Maternal Effect

Pattern offspring is determined by the genotype of the mother, regardless of phenotype.

Maternal effect genes in *Drosophila* development

It was discovered by **Edward Lewis, Christiane Nusslein-Volhard and Eric Weischaus** (Nobel Prize in Physiology or Medicine, 1995). The gene products are deposited in the egg cells of the mother's ovary. These genes set up the first embryonic axes: dorsal-ventral and anterior-posterior and control later zygotic genes.

Sex-linked chromosome

A trait whose gene is located on a sex chromosome (almost always the X-chromosome) and common sex-linked diseases are

- Colorblindness
- Hemophilia
- Duchenne muscular dystrophy

Mutation

Any sudden change occurring in the hereditary material is called a mutation. They may be harmful, beneficial or neutral. In multicellular organisms, there are two broad categories of mutations: Somatic mutations & germ line mutations.

Somatic mutations

Arise in the somatic cells. Passed on to other cells through the process of mitosis. Effect of these mutations depends on the type of the cell in which they occur & the developmental stage of the organism. If occurs early in development, larger the clone of the mutated cells.

Germ line mutations

They occur in the cells that produce gametes. Passed on to future generations. In multicellular organisms, the term mutation is generally used for germ line mutations. The term mutation was given by **Devries** in 1901 while studying evening primrose Oenothera lamarckiana. Most of these were chromosomal variations. Some were point variations. Originally the term mutation was given to both chromosomal as well as point mutations. Recently chromosomal mutations are studied separately. The term mutation is now given only to point mutations.

Types of gene mutation

Number of ways to classify gene mutations. On the basis of the molecular nature of the defect. On the nature of the phenotypic effect-- amino acid sequence of the protein is altered or not. On the basis of the causative agent of the mutation.

Mutations on the basis of the molecular nature of the defect

- Base substitution.
- Insertions & deletions.

Base substitution

It is the simplest type of gene mutation. Involves the alteration of a single nucleotide in the DNA. A base substitution usually leads to base pair substitution. Base substitution is of two types.

Transition

- Purine is replaced with a purine
- A into G
- G into A
- Pyrimidine is replaced with a pyrimidine
- C into T
- T into C

Transversions

- A purine is replaced by a pyrimidine
- A into C
- A into T
- G into T
- G into C
- **or a pyrimidine is replaced by a purine**
- C into G

- C into A
- T into G
- T into A

Insertions & deletions

2nd major class of gene mutation. Addition or the removal, respectively, of one or more nucleotide pairs. Usually changes the reading frame, altering all amino acids encoded by codons following the mutation Also called as frame shift mutations. Additions or deletions in the multiples of three nucleotides will lead to the addition or deletion of one or more amino acids. These mutations are called in-frame insertions and deletions, respectively.

Mutations on the basis of the Phenotypic effects of mutations

The most common phenotype in natural populations of the organism is called as wild-type phenotype. The effect of mutation is considered with reference to wild-type phenotype.

Forward mutation

A mutation that alters the wild type phenotype Reverse mutation (reversion). A mutation that changes a mutant phenotype back into the wild type.

Missense mutation

A base is substituted that alters a codon in the mRNA resulting in a different amino acid in the protein product.

Nonsense mutation

It changes a sense codon into a nonsense codon. Nonsense mutation early in the mRNA sequence produces a greatly shortened & usually nonfunctional protein. Silent mutation: alters a codon but due to degeneracy of the codon, the same amino acid is specified.

Neutral mutation

A mutation that alters the amino acid sequence of the protein but does not change its function as a replaced amino acid is chemically similar or the affected amino acid has little influence on protein function.

Loss of function mutations

Complete or partial loss of the normal function. The structure of protein is so altered that it no longer works correctly. Mutations can occur in the regulatory region that affects transcription, translation or splicing of the protein. Frequently recessive.

Gain of function mutations

It produces an entirely new trait. It causes a trait to appear in inappropriate tissues or at inappropriate times in development. It is frequently dominant.

Conditional mutations

It expressed only under certain conditions.

Lethal mutations

It causes the death of the organism.

Suppressor mutation

Suppresses the effect of other mutation

It occurs at a site different from the site of the original mutation. The organism with a suppressor mutation is a double mutant but exhibits the phenotype of unmutated wild type. It is different from a reverse mutation in which the mutated site is reverted back into the wild type sequence.

On the basis of causative agent of mutation

Spontaneous

Mutations that result from natural changes in DNA.

Induced

The results from changes caused by environmental chemicals & radiations. Any environmental agent that increases the rate of mutation above the spontaneous is called a mutagen such as chemicals & radiations.

Chemical Mutagens

The first discovery of a chemical mutagen was made by **Charlotte Auerbach**.

Base Analogs

Chemicals with structures similar to that of any of the four standard bases of DNA. DNA polymerases cannot distinguish these analogs. They may be incorporated into newly synthesized DNA molecules.

2-amino purine (P)

Base analog of adenine. Normally pairs with thymine. May mispair with cytosine. Causes a transition mutation. Both base analogs produce transition mutations. Mutations by base analogs can be reversed by the treatment with the same analog or different analog.

Alkylating agent

Chemicals that donate alkyl groups e.g. ehylmethanesulfonate (EMS). It adds an ethyl group to guanine and produces 6-ethylguanine, which pairs with thymine and leads to CG: TA transitions. Also adds an ethyl group to thymine to produce 4-ethylthymine, which then pairs with guanine, leading to a TA: CG transition. Mutations produced by EMS can be reversed by additional treatment with EMS. Mustard gas is another alkylating agent. Nitrous acid produces exclusively transition mutations. Both C.G T.A & T.A C.G transitions are produced. Thus mutations can be reversed with nitrous acid.

Hydroxl amine

Specific base modifying mutagen which adds a hydroxyl group to cytosine producing hydroxlamine cytosine which pairs with adenine instead of guanine. This Leads to C.G T.A transitions. It acts only on cytosine thus can not revert the mutation produced.

Oxidative reactions

Reactive forms of oxygen like superoxide radicals, hydrogen peroxide and hydroxyl radicals produced in the course of normal aerobic metabolism or by radiation, ozone, peroxides, and certain drugs cause damage to DNA & induce mutations by chemical changes. Oxidation converts guanine into 8-oxy-7,8-dihydrodeoxyguanine which mispairs with adenine leading to G.C: T.A transversion.

Intercalating agents

Proflavin, acridine orange, ethidium bromide, and dioxin. They are about the same size as a nucleotide. They produce mutations by sandwiching themselves (intercalating) between adjacent bases in DNA. They distort the three-dimensional structure of the helix and cause single-nucleotide insertions and deletions in replication. These insertions and deletions frequently produce frameshift mutations.

Radiations

Ionizing radiations. In 1927, **Herman Muller** demonstrated that mutations could be induced by X-rays. X-rays, gamma rays, and cosmic rays are all capable of penetrating tissues and damaging DNA. They remove electrons from the atoms that they encounter, changing stable molecules into free radicals and reactive ions which then alter the structures of bases and break phosphodiester bonds in DNA. Ionizing radiation also frequently results in double-strand breaks in DNA.

Mutation rates

The frequency with which a gene changes from the wild type to a mutant is referred to as the mutation rate. Expressed as the number of mutations per biological unit i.e. mutations per cell division, per gamete per round of replication e.g. the mutation rate for achondroplasia (hereditary dwarfism) is about 4 mutations per 100,000 gametes.

Mutation frequency

Incidence of a specific type of mutation within a group of individual organisms e.g. for achondroplasia, the mutation frequency in the United States is about $2x10^{-4}$.

22. GENE FREQUENCIES IN POPULATION, HARDY-WEINBERG LAW

Gene pool is the complete set of genetic information in all individuals within a population. Genotype frequency = proportion of individuals in a population with a specific genotype. Genotype frequencies may differ from one population to another. Allele frequency is the proportion of any specific allele in a population. Allele frequencies are estimated from genotype frequencies. Allele frequencies when mating is random.

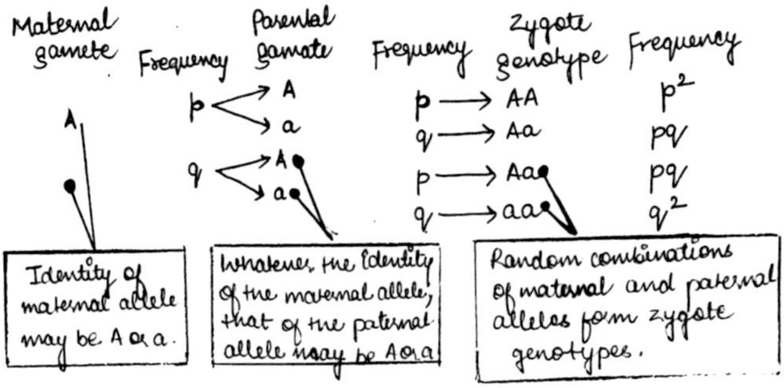

Hardy–Weinberg Law

When gametes containing either of two alleles, *A* or *a*, unite at random to form the next generation, the genotype frequencies among the zygotes are given by the ratio.

$p^2 : 2pq : q^2$ this constitutes the Hardy–Weinberg (HW) Principle

p = frequency of a dominant allele *A*

q = frequency of a recessive allele *a*

$p + q = 1$

One important implication of the HW Principle is that allelic frequencies will remain constant over time if the following conditions are met:

- The population is sufficiently large.
- Mating is random.
- Allelic frequencies are the same in males and females.
- Selection does not occur = all genotypes have equal in viability and fertility.
- Mutation and migration are absent.
- Another important implication is that for a rare allele, there are many more heterozygotes than there are homozygotes for the rare allele.
- HW frequencies can be extended to multiple alleles.
- Frequency of any homozygous genotype = square of allele frequency = p_i^2.
- Frequency of any heterozygous genotype = 2 x product of allele frequencies = $2p_ip_j$.

Allele

Mutation is the origin of new genetic capabilities in populations. They have an ultimate source of genetic variation. Natural selection is the process of evolutionary adaptation. The genotypes best suited to survive and reproduce in a particular environment give rise to a disproportionate share of the offspring. Migration is the movement of organisms among subpopulations. Random genetic drift = the random, undirected changes in allele frequencies, especially in small populations.

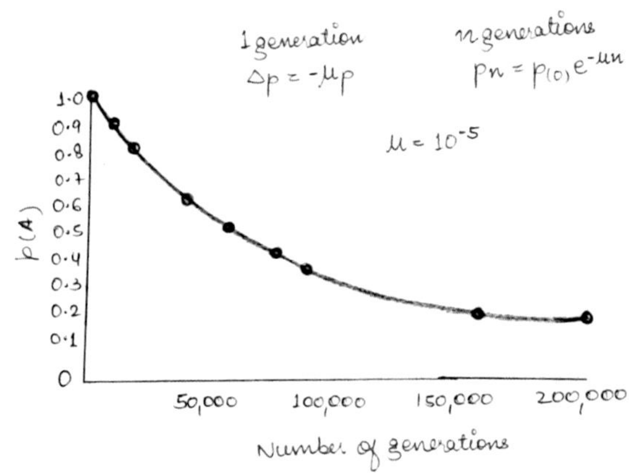

Selection and fitness

Organisms differ in their ability to survive and reproduce, and some of these differences are due to the genotype. Fitness (W) is the relative ability of genotypes to survive and reproduce. Relative fitness measures the comparative contribution of each parental genotype to the pool of offspring genotypes in each generation. The selection coefficient refers to the selective disadvantage of a disfavored genotype.

Selection in diploids

The frequency of the favored dominant allele changes slowly if the allele is common. The frequency of the favored recessive allele changes slowly if the allele is rare. Rare alleles are found most frequently in heterozygotes. When a favored allele is dominant, the recessive allele in heterozygotes is not exposed to natural selection.

Heterozygote superiority

Heterozygote superiority is fitness (measurement of viability and fertility) of a heterozygote is greater than that of both homozygotes. When there is heterozygote superiority, neither allele can be eliminated by selection. In sickle cell anemia, the allele for mutant hemoglobin is maintained in high frequencies in regions of endemic malaria because heterozygotes are more resistant to this disease.

Selection in Diploids

Selection can be balanced by new mutations. New mutations often generate harmful alleles and prevent their elimination from the population by natural selection. Eventually, the population will attain a state of equilibrium in which the new mutations exactly balance the selective elimination.

Migration

The population on an island (i) has migrants (m) from the mainland. Allele frequency in the next generation is the weighted average of the two populations

$$p'_{(i)} = (1\text{-}m)p_{(i)} + (m)p_{(m)}$$

Random genetic drift

Some changes in allele frequency are random due to genetic drift. Random genetic drift comes about because populations are not infinitely large. Only relatively few of the gametes participate in fertilization = sampling. With random genetic drift, the probability of fixation of an allele is equal to its frequency in the original population. The chance that a new mutation will be lost: (2N-1)/(2N). The chance that a new mutation will become fixed in the population:1/(2N).

Inbreeding

Inbreeding means mating between relatives. Inbreeding results in an excess of homozygotes compared with random mating. In most species, inbreeding is harmful due to rare recessive alleles that wouldn't otherwise become homozygous. A convenient measure of the effects of inbreeding is based on the reduction of heterozygosity H_I and is called the inbreeding coefficient F

$$F = (2pq - H_I)/2pq$$

The overall genotype frequencies in the inbred population are

$$f(AA) = p^2(1 - F) + pF$$
$$f(Aa) = 2pq(1 - F)$$
$$f(aa) = q^2(1 - F) + qF$$

Molecular evolution

The discovery that DNA is the genetic material made it possible to compare corresponding genes even in distantly related species. DNA and protein sequences contain information about evolutionary relationships among species. Comparative studies of macromolecules, the study of how and why their sequences change through time constitutes molecular evolution. Accumulation of sequence differences through time is the basis of molecular systematics, which analyses them in order to infer evolutionary relationships. A gene tree is a diagram of the inferred ancestral history of a group of sequences. A gene tree is only an estimate of the true pattern of evolutionary relations. Neighbor joining is one of the ways to estimate a gene tree. Bootstrapping is a common technique for assessing the reliability of a node in a gene tree. Taxon is the source of each sequence. A gene tree does not necessarily coincide with a species tree:

- The sorting of polymorphic alleles in the different lineages.
- Recombination within a gene makes it possible for different parts of the same gene to have different evolutionary histories.

The rate of sequence evolution is the fraction of sites that undergo a change in some designated time interval number of replacements per site per billion years. Rates of evolution can differ dramatically from one protein to another. There are different kinds of nucleotide sites and nucleotide substitutions depending on their position and function in the genome.

Synonymous substitution no change in amino acid sequence = primarily at the third codon position.

Nonsynonymous substitution = amino acid replacement.

- Rates of the evolution of nucleotide sites differ according to their function.
- New genes usually evolve through duplication and divergence.
- Orthologous genes are duplicated as an accompaniment to speciation, retain the same function.

Paralogous genes are duplicated in the genome of the same species, acquire new or more specialized functions. Pseudogenes are duplicated genes that have lost their function.

Population genetics

Population genetics = the application of genetic principles to entire populations of organisms. Population = a group of organisms of the same species living in the same geographical area. Subpopulation = any of the breeding groups within a population among which migration is restricted. Local population = a subpopulation within which most individuals find their mates (www.scribd.com., www.slideshare.com., www.coursehero.com., www. biocyclopedia.com., www. molecular-plant-biotechnology.info., www.slideplayer.com., www.documents.mx., www.docplayer.net., www. bmb.psu.edu.,www.archive.org.,www.mafiadoc.com., www.powershow. com., www.bioscience.jbpub.com., www.outhorstream.com., www. docslide.com., www.msluay.com., www.scienceprofonlue.com., www.csus. edu., www.krishikosh.egrouth.ac.in., www.tankonyvtor.hu., www.silo.pub., www.studyblue.com., www.Preservearticles.com., www.biotechuniverse. blogspot.com).

PART IV

BIOSTATISTICS AND COMPUTERS

23. GRAPHIC AND DIAGRAMMATIC REPRESENTATIONS, CLASSIFICATION AND TABULATION, MEASURES OF CENTRAL TENDENCY AND DISPERSION, SAMPLING THEORY AND ERRORS

Statistics

It is the science which deals with the development and application of the most appropriate methods for the:

- Collection of data.
- Presentation of the collected data.
- Analysis and interpretation of the results.

It is the quantitative information of any data such as birth and death rate, health management, production of medicine, profit and loss of different industries. Another language, it is the collection, compilation, presentation, analysis and interpretation of both qualitative and quantitative information about data for example in one family planning, a number of born children can be collected from married women. The characteristics to be measured depend on the objectives of the study. For example: study the knowledge of child health care, child mortality, child fertility and some other characteristics to be recorded are age of mother, education of mother, number of born children, duration of marriage, breastfeeding period, number of dead children and proper vaccination is done or not in a month.

Other Definitions for "Statistics"

It is frequently used in referral to recorded data. It denotes characteristics calculated for a set of data: sample mean.

Role of Statisticians

To guide the design of an experiment or survey prior to data collection. To analyze data using proper statistical procedures and techniques. To present and interpret the results to researchers and other decision makers. The term is used when the tool is applied to the analysis of data that is come from biological organisms. Statistics of marriage, birth-death rate, migration, family planning, level of education, health care of pregnant women and many more problems that are affecting the welfare of mankind. Definition: Science of collection, presentation, analysis, and reasonable interpretation of data. Statistics presents a rigorous scientific method for gaining insight into data. For example, suppose we measure the weight of 100 patients in a study. With so many measurements, simply looking at the data fails to provide an informative account. However, statistics can give an instant overall picture of data based on a graphical presentation or numerical summarization irrespective of the number of data points. Besides data summarization, another important task of statistics is to make an inference and predict relations of variables Figure 23.1, 23.2 and 23.3.

Taxonomy of Statistics

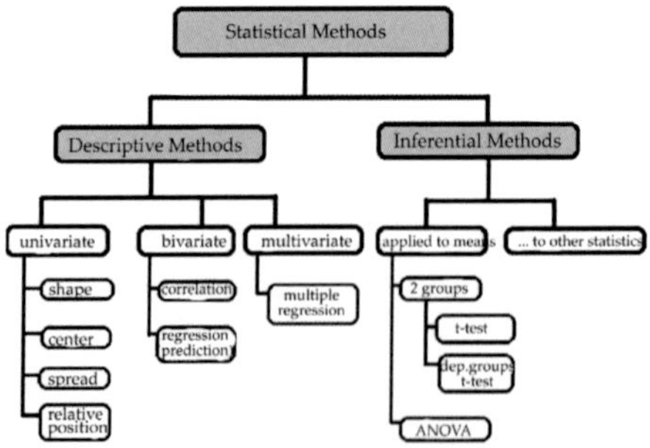

Figure 23.1 Source of data

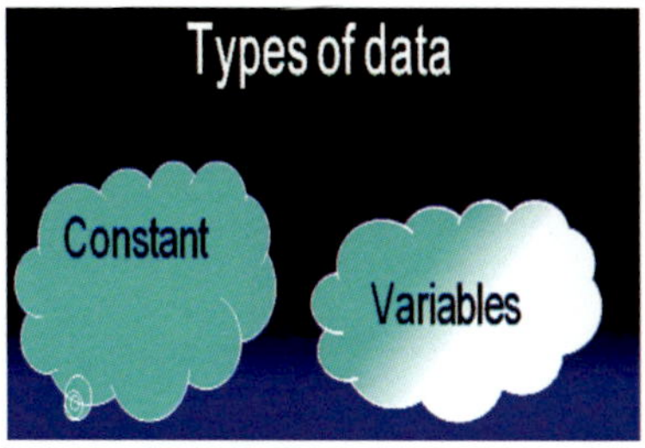

Figure 23.2 Types of data presented

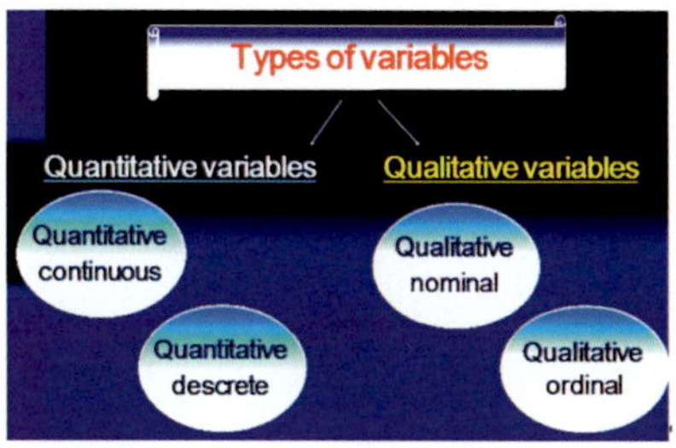

Figure 23.3 Types of variables

Methods of presentation of data

- Numerical presentation
- Graphical presentation
- Mathematical presentation
- Numerical presentation
- Tabular presentation (simple – complex)

Simple frequency distribution Table I

Name of variable (Units of variable)	Frequency	%
- - Categories -		
Total		

Table (I): Distribution of 50 patients at the surgical department of Alexandria hospital in May 2008 according to their ABO blood groups

Blood group	Frequency	%
A	12	24
B	18	36
AB	5	10
O	15	30
Total	50	100

Table (II): Distribution of 50 patients at the surgical department of Alexandria hospital in May 2008 according to their age

Age (years)	Frequency	%
20-<30	12	24
30-	18	36
40-	5	10
50+	15	30
Total	50	100

Complex frequency distribution table

Table (III): Distribution of 20 lung cancer patients at the chest department of Alexandria hospital and 40 controls in May 2008 according to smoking

Smoking	Lung cancer				Total	
	Cases		Control			
	No.	%	No.	%	No.	%
Smoker	15	75%	8	20%	23	38.33
Non smoker	5	25%	32	80%	37	61.67
Total	20	100	40	100	60	100

Table (IV): Distribution of 60 patients at the chest department of Alexandria hospital in May 2008 according to smoking & lung cancer

Smoking	Lung cancer				Total	
	positive		negative			
	No.	%	No.	%	No.	%
Smoker	15	65.2	8	34.8	23	100
Non smoker	5	13.5	32	86.5	37	100
Total	20	33.3	40	66.7	60	100

2- Graphical presentation

Graphs drawn using Cartesian coordinates

- Line graph
- Frequency polygon
- Frequency curve
- Histogram
- Bar graph
- Scatter plot

Pie chart

Statistical maps

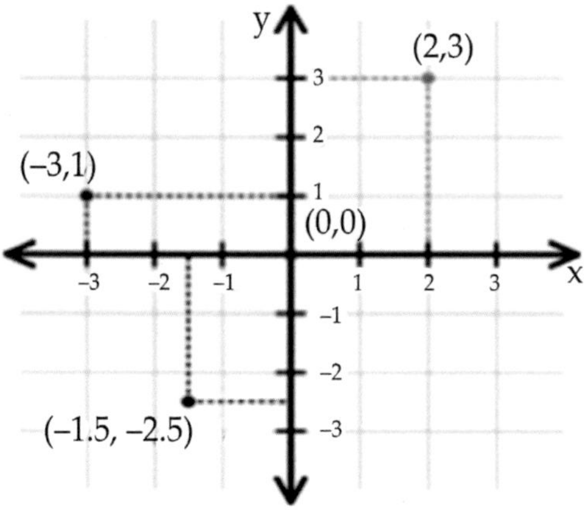

Line Graph

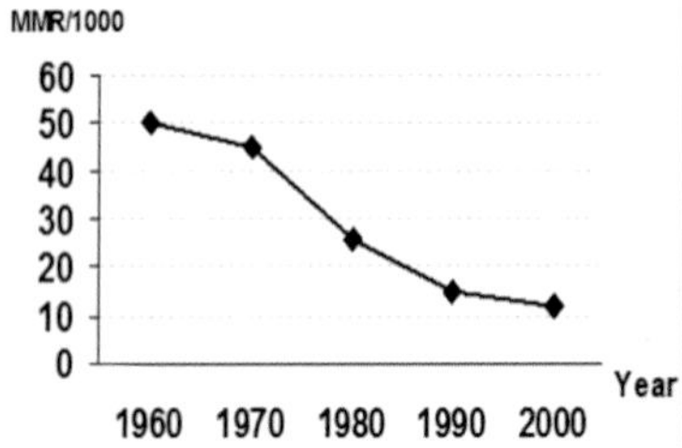

Year	MMR
1960	50
1970	45
1980	26
1990	15
2000	12

Figure 23.4: Maternal mortality rate of India 1960-2000

Frequency polygon

Age (years)	Sex		Mid-point of interval
	Males	Females	
20 -	3 (12%)	2 (10%)	(20+30) / 2 = 25
30 -	9 (36%)	6 (30%)	(30+40) / 2 = 35
40-	7 (8%)	5 (25%)	(40+50) / 2 = 45
50 -	4 (16%)	3 (15%)	(50+60) / 2 = 55
60 - 70	2 (8%)	4 (20%)	(60+70) / 2 = 65
Total	25(100%)	20(100%)	

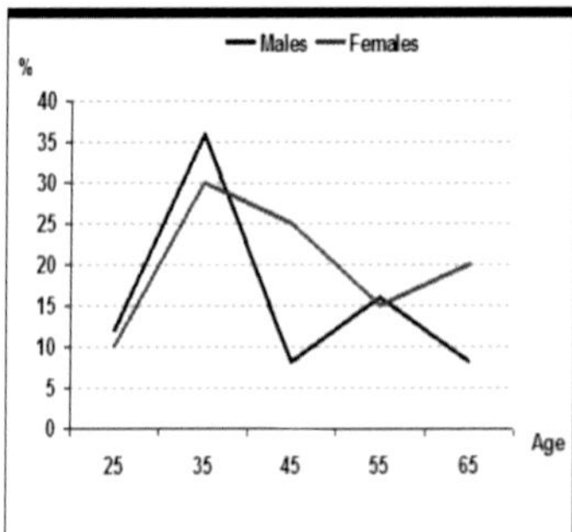

Age	Sex		M-P
	M	F	
20-	(12%)	(10%)	25
30-	(36%)	(30%)	35
40-	(8%)	(25%)	45
50-	(16%)	(15%)	55
60-70	(8%)	(20%)	65

Figure 23.5: Distribution of 45 patients at a place in time by age and sex.

Frequency curve

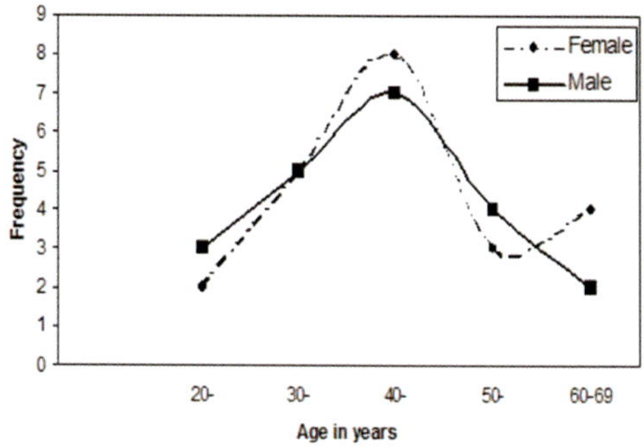

Histogram

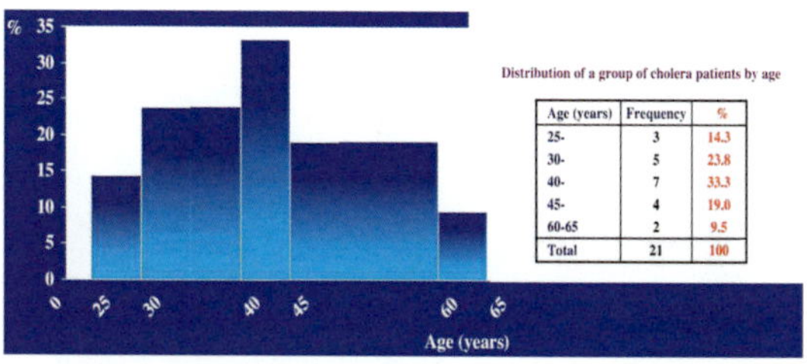

Age (years)	Frequency	%
25-	3	14.3
30-	5	23.8
40-	7	33.3
45-	4	19.0
60-65	2	9.5
Total	21	100

Figure 23.6: Distribution of 100 cholera patients at a place in time by age.

Bar chart

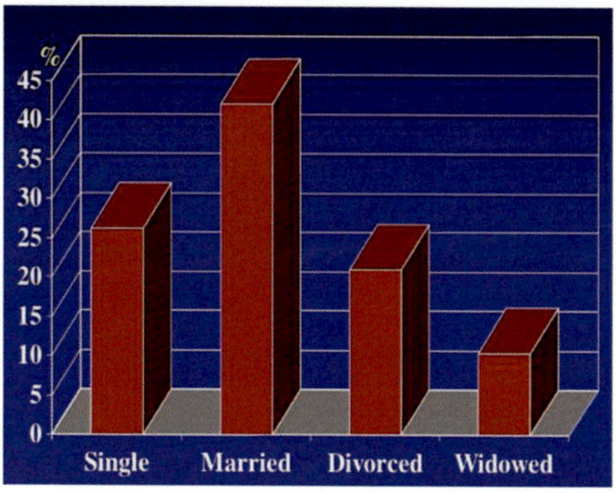

Figure 23.7: Marital Status

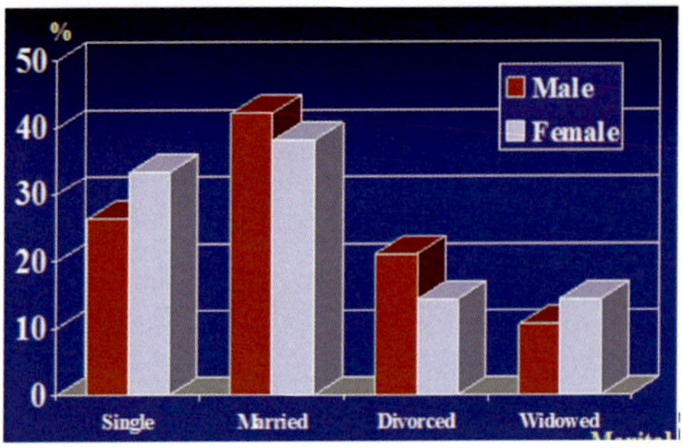

Figure 23.8: Marital Status on the basis of gender.

Pie chart

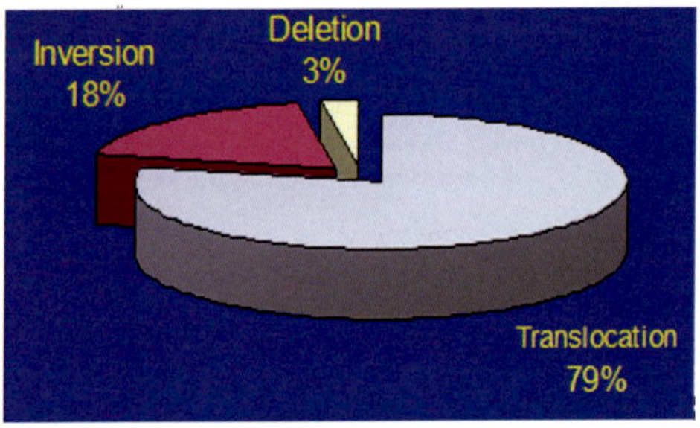

Figure 23.9 Pie chart.

Mathematical presentation

Measures of location

1- Measures of central tendency

2- Measures of non-central locations

(Quartiles, Percentiles)

Measures of dispersion

- Range
- Variance
- Standard deviation
- Semi-interquartile range

- Coefficient of variation
- "Standard error"

Measures of central tendency (averages)

- Midrange
- Smallest observation + Largest observation
- 2

Mode

The value which occurs with the greatest frequency i.e. the most common value

Median

The observation that lies in the middle of the ordered observation.

- Arithmetic mean (mean)
- Sum of all observations
- Number of observations
- Standard deviation SD

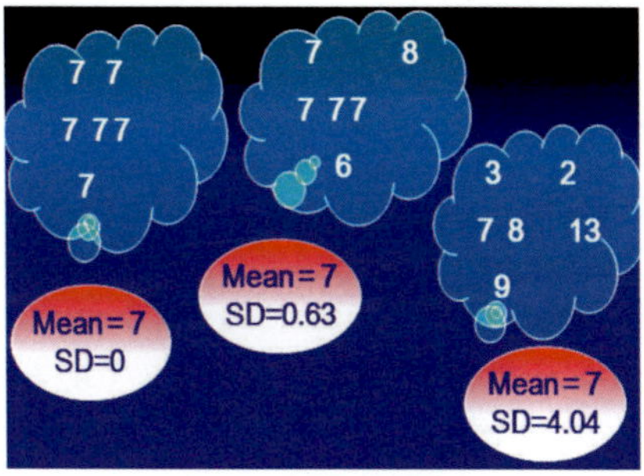

Standard error of mean SE

A measure of variability among means of samples selected from a certain population.

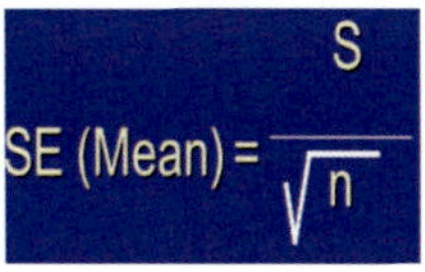

Importance of statistics

It is applied for the aggregate of facts. Statistics are affected by the number of causes by multiplicity. It should be capable of being related to each other, so their effect on the relationship can be induced. This is applied to the standard of accuracy that should be maintained for the analysis of data. Statistics is helpful for the presenting of small to large quantity of data in a simple or very difficult form. This tool is applied for the comparison of data. There are different tools that used in a number of fields viz. railways, bank, army, buildings, plants and animals, etc. It tries to give material for the administrators so as to serve as a guide to shaping or planning in future policies.

Problems restricted in statistics

- It can be used to analyze only collective matters no single events.
- It is applying for only the quantitative data, not for the quality data.
- It is associated with only uncertain samples like biased.
- This is done by only the true sample average run.

Uses of biostatistics

Public health and medicine community

The usefulness of vaccines and sera to be used in the field of the percentage of attacks or deaths among the vaccinated subject is compared with the unvaccinated ones. To measures the affected peoples in publically health related. Differentiated epidemiological studies for the role of causative factors are applied for the evaluation.

Medicine

Identify the symptoms and signs of a syndrome or a disease. Typhoid and cough are found by chance and fever is almost found in every case. The association is done between the attributes such as cancer and smoking. To find out the effect of a drug on the patient or the efficacy of drugs among the people who are taking for the analyses. To check the data or drug to know the percentage of cured and deaths after applying a drug to the patients with control groups for the comparisons.

Anatomy and physiology

Check the correlation between the two factors or variables e.g. X and Y (such as height or weight and color or height) for the observation of normal in a population. To check out the means or average proportions of normal or two places in different modes. To compare the healthy or unhealthy population in data.

Pharmacology

To know the potency of the new drug with respect to the standard drug (a standard drug is taken for the comparisons of tested drugs). Take different drugs or two successive dosages of the same drug for the comparisons of which one is more effective for the prevention of diseases. The drug is given to the animals or humans to view the changes due to the drug effect or due to the off-target effect (means by chance) to know the action of the drug.

In this method, we collect data from some selected units under investigation so that the quality of data can be ensured.

Data

It is a collection of observations expressed in numerical figures. It may be done in two ways.

- Complete enumeration
- Survey method

Types of Data

Statistical data can be divided into two ways

Qualitative

Quantitative

Qualitative data

There is no numerical relation with one another such as

- SEX = Female and Male
- Color of Skin = Fair and Dark
- Color of Eye = Brown and Black

Some important facts

- The results are expresses as ratio or proportion form.
- The persons with same character are counted to form groups.
- It has no magnitude.
- It is always discrete

Quantitative data

- There is a numerical relation with one another and it may be continuous or discrete Such as:
- Discrete = Number of students, Number of Bags, Number of pencil and Number of books
- Continuous = Weight of person and Height of person.
- **Important facts**
- Such data are analyzed through statistical methods such as mean, median, mode and standard deviation, etc.
- It has magnitude.
- It is both continuous and discrete.
- It is arranged by both frequency and character.

In this method, we collect data from all population units under investigation example money and time.

There are two methods of data collection has been discussed.

- Primary data
- Secondary data

Primary data

- The data are collected by personal interview or by mail.
- These data are called as raw data.
- These data are collected directly from the field of enquiry for specific purposes.
- These are raw data that is original in nature and directly collected from the population.

- The primary collection of data may be made through either by complete enumeration or sampling survey methods.

Example

- The amount of fasting blood sugar of some persons (male and female) recorded in the first 10 minutes by a working day.

Primary data collection

- Schedules sent through investigators
- Direct personal observation
- Questionnaires sent by email
- Indirect oral presentation

Questionnaire

- This is the Performa containing a sequence of questions for the statistical enquiry.
- This is used for the collection of the primary data from individual persons through their responses.

Primary data advantages

- It provides details information but information may be suppressed in secondary data.
- It is free from errors.
- It contains information regarding methods of procuring data whereas primary data often included.
- It is cost-effective.
- It is less time-consuming process or suitability.
- It gives accurate results for the data analysis.

Population

- There are a group of peoples or study the elements for measurements having some common fundamentals characteristics.

(a) **Finite:** Population has a fixed value e.g., number of days in a year.

(b) **Infinite:** Population consists of an endless succession of values e.g., number of plants in the ocean.

Secondary data

- The data are collected from official records, from published work or by a second person. Example census reports are known as secondary data.
- These are numerical information that has been collected by some agency for a specific purpose and are subsequently compiled from that source for application in a different connection.

Population

- It is an entire group of people or study elements-persons, things or measurements having some common fundamentals characteristics.

Infinite

- If a population consists of an endless succession of values, it is said to be infinite e.g., number of animals in the ocean.

Finite

- If a population consist of a fixed number of value, it is said to be finite e.g., number of days in a week.

Statistical Error

- There are errors during the collection of data or error shows the extent to which the observed value of a quantity exceeds the true value.

Error = Observed value-True value

Types

- It is classified as:
- **Biased errors:** due to personal prejudices or bias of investigator.
- **Unbiased errors:** statistical enquiry due to chance causes.

ARRAY

- Presenting data in ascending order of magnitude is known as array.

Tally

- A value occurs more than 4 times the fifth occurrence is denoted by a cross (\) tally mark and running diagonally across the four tally marks.

Examples: Form a frequency table for the different variables.

1, 4, 7, 8, 5, 8, 4, 9, 23, 56, 23, 45, 67, 78, 25, 4, 8. 9, 4, 1, 23, 78, 56, 9, 45, 78, 25, 67, 25, 78, 45, 23.

Solution:

Variables	Tally	Frequency
1	II	6
4	IIII	7
7	I	4
8	III	6
9	III	7
11	III	3
23	IIII	5
55	IIIII	3
56	II	5
45	III	7
67	II	7
78	IIII	3
25	IIII	5
78	IIII	6

The data are collected by a survey is called raw data. The raw data are not always suitable for proper statistical analysis. So, we have to classified, tabulated and presented by graphs and diagrams.

Different graphs and diagram

There are many important diagrams and graphs used in biostatistics. The representation of data through charts and diagrams is known as the graphical representation of data.

Advantages of graphical representations

- It is easily understood.
- The data is presented in a more attractive form.
- It shows the tendency and trends of values of the variable.
- It is useful to detect mistakes.
- It shows easily the relationship between two data sets.
- It is universally applicable.
- It is helpful for the assimilation of data quickly.

Limitations of graphical representation data

- It does not show all the facts.
- It can reveal only the approximate position.
- It can take lot of time.

Different types of diagrammatic representations

There are different types of graphs in the form of diagrams and charts such as:

- Line diagram
- Bar diagram
- Pie diagrams
  - These three are used for the qualitative presentation of data
- Stem-and leaf plot
- Histogram
- Ogive (cumulative frequency polygon)
- Scatter diagram
- Frequency polygon
  - These five are used for the quantitative presentation of data.
- Statistics is an important part for all students from graduation to Ph.D. level.

- Here we are going to discuss stats via using some important software that make them very easy as well as less time-consuming.
- Excel is basic stat software that should be known to each and every student of any discipline. After excel, some other advanced stats tools are also discussed here in this chapter such as Origin Pro software.
- The data in the form of raw scores is known as ungrouped data and when it is organized into frequency distribution then it is referred to as grouped data.
- Separate modes and methods are used to represent these two types of data ungrouped and grouped.

Line diagram

- The data is usually represented as lines. The plotted data on the Y-axis is connected to the X-axis by a line. The resultant diagram is called the Line diagram.

Mean

- The sum of a collected data and divide it by the number of the set of data. It is also called average. You can see how to calculate the mean or average. **For example**

No. of boys participate in cricket games and get marks.

3, 3, 4, 5, 3, 6, 2, 1, 7, 8 and 10 (N = 11)

Calculate the mean = 3+3+4+5+3+6+2+1+7+8+10= 52/11= 4.72

Merits and demerits of mean

Merits

- It has the simplest formula that is understandable easily and easy to compute.
- It is defined rigidly by a mathematical formula that gets the same results for a single problem.
- It is based on all the observation calculations.
- It is based on the simplest fluctuations.
- The mean is typical i.e. it balance the value on either side.

- It is the best for the two series measure for the comparison.
- It is calculated using the value.
- It does not depend upon any position.

Demerits

- If the values are not known, it cannot be calculated.
- The value is affected by the extreme mean.
- It is not determined by the qualitative data such as honesty, love, beauty ,etc.

Uses

- It is always obtained by the mean.
- It is used in practical statistics.
- The common people use the mean for calculating average marks obtained by students.

Mean deviation

- The mean deviation is called the average deviation. It is the average difference between the items in a distribution and the median and mean of that series.

MEAN DEVIATION

- It is about the mean.
- It is about the median

Co-efficient of Mean Deviation

It is the ratio of mean deviation to its arithmetic mean or median multiplied by 100.

C.M.D = MD× 100

Mean/ Median

Calculation of Mean Deviation

Ungrouped data

- It can be calculated mean and median. Calculate the deviation from mean denoting by "D" and ignoring the sign positive (+) or negative (-).

- Where "n" is the total number of observation items.
- Divide the sum of the $\sum D$ by the total number of items.

Mean Deviation (MD) = $\frac{\sum |X - X|}{n} = \frac{\sum |D|}{n}$
Mean Deviation about the mean MD = $1/N \sum |X - X| f$
Mean Deviation about the median MD = $1/N \sum f(x - \text{median})$

Median

- It is the value in given data that divides into two equal parts such that half of the observation is below or another half are above it.
- It is the middle most point or the central value of the variable in a set of observations when observations are arranged either in ascending or in descending order of their magnitudes.

"It is the value of that in a series which decides the series into two equal parts, one part consisting of all values greater than it". **(Prof. Ghosh & Chowdhury).**

- Simple series (ungrouped data)

The arrangement of data is either descending or ascending order. If the number of observation is odd, the value of the middle is most of the median. However, if the number be even, the arithmetic mean of the two middle most items is taken as the median.

When "n" is odd. In this case, the (n+1/2)th value is the median.

= M = (n+1/2)th term.

When "n" is even then there are two middle terms n/2 and (n+1/2)th. The median is the average of these two terms.

$$M = \frac{n/2 + (n/2 + 1)}{2}$$

Example1

Find the median of the following numbers

- 21, 12, 49, 37, 88, 46, 55, 74, 63
- 88, 72, 33, 29, 70, 86, 54, 91, 61, 57.

Solution

(a) Let arrange the data in order: 12, 21, 37, 46, 49, 55, 63, 74, 88.

In this data the number of item is n = 9 (odd)

Median

= average of (n/2)th + (n/2+1) terms.

Average of (10/2)th and (10/2) th +1 terms.

= average of 5th and 6th terms

M = 61+70/2 = 131/2 = 65.2

Median is 65.2

Merits and demerits of the median

Merits

- It is no popular as the mean and it is easily understood.
- It is not affected by the variation in the magnitude of the extreme values.
- The value of the median can be graphically ascertained to gives.
- It is the best measure for qualitative data such as intelligence and beauty , etc.
- It can be determined even by inspection in many cases.

Demerits

- Data must be arranged for the calculation of the median.
- The positional average cannot be dependent on each and every observation.
- It is subjected to algebraic treatment.
- It is more affected by sampling fluctuations than the arithmetic mean.

Mode

There is a given value of the data set which occurs most frequently. For example, let us consider the age of boys, where ages are **8, 5, 10, 9, 5, 8, 10, and 8**. Here 8 is repeated most of the time or we can say occurs the maximum time. Therefore, the mode is 8, because 8 occurs most frequently in the given data set value. The data follows the description given in the median and follow step same as in the median and get the values of mode. According to the CRAFT and COWDEN, "The mode of the distribution of the value at the point around which the items tend to be most heavily concentrated. It may be regarded as the most typical value".

- Ungrouped data (simple series): Mode can be determined by locating that value that occurs the maximum number of times. It is that value of the variable which corresponds to the largest frequency.

Example 1

Find out the Mode of the following data.

1,3, 1, 3, 3, 5, 3, 3, 1, 5, 3, 3, 4, 5, 4, 2, 3, 2, 3, 7, 6, 3, 2, 5, 2, 3, 3, 2, 6, 2, 3, 2, 4, 2, 3.

Solution

Let prepare the table

Values	Number of items (f)
1	3
2	8
3	14
4	3
5	4
6	2
7	1

Here, 3 is repeated 14 times and is the most frequent; so the mode is 3.

The word average denotes representatives of a whole set of observations. It is a single figure that describes the entire series of observations with their varying sizes. It is the value that occupies the central position where some observation is larger and some are smaller than it. The values of variables tend to concentrate around the central value is called as the central tendency.

Characteristics of central tendency

- It is rigidly defined.
- It is based on all the items.

- It is affected by extreme values.
- It should be least affected by the fluctuations of the sampling.
- It should be easy to interpret.
- It is easy to be subjected to further mathematical calculations.

Measures of central tendency

- Mean
- Median and
- Mode

Here we describe some of the basic statistics, which are divided into four sections (1) Descriptive statistics (2) Graphing data (3) Association statistics (4) Comparative statistics

Descriptive statistics

There is a use of simple graphics as well as simple calculations to estimate the collected data. It involves different parameter values like mean, median, mode, range, variance, standard deviation, standard error ,etc.

24. INTRODUCTION TO PROBABILITY, AND DISTRIBUTION, CORELATION, REGRESSION, SKEWNESS AND KURTOSIS

Theoretical distribution

Normal distribution was first discovered in 1733 by mathematician De Movire. He obtained this continuous distribution as a limiting case of the binomial distribution. So the normal distribution is also called the Gaussian distribution named after Karl Friedrich Gauss, who used this normal curve to describe the theory of accidental errors of measurements involved in the calculation of orbits of heavenly bodies. It is a continuous probability distribution that is bell-shaped unimodal and symmetrical. The normal distribution of a variable when represented graphically takes the shape of a symmetrical curve is called the Normal curve.

Properties of the normal distribution and normal probability curve

- It has two parameters viz. mean (μ) and standard deviation (σ) It is bell-shaped and symmetrical about the line x = μ.
- It has only one mode occurring at μ i.e., it is unimodal.
- The mean, median and mode value will coincide at the center (x = μ) because of the symmetrical and single-peaked.
- Mean = Median = Mode = μ
- The normal curve has asymptotic tails i.e. progressively nearing the abscissa or x-axis.
- The range is unlimited in both directions but as the distance from μ increases, the curve approaches the horizontal axis more and more closely and never touches the horizontal axis (X).
- The two points of inflections, the point where the change in curvature occurs at a distance σ on either side mean.
- The points of inflections of the normal curve are at x = μ ± σ.
- The curve changes from concave to convex and vice versa.
- The quartiles of the first (Q_1) and the third (Q_3) are equidistant from μ.

$Q_3 - \mu = \mu - Q_1$

Quartile deviation = $Q_3 - Q_1/2$

= 0.67458

$Q_1 = \mu - 0.6745$

$Q_3 = \mu + 0.6745$

- Thus the range of the normal distribution from -0.6745 to 0.6745.
- The normal distribution is bilaterally symmetrical so, it is free from skewness so its coefficient of skewness amounts to zero.

Skewness = 0

Kurtosis = 0

- The ordinates maximum (y) lies at the mean i.e., at x = μ.
- Its value is y = 1/ σ √ 2x. When S.D increases the maximum ordinate decreases.
- Since the mean = median = μ, the coordinates at x = μ (or Z = 0) divides the whole area into two equal parts.

- The area right to the ordinate and left to the ordinates at x = μ (Z = 0) is 0.5.
- The mathematical equation is completely determined if the mean (μ) and standard deviation (σ) of the variable are completely known to fix the curve.
- The curve will always remain symmetrical about the maximum ordinate. But the change in the value of mean (μ) and the standard deviation (σ) or in both will alter the shape of the curve.
- The maximum ordinate is at the mean (μ) and at various standard deviation (σ) the distances are in a fixed proportion to the ordinate at the mean when the total mean is under the normal curve is equal to unity.
- There is the following table given under the normal probability curve for some important values of Z.

Distance from the eman ordinate in terms of ± σ	***Area under the curve***
Z = ± 0.6745	50% = 0.50
Z = ±1.0	68.27% = 0.6827
Z = ±1.96	95% = 0.95
Z = ±2.00	95.45% = 0.9545
Z = ±2.58	99% = 0.99
Z = ±3.0	99.73% = 0.9973

Uses of normal distribution

- It is applied in sampling theory.
- It can be used to approximate the binomial and Poisson distribution.
- It has considerable application in statistical quality control.
- It is mostly used in statistical hypotheses.
- It is used to test the significance value.
- It has different mathematical properties that make it popular and comparatively easy to manipulate for use in social natural sciences.

Best fitting normal distribution

- It is the normal distribution that fits best with an observed distribution and has the same mean (X), the same standard deviation (σ) and the same sample size (n) as the latter.
- The mean value, the standard deviation is computed.
- So, the X_m is transformed into a Z score.
- $\mathbf{Z = X_m - X/\sigma}$

Example 1

The best fitting of the normal distribution for the following blood glucose concentration of 80 patients of the Faridabad district hospital. Calculate the mean and the standard deviation of the given data.

Class Interval	**100-109**	**110-119**	**120-129**	**130-139**	**140-149**	**150-159**	**160-169**
Frequency	6	11	10	17	16	13	7

Solution

Let assume the mean 134.5

Class Interval	*Mid value (X_m)*	*Frequency (f)*	*Assumed mean A = 134.5 X_m - A*	*X_m – A/ C = d*	*fd*	*fd^2*
100-109	104.5	6	104.5-134.5 = -30	-3	-18	54
110-119	114.5	11	114.5-134.5 = -20	-2	-22	44
120-129	124.5	10	124.5-134.5 = -10	-1	-10	10
130-139	134.5	17	134.5-134.5 = 0	0	0	0
140-149	144.5	16	144.5-134.5 = +10	+1	+16	16
150-159	154.5	13	154.5-134.5 = +20	+2	+26	52
160-169	164.5	7	164.5-134.5 = +30	+3	+21	63
		$\sum f$ = 80			*$\sum fd$ = 63-50 = 13*	*239*

Mean = A + ∑ fd/N × i

X = 134.5 + 13/80 × 10

= 134.5 + 1.625

= 136.125

= 136.1

Standard Deviation (SD) $= \sqrt{\sum fd^2/N - (\sum fd/N)^2} \times i$

$= \sqrt{239/80 - (13/80)^2} \times 10$

$= \sqrt{2.988 - 0.02656} \times 10$

$= 10\sqrt{2.988 - 0.0266}$

$= 10\sqrt{2.96}$

$= 10 \times 1.72$

$= 17.2$

- The deviation of each (X_m) from (X) is then transformed into Z score which entered in the following table.

Class Inter-val	*Frequen-cy (f)*	*Mid Value (x)*	$X_m - X$	*Z = Xm – X/ S.D*	*Y*	*Y*
100-109	6	104.5	104.5 – 136.1= -31.6	-31.6/17.2 = -1.84	0.0734	3.4
110-119	11	114.5	114.5 – 136.1 = -21.6	-21.6/17.2 = -1.26	0.1804	8.4
120-129	10	124.5	124.5 – 136.1 = -11.6	-11.6/17.2 = -0.67	0.3187	14.8
130-139	17	134.5	134.5 – 136.1 = -1.6	-1.6/17.2 = -0.09	0.3973	18.5
140-149	16	144.5	144.5 – 136.1 = +8.4	8.4/17.2 = 0.49	0.3538	16.5
150-159	13	154.5	154.5– 136.1 = 18.4	18.4/17.2 = 1.07	0.2251	10.5
160-169	7	164.5	164.5– 136.1 = 28.4	28.4/17.2 = 1.65	0.1023	4.8
	$\sum f = 80$					

Neglecting the algebraic sign of the Z score, the height y of the ordinate at each Z score is then recorded from the unit normal curve table.

The Y ordinate height of the best fitting normal distribution is computed for each Z score by multiplying its y score with in/ SD (I = class interval , n = total frequency, SD = standard deviation); for example:

$Y = y \times in/ SD$

$= 0.1804 \times 10 \times 80/17.2$

$= 0.1804 \times 46.51$

$= 8.39$

$= 8.4$

Binomial distribution

Binomial distribution was discovered by Swiss mathematician James Bernoulli (1654-1705). It is derived from the process is called the Bernoulli trial. It is the discrete probability distribution that is obtained when the probability (P) of the happening of an event is the same in all the trials and there are only two events in each trial. It is a theoretical probability distribution because it can be worked out theoretically using the series of the terms of the binomial equation.

Condition under binomial distribution

The trials or events must be repeated under the conditions. The number of trials should be finite and also small. The events or trials must be independent it means the happening of one event or trials must not affect the happening of other events. The variable should be discrete: it means the values of X should be 1, 2, 3, 4, or 5 ,etc. It never be 1.5, 2.5, 3.7, 5.5 ,etc. It should be either a success or failure so that dichotomy exists, i.e., the happening of an event has two possible outcomes.

Properties of binomial distribution

- It is presented the discrete probability distribution.
- It has two parameters "p" or "q" the probability of success or failure and "n" the number of trials.
- The parameters "n" is always an integer.
- The mean (μ), SD (σ) variance (σ^2) and the coefficient of dispersion (C.D) of a binomial distribution of cases of the class having the proportion "p" in the population are obtained from the sample size (n) and the proportions (p and q) of the cases in the two classes.
- Mean (μ) = $\boldsymbol{np}$
- Standard deviation (σ) = $\boldsymbol{\sqrt{npq}}$
- Coefficient of dispersion (CD) = $\boldsymbol{npq/np = q}$
- Variance (σ^2) = $\boldsymbol{npq}$

Kurtosis and Skewness of binomial distribution depend on the proportion of p and q in the population.

- Skewness (Y_1) = $(q - p)^2/npq$
- Kurtosis (Y_2) = $3 + 1 - 6\ pq/npq$
- It is symmetrical of $p = q = 0.5$
- It is positively skewed if $p < 0.5$

- It is negatively skewed if $p > 0.5$
- The binomial coefficients are given by **Pascal's triangle**.

Computational binomial probabilities

Multinomial distribution

It may be regarded as a generalization of the binomial distribution for accommodating any number of variables when they are more than two mutually exclusive outcomes of a trial the observations lead to multinomial distribution.

Suppose that

E_1, E_2, E_3 E_4……E_K are k mutually exclusive and exhaustive outcomes of an event trial with respective probabilities p_1, p_2, p_3, p_k will occur k_1, k_2, k_3, k_4…k_n times respectively is

$\mathbf{C = N/ k_1 | k_2 | k_3 | Kn \times P_1^{k1} P_2^{k2} \ldots\ldots P^{kn}}$

Where $\mathbf{k_1 + k_2 + k_3 + kn = N}$

C = the number of permutation of the events E_1, E_2….E_k

Single term of the expansion

It represents the number of ways in which the conditions of each term may be satisfied. The number of combinations (C) of n different taken k at a time is expressed by

$n^{C}k = n/ k\ n-k$

Bernoulli expansion

It is having total number of classes, events with probability of the Bernoulli expansion:

n = total number of events

p = classes

q = classes

X = number of classes in p classes

n-X = number of cases in q classes

Probability of P (X) is expressed by Bernoulli expansion

$\mathbf{P(X) = n\ p^x\ q^{n-x} / X\ |N - X|}$

Binomial expansion

It is having a total number of events and trials with a probability of occurrence of success or failure.

n = total number of events/ trials

p = probability of occurrence of success

q = probability of occurrence of failure.

p + q = 1 or q = 1- p

The general expression of Binomial expansion

$(p+q)^n = p^n + np^{n-1} + n\,(n-1)/\,1\times2\,p^{n-2}q^2 + n\,(n-1)\,(n-2)/\,1\times2\times3\,p^{2-3}q^3 + ...+ n\,(n-1)\,(n-2)\times3\times2/\,1\times2\times3\times.........\times(n-1)\times pg^{n-1} + q^n$

Probability distribution of the binomial expansion for r success in n trials is given by:

$P(r) = {}^nC_r\, q^n - r\, p^r$

It has mean = ***np*** and variance = ***npq***.

Mean, Variance and Standard deviation of Binomial distribution

The probability of distribution of binomial distribution for r success in n events and trials is given by

$P(r) = {}^nC_r\, q^n - r\, p^r$

So, mean (μ) = np
Variance (σ^2) = npq
Standard deviation (σ) = √npq
n =number of independent events
p = probability of success
q = probability of failures

Example 1

Calculate the probability of getting head three times when a coin is tossed 5 times.

Solution

When a coin is tossed there are only two outcomes either head or tail.

Therefore it shows the binomial distribution. If the probability of getting a head then it is "p" and of tail, it is "q". The probability of getting a head 3 times can be calculated in the following way:

n =5, X = 3, P = ½ = 0.5, Q ½ = 0.5

$$\mathbf{P (X) = n\, p^x q^{n-x} / X (n-X)}$$

$= 5 (0.5)^3 (0.5)^{5-3} / 3 (5-3)$

$= 5.4.3.2 \times (0.5)^3 \times (0.5)^2 / 3.2.2$

$= 5 \times (0.5)^5$

$= 10 \times 0.5 \times 0.5 \times 0.5 \times 0.5 \times 0.5$

$= 5 \times 0.03125$

$= 0.3125$

Example 2

The two male and female children in a family of four children by applying the binomial theorem.

Solution

Total number of children = 2 + 2 = 4

Male = 2

Female = 2

Initial probabilities of male child (p) = 2/4 = ½ = 0.5

Initial probabilities of female child (q) = 2/4 = ½ = 0.5

According to the binomial equation

$$(p + q)^4 = p^4 + 4p^3q + 6\, p^2\, q^2 + 4pq^3 + q^4$$

p = 6p2q2

= 6. (0.5)2 (0.5)2

$= 6 \times 0.5 \times 0.5 \times 0.5 \times 0.5$

$= 6 \times 0.0625$

$= 0.375$

So, the probability of two boys and two girls in a family is 0.375.

Example 3

In two hundred families with three children, a population of the Arambagh subdivision is sampled at random. How many families do we expect to have

(a) no girls (b) one girl (c) two girls? Assume the sex ratio to be 1:1?

Solution

Probabilities for girls and boys = ½

g for girls and b for boys.

Now we expand the binomial (g and b), n = 3

$(g + b)^3 = g^3 + 3g^2b + 3gb^2 + b^3$

No girls relate to b3 term

(1) ½ × ½ × ½ = 1/8 = 1/8 ×200 = 25 (200 = families)

(2) One girls relates to $3gb^2$ term

3 × ½ (1/2)2 = 3 ×1/2 × ¼ = 3/8 = 3/8 ×200 = 75

(3) Two girls relate to $3g^2b$ term

3 × (1/2)2 ½ = 3× ¼ ×1/2 = 3/8 = 3/8 × 200 = 75

Example 4

A plant breeder has 45 different inbred strains of pea plants. How many different hybrids can be obtained from a total 45 plants?

Solution

Hybrid has two genes

N = 45 r = 2

According to the formula:

nCr = n/r (n –r)

= 45/ 2 (45 -2)

= 45 × 44 ×43/ 2.1 ×4 ×43

= 45 ×44/2

45 × 22 = 990

Example 5

In a family with two children in Serampore subdivision where both parents are heterozygous for albinism. What proportion of this family would be expected to have (a) neither child with albinism (b) one child with albinism, (c) both children with albinism?

Solution

Let the symbol "a" for albinism and "A" for normal

Expand the binomial expansion:

$(A + a)^2 = A^2 + 2Aa + a^2$

The parents are heterozygous so therefore probabilities of normal ¾ and albino ¼

Two children in a family both are normal i.e., A2 = (3/4)2 = 9/16

Among the two children, one is with albinism i.e., 2Aa = 2Aa = 2 × ¾ × ¼ = 6/16

Both children with albinism i.e . $a^2 = (1/4)^2 = 1/16$.

Example 5: Consider the parents of a Sinha Roy family in which both of them heterozygous for a severe genetic syndrome, that is autosomal recessive. Of their six children, five of them have this particular syndrome. How unlucky is this family?

Solution

n = 6

p = ¾

q = ¼

Probabilities for diseases (t) = 5

Normal (s) = 1

$P = n \times p^{i} q^{5} /s \times t$

= 6 × (3/4)1 (1/5)5

= 6 ×5/ 5 ×3/4 ×1/1024 / 1 × 5

= 18/4096

= 0.00439

= 0.0044

Example 6

A couple of heterozygous for albinism (Aa). What is the probability that (a) 4 out of 6 children born to them are normal: (b) 4 normal and 2 albino out of 6 children?

Solution

Let “a” = allele for albinism

A = allele for normal skin color

Generally, heterozygous parents have ¾ normal children and ¼ albino children.

$Pa = 3/4$

$q\ (a) = 1/4$ and n = 6

Probability of 4 children being normal

$P_A = (3/4)^4 = 3/4 \times 3/4 \times 3/4 \times 3/4 = 81/256 = 0.316$

n = 6, s = 4 and I = 2 (4 normal and 2 albino)

$P = n \times (p)^4 \times (q)^2 / s \times t$

$= 6 \times (3/4)4 \times (1/4)2$

$= 6 \times 5 \times 4 \times 3/4 \times 3/4 \times 3/4 \times 3/4 \times 1/4 \times 1/4\ /\ 2 \times 4$

$= 15 \times 81/256 \times 1/16 = 1215/4096 = 0.2966$

$= 0.297$

Through binomial expansion:

$(p+q)^6 = p^6 + 6p^5q + 15\ p^4 q^2 + 20\ p^3 q^3 + 15\ p^2q^4 + 6\ pq^5 + q^6$

$P = 15p^4q^2$

$= 15 \times (3/4)4 \times (1/4)2$

$= 15 \times 81/256 \times 1/16$

$= 1215/4096$

$= 0.2966$

$= 0.297$

Example 6

Four babies were born in Aligarh general hospital (a) what was the chance that two will be boys and two girls (b) what was the chance that all four would be girls?

Solution

The probabilities of boys and girls were ½ = 0.5

P $= \frac{1}{2}$

$q = \frac{1}{2}$

(a) n = 4. s = 2 and t = 2
(b) (s) for girls and
(c) (t) for boys
(d) $P = n \times p^s \times q^t / s \times t$
(e) $= 4 \times (1/2)^2 \times (1/2)^2 / 2\times2$
(f) $= 4.3.\ (2) \times 1/4 \times 1/4 / 2.1(2)$
(g) $= 6 \times \frac{1}{4} \times \frac{1}{4}$
(h) = 3/8

Probabilities was 3/8

(a) $(p+q)^4 = p^4 + 4\, p^3q + 6\, p^2q^2 + 4\, pq^3 + q^4$
(b) $p = q^4\ (1/2)^4 = 1/8$

Example 7

There are eight children in a family, where both parents are heterozygous for albinism what mathematical expression predicts the probability that six are normal and two are albinos?

Solution

As both parents are heterozygous the probabilities of normal is ¾ and albinos ¼

i.e., $p = \frac{3}{4}$

$q = \frac{1}{4}$

n= 8

s = 6 and t = 2

$P = n \times p^s \times q^t / s \times t$

$= 8 \times 3/4 \times (1/4)2 / 6\ (2)$

$8.7\ (6) \times (3/4)6 \times (1/4)2$

$= 28 \times (3/4)^6 \times (1/4)^2$

$= 28 \times 3/4 \times 3/4 \times 3/4 \times 3/4 \times 3/4 \times 3/4 \times 1/4 \times 1/4$

$= 7 \times 729/16384$

= 5103/16384

= 0.31146

Example 17

A multiple allelic system is known to consist of seven alleles. Assuming that this is a diploid species, how many different genotypes could exist in the population?

Solution

Number of possible genotypes = Number of different allelic combination (heterozygotes) × number of genotypes with two same allele (homozygotes)

= n + n /k (n – k)

n = 7

k = 2 (heterozygotes)

= (7)/ 2 (7-2) +7 = 7.6 (5) / 2.5 +7 = 21 +7

= 28 genotypes

Poisson distribution

- The poison distribution was derived by French mathematician Simeon Denis Poisson (1837) and is known as Poisson distribution.
- It represents the Poisson distribution of discrete, random variables of rare events whose probability occurrence is very small but the number of events / trials is very large and may approach infinity. It is basically applicable to such events where the binomial expression formula can be used in determining theoretical probabilities.
- So the definition of Poisson distribution is a discrete probability distribution of rare events and the mean and the variance is equal.

Characteristics

- The events are independent and random.
- It is a limited form of the binomial distribution.
- It may be expected in cases where the chance of any individual event being a success is small.
- It has a single parameter, the mean of the distribution.
- The mean and the variance are equal.
- It is a discrete probability distribution because it is a probability distribution of whole numbers (0, 1, 2, 3…..n) of events.

- It is positively skewed and declines with the rise of the mean.
- It is leptokurtic which decreases with the increase in the mean.

Conditions when the Poisson distribution is used

- It is applicable when the observation or number of events is very large but the probability of success is very small.
- The random variables should be discrete.
- A dichotomy exists i.e., the happening of the events must be divided into two classes viz., Success or failure occurrence or nonoccurrence.
- P should be small or zero.
- It is independent it means the happening of one event does not affect the happening of other events.

Computation of Poisson distribution

- The random variable(X) is applicable for the probability of distribution so it is said to have Poisson distribution.

$P(X) = e^{-m} m^x / x$

P = Probability of success

x = variables (such as 0, 1, 2, 3 and n)

e = constant 2.7183 (base of natural logarithm

m = finite positive constant and is known as parameter of Poisson distribution.

Number of success ***(X)***	**0**	**1**	**2**	**3**	**R**	**N**	**Total**
Probabilities ***P(X)***	e^{-m}	$e^{-m} m/1$	e-m $m^2/2$	$e^{-m} m^3/3$	$e^{-m} m^r/r$	$e^{-m} m^n/n$	1

$P(0) + p(1) + p(2) + p(3) + p(r) + p(n) + \ldots\ldots\ldots\alpha$

$= e^{-m} + e^{-m} m/1 + e\text{-}m\ m^2/2 + e^{-m} m^3/3 + e^{-m} m^r/r + e^{-m} m^n/n + \ldots\ldots\ldots\alpha$

$= e^{-m} (1 + m/1 + m^2/2 + m^3/3 + m^r/r + m^n/n + \ldots\ldots\ldots\alpha$

$= e^{-m} e^{m} = e^{0} 1 = 1$

Mean = m = p

Standard deviation = $\sqrt{m}$

Skewness given by (β_1) = 1/m

Kurtosis given by (β_2) = 3 + 1/m

Variance = m

Examples of Poisson distribution

- The number of fish death in tanks in one week due to some pesticide treatment in water.
- The number of bacterial colonies in a given culture per unit area on a microscopic slide has been seen under a microscope.
- The emission of radioactive particles **(alpha = α)**
- The number of mistakes committed by a good typist per page.
- The numbers of buses passing through a certain road (M.G Road, Agra).
- The number of diseases or death by cancer or heart attack in any city like in (Agra) hospitals in a year.

Example 1

In a Biotechnology book of 520 pages, 390 typological error occurs. Assuming Poission's law for the number of pages error per page, find probabilities that random sample of 5 pages will contain no error?

Solution

Here n = 5, Book has 520 pages

Typological errors 390 pages

Therefore probabilities (P) = 390/ 520

= 0.75

Mean np = 5 × 0.75

= 3.75

Using Poisson probabilities law

$\boldsymbol{P(r) = e\text{-}^{m}\, m^{r}/\, r = e^{-0.75}\, m^{r}/r}$

Probabilities error zero,

Therefore $P(0) = e^{-0.75}\,(3.75)^{0}/0$

$\boldsymbol{= e^{-0.75} 3.75}$

Example 1

The Biostatistics book with 585 pages contains 43 typological errors. If these errors are randomly distributed throughout the book, what is the probability that 10 pages, selected at random will be free from errors?

Solution

Here n = 10

Book has 585 pages

Typological errors 43 pages

Therefore probabilities $P = 43/585 = 0.0735$

Mean (m) $np = 10 \times 0.0735$

$= 0.735$

Poisson distribution $(Pr) = e^{-m}\ m^r/r$

$= -0.735 \times (0.735)^r/\ r$

Probability zero error $P\ (0) = e^{-0.735} \times (0.735)^0\ /\ 0$

$= e^{-0.735} \times 1$

$= e^{-0.735}$

= 0.4795

Example 3

If the mean of the Poisson distribution is 4 then find the S.D, $ß_1$, $ß_2$, μ_3, μ_4

Solution

Here m = 4

Variance = m = 4

S.D = √4

= 2

Skewness ($ß_1$) = 1/m

=1/4

=0.25

Kurtosis ($ß_2$)

= 3 +1/m

$= 3 + ¼$

0.325

$\mu_3 = m$

= 4

$\mu_4 = m + 3m^2$

$= 4 + 3\,(4)^2$

$= 4 + 3.16$

$= 4 + 48 = 52$

Example 4

The following data are obtained from vector cytogenetics research laboratory of Dr. M.P.S College Agra.

Events (x_1)	*Frequency (f)*
0	20
1	26
2	16
3	4
4	2

1. What is the mean?
2. What is e-m?
3. What is P(0)?

Solution

Mean (X) = ∑ fx/ ∑ f

$= 0+26+32+12+18 \,/\, 20+26+16+4+2$

$= 78/68$

1.147

$=1.15$

(X) = m = 1.15

$e^{-1.15} = 0.32$

$P(0) = e^{-m}\,(1.15)^0/0$

$= 0.32 \times 1 /1 = 0.32$

Example 5

In a family of 8 children where both parents are heterozygous for albinism, what mathematical expression predicts the probability that six are normal and two are albino?

Solution

Both parents are heterozygous, the probabilities of normal is 3/4 and albinos is 1/4.

i.e $p = 3/4$

$q = 1/4$

$n = 8$

$s = 6$ *and* $t = 2$

$P = n \times p^s\, q^t / s(t)$

$= 8/6\,(2) \times (3/4)^6 \times (1/4)^2$

= 8.7 (6)/ 2.1 (6) × (3/4)6 × (1/4)2

$= 28 \times (3/4)^6 \times (1/4)^2$

$= 28 \times 3/4\times3/4\times3/4\times3/4\times3/4\times3/4 \times1/4\times1/4$

$= 7\times729/ 1638$

$= 5103/16384$

$= 0.31146$

Correlation

- Select a new column for the correlation and then click on cell type CORREL and choose option and then enter.
- There are two type of correlations +1 (perfect correlation) 0 (no correlation) to –1 (perfect negative correlation).
- After calculating the correlation we can determine the probability of observed correlation occurred by chance means we can conduct a significance test. Most often we are using this to determine whether our hypothesis is a real one and not a chance occurrence. There are two hypotheses **(1) Null hypothesis and (ii) Alternative hypothesis**

Null hypothesis

- The hypothesis is usually verified for possible rejection by test, even it is true known as the null hypothesis. It is denoted by H_0. The formula is given as follows:
- $H_0 : \mu = \mu_0$

Alternative hypothesis

- The statement against the null hypothesis is known as the alternative hypothesis. Thus, if the null hypothesis is $H_0 : \mu = \mu_0$ then alternative hypothesis may be HA : $\mu \neq \mu_0$ or HA : $\mu < \mu_0$ or HA : $\mu > \mu_0$. The commonly are used significance level 0.05 for agriculture and 0.01 for the pharmaceutical industry. For example, if we get a correlation value r = 0.972, which is more than 0 means our value accepts the alternative hypothesis and rejects the null hypothesis.
- It is used to denote estimation or prediction of the average value of one variable for a specified value of the other variable. One of the variables is known as independent or the explained variable and the other is called dependent or the explaining variable.
- (*It is the measure of the average relationship between two or more variables in terms of the original units of the data.* ***M.M. Blair).***

Regression lines

- The bivariate data are plotted as points on graph paper it will be found that the concentrations point follows a certain pattern showing the relationship between the variables.
- When the trends points are found to be linear, we determine the best fitting straight line by Least Square Method.
- Such straight lines are used to obtain best estimates of one variable for given values of the other are called regression lines.
- If two variables are linearly related then that relationship can be expressed as Y = bx +a.
- Where: "b" is the slope of the line relating Y to X and "a" is the "Y" intercept of that line.
- ***"A line of regression is the straight line gives the best fit in the least square sense to given sets of data".***

Regression coefficient

- It is expression of how much one dependent variable (Y) may be expected to change per unit change in some other independent variable (X) is known as regression coefficient (b).

Types of regression: 4 types

- Simple regression

The ependent variable is a function of a single independent variable.

- Multiple regression

The dependent variable is a function of two or more variables.

- Linear regression

The dependent variable is linearly correlated with the predictor (independent variable). It forms a straight line.

- Nonlinear regression

The dependent variable has a nonlinear correlation with the independent variable. It forms a sigmoid or hyperbolic curve.

Properties regression

- The expression of the dependent variable is applied as a function of an independent variable.
- It works out only when there is a significant correlation between the dependent and independent variables.
- Regression predicts
- It is worked out using statistics for regression coefficient.

Methods of studying regression

There are two types of methods such as:

- Graphic Method
- Algebraic Method

Useful properties of regression

- It is a mathematical measure showing the average relationship between two variables.
- It is random variables (x) and y is a fixed variable.

- Sometimes both variables are randomly collect.
- It causes and affects the relationship between the variables.
- It is used for the prediction of one value in respect to the other given values.
- It is an absolute figure.
- It has wide application as it studies linear and nonlinear relationships between the variables.
- It explains that the decrease in one variable is associated with the increase in the other variable.

Skewness

- The distribution is said to be symmetrical when mean, median and coincide. It has three parts left tail and middle part also. It has also right and left tails that are equal in length.
- It is used to denote the extent of symmetry in the data.
- When the frequency distribution is not symmetrical it is said to be skewed.
- The meaning of skewness is "lack of symmetry". A symmetrical distribution has therefore zero skewness.

Characteristics

- It may be positive or negative.
  - Positive skewness:
- The curve of the distribution has a longer tail toward the right it means the higher values of the variable.
- Mean > Median > Mode
- If the curve of the distribution has a longer tail towards the left i.e. the lower values of the variable.
- Mean < Median < Mode
- Here Mean, Median and mode are failed to coincide. Both median and mean are displaced from the mode towards the skewed tail.

Mean > Median > Mode (Positive Skewed)

Mean < Median < Mode (Negatively Skewed)

Here the first quartile is displaced toward the skewed tail.

Therefore,

$Q_3 - Q_2 > Q_2 - Q_1$ (Positively Skewed)

$Q_2 - Q_1 > Q_3 - Q_2$ (Negatively charged).

Measures of skewness

It indicates not only the extent of skewness in numerical expression but also the direction i.e the number in which the deviations are distributed. It is normally measures of symmetry are called measures of skewness. The absolute measures are known as measures of skewness. It tells us the extent of symmetry whether it is positive or negative.

Absolute skewness = Mean – Mode

Mean > Mode (Positive skewness)

Mode > Mean (Negative skewness)

It is also known as the coefficient of skewness and it is given by $\beta_1 = \mu^2{}_3 / \mu^3{}_2$

μ_3 = 3rd moment and μ_2 2nd moment.

There are important measures of relative skewness.

Karl Pearson's coefficient of skewness

S_k = Mean – Mode/ Standard deviation

S_k = 3 (Mean – Median) / Standard deviation

Bowley's quartile coefficient of skewness

$Sk = Q_3 - 2Q2 + Q_1 / Q_3 - Q_1$ (Q_1 = First quartile; Q_2 = Second quartile)

Kelly's coefficient of skewness

$Sk = P_{90} + P_{10} - 2$ Median / $P_{90} - P_{10}$ (P_{10} = 10th Percentiles; P_{90} = 90 th Percentiles

Kurtosis

It is used to describe the degree of peakedness of a frequency distribution compared to that of normal distribution.

Characteristics

1. It measures the peakedness of a normal curve.
2. It is also called as measures of convexity of the curve.

3. It introduced the three broad patterns.
4. If the peakedness viz., (Leptokurtic, Mesokurtic and Platykurtic).

✓ The curve is neither flat nor peaked is known as mesokurtic i.e., normal curve.

✓ The curve which has higher and sharper peaked is called as mesokurtic; it means the normal curve.

✓ The curve is higher and sharper peaked and narrow-bodied, then the normal curve is known as Leptokurtic.

✓ The curve is flatter and its center, broader in the body and thinner at tails than normal curve is called as Platykurtic.

Measures of kurtosis

The frequency of distribution is based upon the fourth moment about the mean of the distribution.

$ß_2 = \mu_4/\mu^2_2$

$= \mu_4/ \sigma_4$

$ß_2 = 3$ Mesokurtic
$ß_2 < 3$ Platykurtic
$ß_2 > 3$ leptokurtic

Importance of skewness

- It tells the direction and extent of asymmetry in a series.
- It provides us an idea about the nature and degree of concentration of items.

Dispersion	Skewness
It spreads the individual's values about the mean it means central value.	It shows the departure from symmetry i.e. direction of variation.
It shows the degree of variability.	It shows the value is higher or the lower concentration.
It is the averages of deviation-average of the second order.	It is not the average but it is measured by the use of mean, median and mode.
It judges the truthfulness of the central tendencies.	It judges the truthfulness of the central tendencies.

Significance

It tells us the extent to which a distribution is more peaked or more flat-topped than the normal curve.

It also depends the shape of the top of a frequency curve.

Moments

It is used in mechanics, physics ,etc. It is also applied in statistics, it describes the various characteristics of frequency distribution viz., central tendency, dispersions and skewness and kurtosis. It can be defined also as the arithmetic mean of various powers of deviations taken from the mean of the distribution.

Role Moments

First moments (μ_1) of frequency distribution = $\sum$ X – X/ N	**It is always zero i.e μ = 0**	**It measures mean of the distribution μ_1 = X = 0**
Second Moments (μ_2) of frequency distribution about the mean is the variance of the distribution = $\sum$ (X –X)^2/N	$\mu_2 = \sigma_2$	It measures the variance i.e the spread of the different terms in a distribution.
Third Moments (μ_3) = $\sum$ (X-X)^3/N	It deals with skewness	It gives an idea about the degree of skewness present
Fourth Moments (μ_4) = $\sum$ (x – x)^4/N	It highlights the height of frequency distribution whether it is more peaked or flat-topped than normal	It measures kurtosis.

Example 1

The coefficient of skewness = 3, Mean = 90, Median = 80 Find the value of S.D.

Solution

S_k = 3 (Mean – Median) / S.D

S.D = 3 (Mean – Median) / Sk

S.D = 3 (90 -80)/ 3

= 3 × 10/ 3

= 10

Tests of significance such as Chi- square and F-test, Z, and t-test

In this, we compare the variables. We can find out variables are basically the same means they could originate from the same population or if they are significantly different means they have a different origin. So, there are two tests ANOVA (analysis of variance) and t-test (Figure 17.1).

Chi-square Test

The calculation of the quantity is used to compare an "observe" ration with an "expected" or "theoretical" ration and to determine how closely the former fits the letter so it involves the entire statistical test for the calculation or observation of the hypothesis to be significant or not significant value. In scientific research, first we make a hypothesis and do experiments after completion of the experiment we analyze the data that is correct or not so chi-square test for the analysis of our observation. So the measure of chi-square enables us to find out the degree of discrepancy between "observed" frequencies and " theoretical or "expected" frequencies and thus to determine where the observed and theoretical frequencies is due to the error of sampling or due to the chance. It is computed on the basis of frequencies in a simple and thus the value of chi- square so obtained is a statistic. Chi-square is not a parametric test as its value is not derived from the observations in a population. Hence, the chi-square test is a Non Parametric test.

Chi-square (X^2) test

A statistical test to determine the number of observations which is derived from those "expected" or "theoretical" number under a particular hypothesis.

$X^2 = i = 1 \sum (1\ (O - E)\ 1\ -1/2)^2 / E$

Where ½ = 0.5 Yates correction

O = observed frequencies

E = Expected Frequencies

Important characteristics of the chi- square test

- Chi-square will be zero if each pair is zero and it might assume any value extending to infinity when the differences between the observed frequency and expected frequency in each pair are unequal.
- Thus, chi-square lies between O and ∞
- It is a statistic not a parameter.
- It is always positive as each pair is squared up.

Useful points

Hypothesis test

- It is the procedure that specifies a set of "rue for decision" whether to "accept" or "reject" the hypothesis under consideration (i.e. null hypothesis).

Null hypothesis

- A hypothesis is set up and the validity is tested for possible rejection on the basis of sample observations is known as the Null Hypothesis.
- It is denoted by H_0
- It is tested against the alternatives.
- "**Null hypothesis** is the hypothesis which is to be tested for possible rejection under the assumption it is true" (**Prof R.A Fisher remarked**).

Alternative hypothesis

- The negation of the null hypothesis is called the alternative hypothesis.
- It is denoted by H1 and H∞.
- It is not tested, but its acceptance (rejection) depends on the rejection (acceptance) of the null hypothesis.
- The alternative hypothesis contradicts depends on the rejection of the null hypothesis.

Statistical hypothesis

- The statement or assertion about the statistical population or the value of its parameters is called a statistical hypothesis.
- It is two types of hypothesis

1. **Simple**
2. **Composite**

Simple hypothesis

- The hypothesis that specifies the population completely is called simple hypothesis.

Composite hypothesis

- The hypothesis that does not specify the population completely is called the composite hypothesis.

Rejection region

- The set values of the statistics which lead to rejection of the null hypothesis is called the rejection region of the test. The probability of the null hypothesis is rejected by the test is often referred to as "size" of the critical region.
- On the other hand the acceptance of the null hypothesis which gives us a region is called as "Acceptance region".

Statistics test

- After the arrangement of the null hypothesis and alternative hypothesis, the test of statistics is computed and it is based on the probability distribution. It is used to test whether the null hypothesis set up should be accepted or rejected.

Significance level

- The probability is maximum with which a true null hypothesis is rejected is known as the level of significance of the test.
- It is denoted by α.
- The level of significance is arbitrarily chosen in advance depending on the consequence of the statistical decision for the farming decision rules.
- The level of significance generally 5% and 1% is taken other levels such as 2% or ½ (0.05) % are also used.

Important facts

Hypothesis testing methodology

- Z- Test for the Mean (s Known)
- p-Value Approach to Hypothesis Testing
- Connection to Confidence Interval Estimation
- One Tail Test

- T-test of Hypothesis for the Mean
- Z-Test of Hypothesis for the Proportion

What is a hypothesis?

A hypothesis is an assumption about the population parameter.

A parameter is a Population mean or proportion

The parameter must be identified before analysis.

I assume the mean GPA of this class is 3.5

The Null Hypothesis, H_0

- States the Assumption (numerical) to be tested
- e.g. The average # TV sets in US homes is at least 3 (H_0: m ³ 3)
- Begin with the assumption that the null hypothesis is TRUE.
- (Similar to the notion of innocent until proven guilty)
- Refers to the Status Quo
- Always contains the ‘ = ‘ sign
- The Null Hypothesis may or may not be rejected.
- The Alternative Hypothesis, H_1
- Is the opposite of the null hypothesis e.g. The average # TV sets in US homes is less than 3 (H_1: m < 3)
- Challenges the Status Quo
- Never contains the ‘=‘ sign
- The Alternative Hypothesis may or may not be accepted

Identify the problem

Steps:

State the Null Hypothesis (H_0: m ³ 3)

State its opposite, the Alternative Hypothesis (H_1: m < 3)

- Hypotheses are mutually exclusive & exhaustive
- Sometimes it is easier to form the alternative hypothesis first.

Hypothesis testing process

Assume the mean age of the population is 50 (Null Hypothesis)

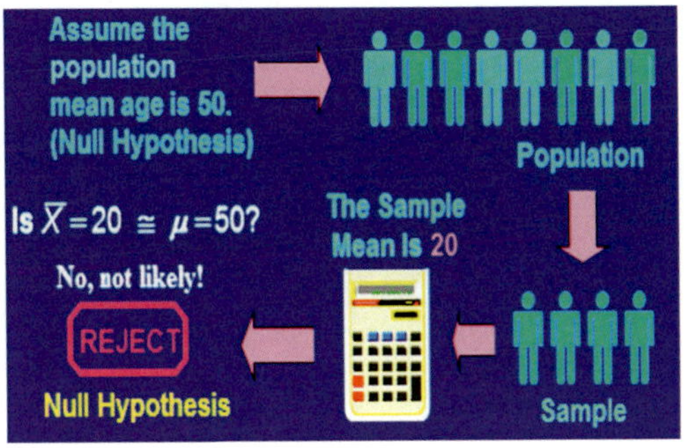

Reason for rejecting H_0

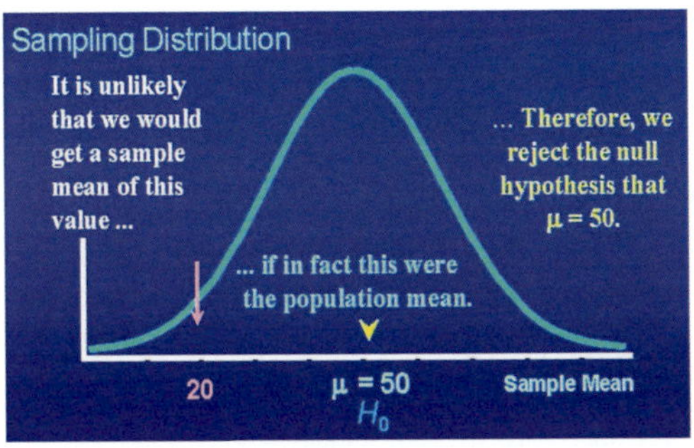

Level of significance, *a*

- Defines Unlikely Values of Sample Statistic if Null Hypothesis Is True
- Called Rejection Region of Sampling

Distribution

- Designated *a* (alpha)
- Typical values are 0.01, 0.05, 0.10
- Selected by the Researcher at the Start
- Provides the Critical Value(s) of the Test
- Level of Significance, *a* and the Rejection Region

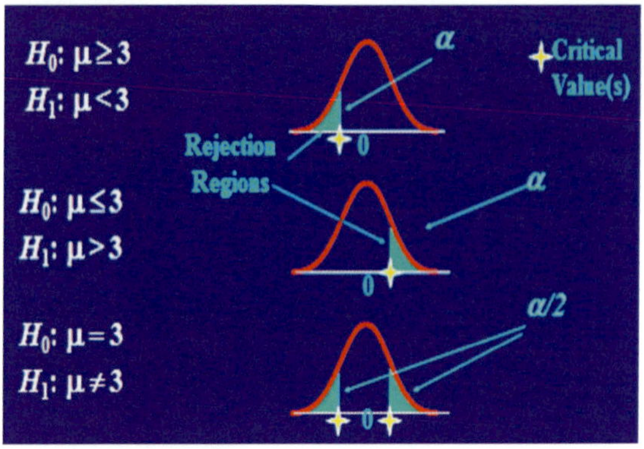

Errors in making decisions

Type I Error

- Reject True Null Hypothesis
- Serious Consequences
- Has Probability of Type I Error Is *a*
- Called Level of Significance

Type II Error

- Do Not Reject False Null Hypothesis
- Probability of Type II Error Is *b* (Beta)
- Result Possibilities

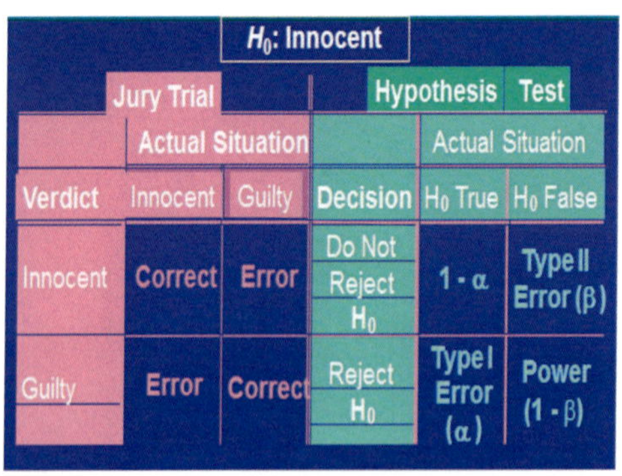

It has *a* & *b* an inverse relationship

25. TEST OF SIGNIFICANCE

Z-test statistics

It converts sample statistic (e.g., to standardized *z* variable. It compares to critical *Z* value(s). If the *Z* test Statistic falls in Critical Region, Reject H_0; Otherwise Do Not Reject H_0.

- value test

- Probability of Obtaining a Test Statistic More Extreme (£ or ³) than Actual Sample Value Given H_0 Is True
- Called Observed Level of Significance
- Smallest Value of a H_0 Can Be Rejected
- Used to Make Rejection Decision
- If *p*-value ³ a, Do Not Reject H_0
- If *p*-value < a, Reject H_0

Hypothesis testing: steps

- Test the Assumption that the true mean # of TV sets in US homes is at least 3.

Test the Assumption that the true mean # of TV sets in US homes is at least 3.		
1.	State H_0	$H_0: \mu \geq 3$
2.	State H_1	$H_1: \mu < 3$
3.	Choose α	$\alpha = .05$
4.	Choose n	n = 100
5.	Choose Test:	*Z Test (or p Value)*

Test the Assumption that the average # of TV sets in US homes is at least 3.

6. Set Up Critical Value(s)	*Z = -1.645*
7. Collect Data	*100 households surveyed*
8. Compute Test Statistic	*Computed Test Stat.= -2*
9. Make Statistical Decision	*Reject Null Hypothesis*
10. Express Decision	*The true mean # of TV set is less than 3 in the US households.*

One-Tail Z Test for Mean (s Known)

Assumptions

- Population Is Normally Distributed
- If Not Normal, use large samples
- Null Hypothesis Has £ or ³ Sign Only
- *Z* Test Statistic:

$$z = \frac{\bar{X} - \mu_{\bar{X}}}{\sigma_{\bar{X}}} = \frac{\bar{X} - \mu}{\sigma / \sqrt{n}}$$

Rejection region

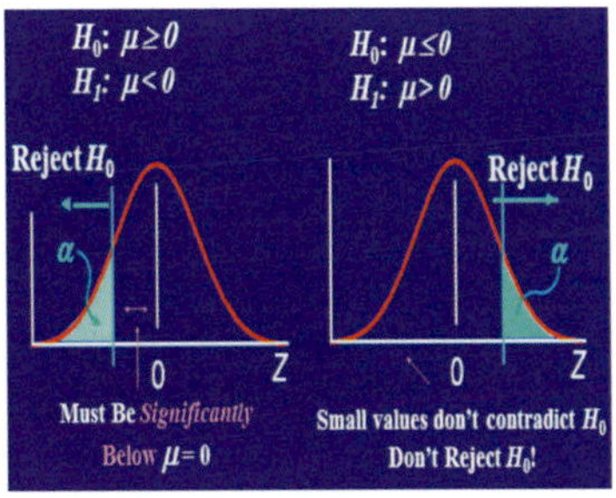

Example: one tail test

Does an average box of cereal contain **more than 368** grams of cereal? A random sample of **25** boxes showed **X = 372.5**. The company has specified **s** to be **15** grams. Test at the α=**0.05** level.

Finding critical values: one tail

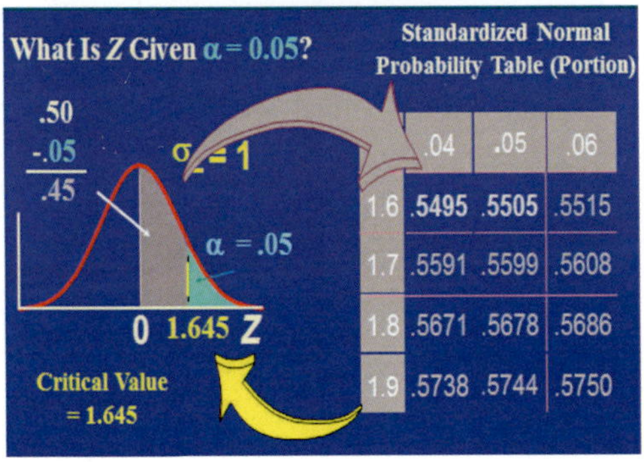

Example solution: one tail

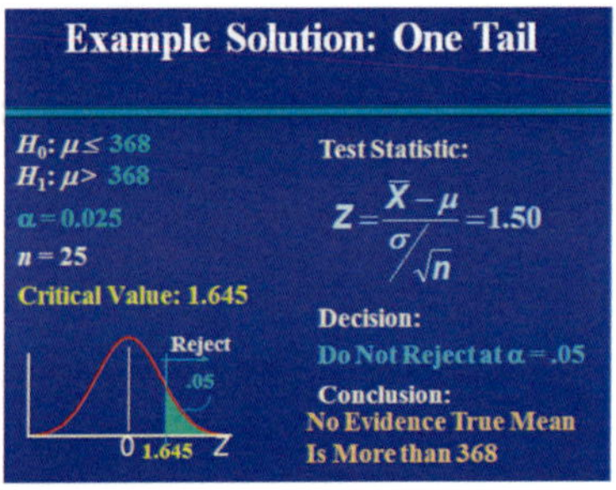

p value solution

p value is $P(Z ^3 1.50) = 0.0668$

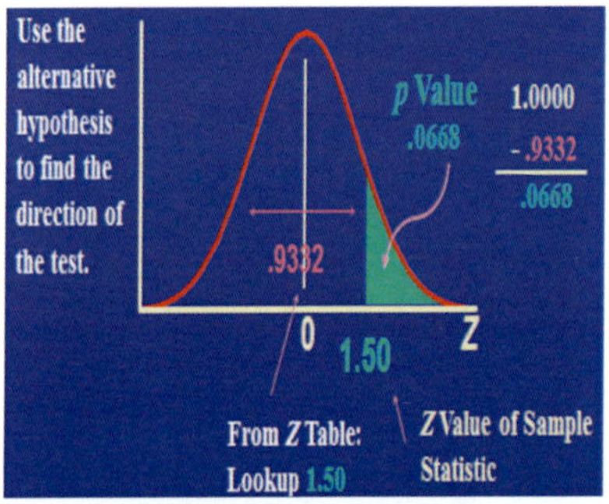

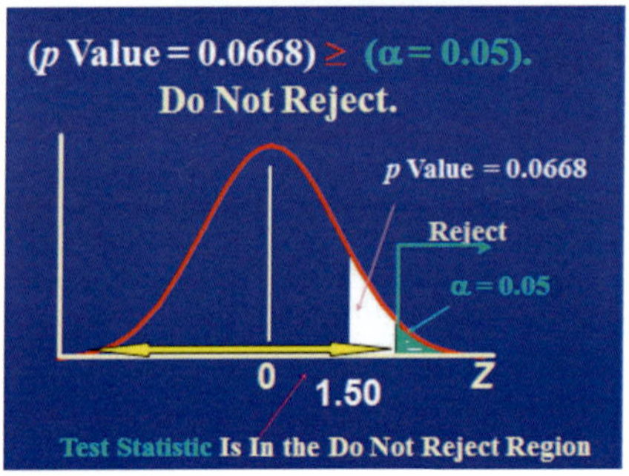

Example: two-tail test

Does an average box of cereal contain **368** grams of cereal? A random sample of **25** boxes showed **X = 372.5**. The company has specified **s** to be **15** grams. Test at the a=**0.05** level.

Example Solution: two-tail

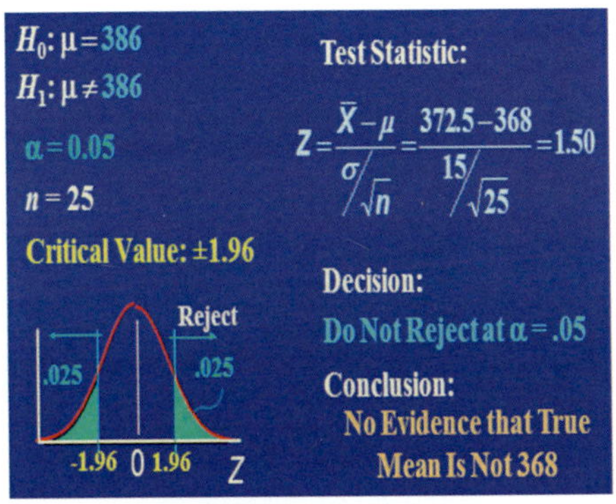

Connection to confidence intervals

For X = 372.5oz, s = 15 and n = 25,

The 95% Confidence Interval is:

372.5 - (1.96) 15/ 25 to 372.5 + (1.96) 15/ 25

or

366.62 £ m £ 378.38

If this interval contains the Hypothesized mean (368), we do not reject the null hypothesis.

It does. Do not reject.

t-Test: s Unknown

Assumptions

The population is normally distributed

If not normal, only slightly skewed & a large sample taken

Parametric test procedure

t-test statistic

$$t = \frac{\bar{X} - \mu}{S/\sqrt{n}}$$

Example: one tail t-Test

Does an average box of cereal contain **more than 368** grams of cereal? A random sample of **36** boxes showed **X = 372.5**, and s = 15. Test at the a=**0.01** level.

σ is not given,

Example: Z test for proportion

- Problem: A marketing company claims that it receives 4% responses from its Mailing.
- Approach: To test this claim, a random sample of 500 was surveyed with 25 responses.
- Solution: Test at a = 0.05 significance level.
- Z Test for Proportion: Solution

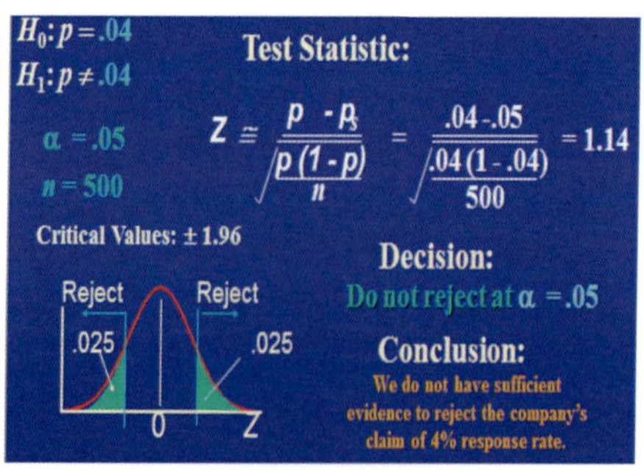

Degree of freedoms

The sample that is freely variable without affecting the mean or it is an integer used to determine whether a chi-square value is statistically significant.

There is the number of data which are given in the form of a series of variables in a row or column or the number of frequencies that are put in cells in a contingency table that can be calculated independently is called the degrees of freedom and is denoted by ***df.***

Calculation for degree freedom

The data is given in the form of series of variables in a row or column then the degrees of freedom (df) = (the number of items in series) -1 i.e. df = n-1. Where "n" is the number of observations.

The number of frequencies are put in cells in a contingency table the degree of freedom will be the product (number of rows less one) and the (number of column less one) i.e.,

df = (R-1) (C-1)

Where **"R"** is the number of rows.

"C" = is the number of columns.

Condition for using the chi-square test

- It should be on a random basis of sampling.
- It should be absolute, not relative terms.
- It is dependent degree of freedom.
- It should be independent of each other for making the sample of each observation.
- If X^2 test is applied in a fourfold table then it will not give a reliable result with one degree of freedom if the expected value in any cell is less than 5.
- In such cases to apply the X^2 test: Yates correction necessary.
- In this test, the total number of observation should be large it mean $n \geq 50$.

Types of chi-square test

- Homogeneity Chi-Square.
- Contingency Chi-Square.

- Goodness of fit.

Goodness of fit (Pearsonian – x^2)

The chi-square test is also applied for the test of "goodness of fit" to determine whether the actual or (observed) numbers or frequencies are similar or in "good agreement" with the expected or (theoretical) number of frequencies. The test is called "goodness of fit"

$X2 = \sum (|O-E| -1/2)^2 / E$

Where "E" Expected.

"O" Observed

½ Yates correction.

Example

Mendel reported the results of the garden pea tests each for goodness of fit for the following tested.

Solution

Cross	Progeny	Hypothesis
Green × Yellow Pods	(F2)428 : 152	3 : 1
Violet red × White Flower	(F1) 47 : 40	1: 1
Round yellow × Wrinkled green	(F1) 31 : 26 : 27 : 26	1: 1: 1: 1

Solution

1. **Null Hypothesis = 3 : 1**
2. **Alternative Hypothesis = 1 : 1**
3. **Calculation =**

Observed (O)	Expected (E)	(O-E)	$(O-E)^2$	$(O-E)^2/E$
428	¾ × 580 = 435	**428 – 435 = -7**	**49**	**49/435 = 0. 113**
152	¼ × 580 = 145	**152 – 145 = 7**	**49**	**49/145 = 0. 338**
Total = 580				**x^2 = 0. 451**

- The critical value of X^2 at 0.05 and for 2-1 = 1 degree of freedom is 3.84.
- The decision; the calculated value of chi – square (x^2) = 0.451 < critical value of x^2 for *df* = 3.84 so the null hypothesis is accepted

i.e. there is no significant variation with the data. So it is the result of the F2 monohybrid cross.

Solution (b)

1. **Null Hypothesis = 1 : 1**
2. **Alternative Hypothesis = 3 : 1**
3. **Calculation =**

Observed (O)	Expected (E)	(O-E)	$(O-E)^2$	$(O-E)^2/E$
47	87/2 = 43.5	47 – 43.5 = 3.5	12.25	12.25/43.5 = 0.281
40	87/2 = 43.5	40 – 43.5 = -3.5	12.25	12.25/43.5 = 0.281
Total 87				X^2 = 0.562

- Critical value: The control value of chi-square at 0.05 and for 2-1 = 1 degree of freedom is 3.84.
- Decision: the calculated value of the chi square x^2 = 0.562 < critical value of x^2 for 1 ***df*** = 3.84 so the null hypothesis is accepted i.e. the variation is non-significant.

Solution (c)

1. **Null Hypothesis = 1 : 1**
2. **Alternative Hypothesis = 3 : 1**
3. **Calculation =**

Observed (O)	Expected (E)	(O-E)	$(O-E)^2$	$(O-E)^2/E$
31	¼ × 110 = 27.5	31 – 27.5 = 3.5	12.25	12.25/27.5 = 0.445
26	¼ × 110 = 27.5	26– 27.5 = 1.5	2.25	2.25/27.5 = 0.082
27	¼ × 110 = 27.5	27– 27.5 = -0.5	0.25	0.25/27.5 = 0.009
26	¼ × 110 = 27.5	26 – 27.5 = -1.5	2.25	2.25/27.5 = 0.082
Total = 110				x^2= 0.618

Critical value

The chi square value of chi-square at 0.05 and 4-1 = 3 so, ***df*** = is 7.82

Decision

Calculated chi-square value (x^2) = 0.618 < critical value of X^2 of 3 ***df*** = 7.82 so the null hypothesis is accepted and the variance is non-significant.

Uses of chi-square test

- Test of homogeneity.
- Test of independent of attributes.
- Test of Goodness of Fit.

Student *"t"* distribution test

- The *t*- test is used with a small sample (n < 30) and it was worked out by W.S Gossett whose pen name was "student". However, the test is called as student t- test distribution.
- It may be defined as quantity representing the difference between the sample mean and true mean or population mean expressed in terms of the standard error.
- T = Difference between sample mean.
- Standard error of the difference between means.

$T = |X_1 - X_2| / SE$

Where X_1 and X_2 = Mean

SE = Standard error

TYPE

✓ Paired *t-Test*
✓ Unpaired *t-Test*

- **Condition for applying *"t"* test**
- Random sample are collected from the normal population.
- The population variances are regarded as equal for the testing the equality of two population means.
- Samples are less than 30.
- Some adjustments in degrees of freedom for ***"t"*** are made in case of two samples.

"t" Distribution properties

- The ***"t"*** distribution curve varies with the degrees of freedom.
- It is a symmetrical distribution with a mean of zero.
- It is asymptotic to X-axis i.e. it extends to infinity on either side.
- The graph is similar to that of normal distribution.
- It has a greater spread than the normal distribution.

- The larger the number of degrees of freedom the more closely *"t"* distribution resembles a standard normal distribution.
- ***"t"* Distribution applications**
- It observed sample correlation coefficient or differences between the mean of two samples.
- The difference between two sample mean when the population variance being equal and unknown.
- The single mean when the population variance is unknown.
- Calculate the t- value:

$T = |X_1 - X_2| / SE$

X_1 and X_2 = Mean

Sx_1 and Sx_2 = SD

n_1 and n_2 = size of sample

SE of $(X_1 - X_2)$ = SD $\sqrt{1/n_1 + 1/n_2}$

SD = $\sqrt{\sum (X_1 - X_2)^2 + \sum (X_2 - X_2)^2 / N1+N2-2}$

Determine the pooled degree of freedom from the formula

df **= (n-1) + (n_2 -1)**

= n_1+n_2-2

Compare the calculated value with the table value at particular degrees of freedom.

Example

There are 13 children who were given a usual diet plus vitamins "A" and "D" tables. While the second comparable group of 12 children was taking the usual diet. After 12 months, the gain in weight in pounds was noted as given in the table. Can you say that Vitamins A and D were responsible for the difference?

A	5	3	4	3	2	6	3	2	3	6	7	5	3
B	1	3	2	4	2	1	3	4	3	2	2	3	–

Solution

Null hypothesis

Vitamins (A and D) are responsible for the gain weight difference

Alternative hypothesis

Vitamins are not responsible for the gain weight differences.

S.no	Gr A (x)	(X –X) = D	$(X-X)^2 = D^2_1$	Gr B (Y)	$(Y-Y) = D_2$	$(Y-Y)^2 = D^2_2$
1	5	5– 4 = 1	1	1	1 – 2.5 =- 1.5	2.25
2	3	3– 4 = -1	1	3	3 – 2.5 = 0.5	0.25
3	4	4– 4 = 0	0	2	2 – 2.5 = -0.5	0.25
4	3	3– 4 = -1	1	4	4 – 2.5 = 1.5	2.25
5	2	2– 4 = -2	4	2	2 – 2.5 = 0.5	0.25
6	6	6– 4 = 2	4	1	1 – 2.5 = -1.5	2.25
7	3	3– 4 = -1	1	3	3 – 2.5 = 0.5	0.25
8	2	2– 4 =-2	4	4	4 – 2.5 = 1.5	2.25
9	3	3– 4 = -1	1	3	3 – 2.5 = 0.5	0.25
10	6	6– 4 = 2	4	2	2 – 2.5 = -0.5	2.25
11	7	7– 4 = 3	9	2	2 – 2.5 = -0.5	0.25
12	5	5– 4 = 1	1	3	3 – 2.5 = + 0.5	0.25
13	3	3– 4 = -1	1	—		0.25
	∑ 52		$\sum D^2_1 = 32$	∑ 30		$\sum D^2_2 = 11$

GrA	GrB
n = 13	n = 12
X = 14	Y = 2.5
$D^2_1 = 32$	$D^2_2 = 11$

SD (A –B) = $\sqrt{\sum}$ (X –X)^2 + ∑ (Y-Y)^2/ $n_1 + n_2$ - 2

= √ d21 + d22/ 13 +12 -2

= √ 32 +11/23

= √43/23

SE = SD √ 1/ n1 + 1/ n2

=√ 43/23 ×b√ 1/13 + 1/12

= √43/23 × √12 +13 / 156

$= \sqrt{43/23} \times \sqrt{25/156}$

$= \sqrt{1.87} \times \sqrt{0.160}$

$= 1.37 \times 0.4$

$= 0.548 = 0.55$

$t = X – Y/ SE$

4 -2.5/0.55

1.5/0.55 = 2.72

5% level of significance it means 0.05.

The critical value is calculated for df 23 is 2.07.

Decision (t) = 2.72

$|t| = 2.72 > t0_{.05, 23} = 2.07.$

100, t approaches Z.

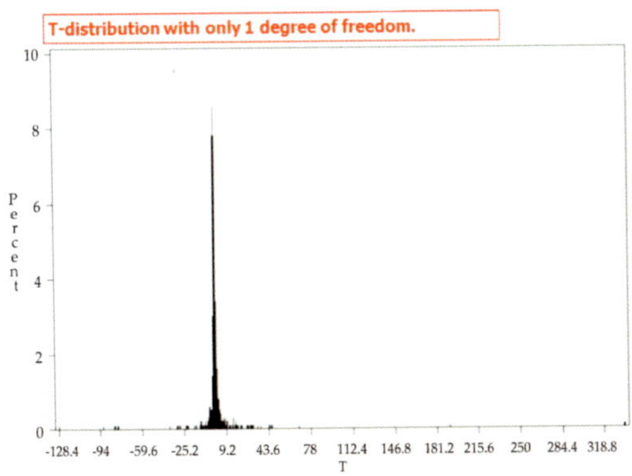

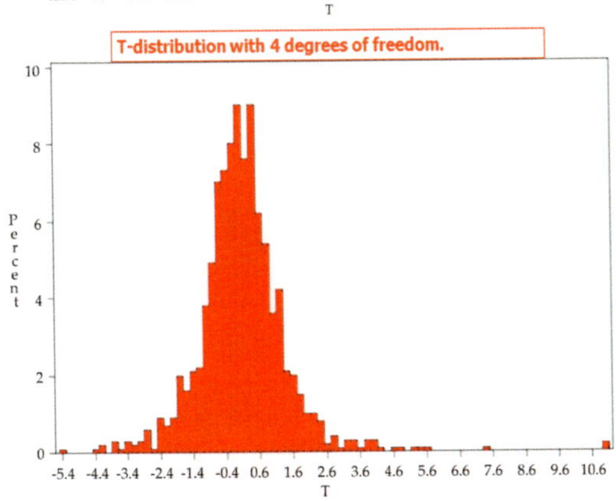

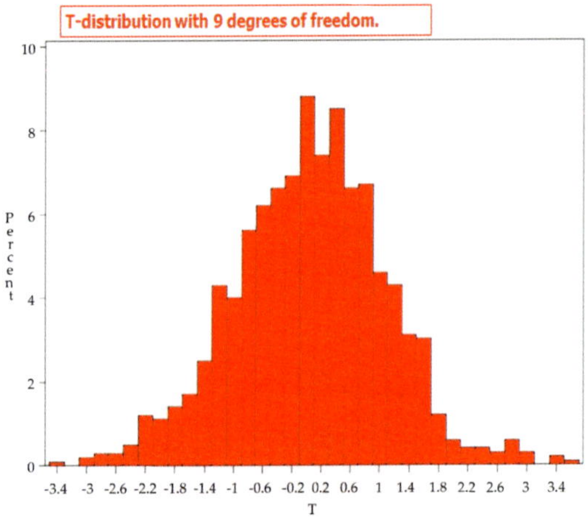
T-distribution with 9 degrees of freedom.
Percent
T

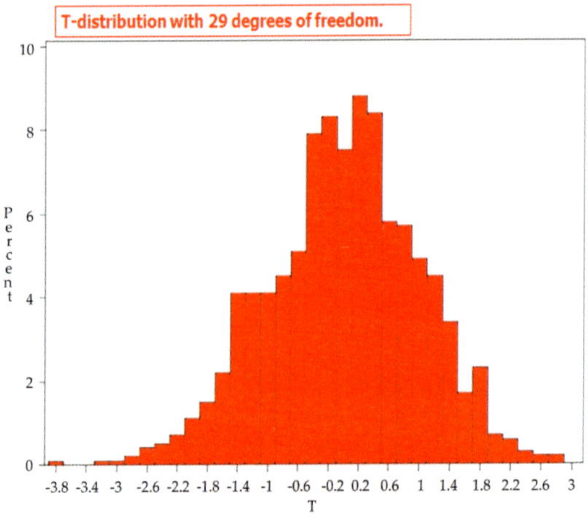
T-distribution with 29 degrees of freedom.
Percent
T

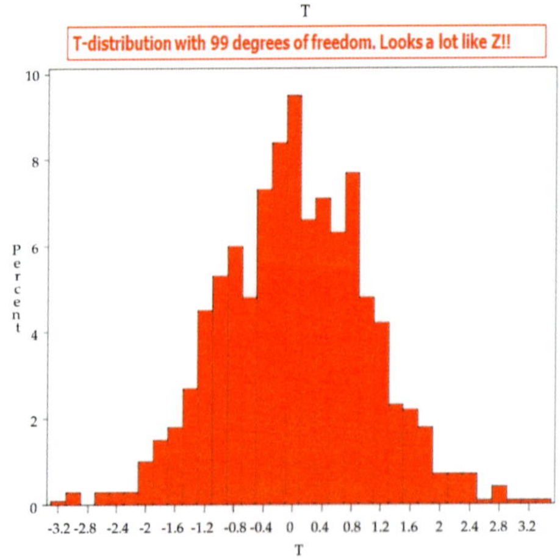
T-distribution with 99 degrees of freedom. Looks a lot like Z!!
Percent
T

Z- test

The deviation from the mean in a normal distribution r curve is called relative or standard normal deviation and is given the symbol "Z'. It is measured in terms of SD and indicated how many observations is bigger or smaller than the mean in units of SD. So "Z" will be ration.

Z = Observations – Mean / SD = X-X/SD

It is applying to the sampling variability and the difference between a sample estimate and that of the population is expressed in terms of SE instead of SD. The score of the value ration between the observed difference and SE is called "Z".

Condition for "Z"

- It must be quantitative.
- It should be assumed to follow a normal distribution.
- It must be randomly collected the data.
- The sample size must be larger than 30.

Test of significance of differences between a sample mean (μ) and population mean (X)

$Z = X - \mu / SE (X)$

Test of significance and difference between the two sample means

$Z = X_1 - X_2 / SE (X_1 - X_2)$

One-tailed t-test

- The statistical hypothesis where either alternative hypothesis is one-sided is called one-tailed test or one-sided test.
- It may be right or left tailed test.
- If we want to know one particular drug is than the other.
- It will be one-tailed test.

Two – tailed test

It is a test of statistical hypothesis based on rejected region represented by both sides of the standard normal curve.

Example

The nourished children or a healthy child is different from that of unnourished or unhealthy children.

Example

The sample of the mean of 1600 IQ level children was 99. It is likely that this was a random sample from a population with mean I.Q 100 and a standard deviation is 15.

Solution

Null hypothesis: the sample has not been drawn from the population with mean I.Q 100.

Alternative hypothesis: sample has been drawn from the population hypothesis.

Here n = 1600

$X = 99$

$\mu = 100$

$SD = 15$

$SE = SD / \sqrt{n} = 15/ \sqrt{1600} = 15/40 = 0.375$

$|Z|\ |X - \mu| / SE\ (X)$

$= |99 - 100|\ / 0.375$

$= |\mathbf{2.67}|$

Critical value is not given so we take $\alpha = 0.05$.

Decision: calculated value of Z is 2.67 since $|Z|$ is $2.67 > Z\alpha\ 1.96$

So the null hypothesis is rejected, the sample has not been drawn from a population with a mean of 100 and SD is 15.

F-Test or Fisher's F test

- The F – test was originated by RA Fisher. It is known as F-test. It is also called as variance ratio test as a comparison of sample variance involves in this test.
- F –Test: hypothesis variances derived from two samples.
- $F = \sigma^2_1 / \sigma^2_2$

Assumption of F-test

- It should be equal for all groups.
- It should be independent of each value.
- It should be normally distributed.

Uses of F-Test

- The two independent estimates of the population variances are homogeneous.
- Two independent samples have been drawn from the normal populations with the same variances (σ^2).
- It is an equality of population variances.
- It gives variance ration at different levels of significance at ***df*** = (n_1-1).
- F- ratio value is smaller than the t- value so the null hypothesis is accepted.
- It indicates that the samples are drawn from the same population.
- It is used to calculate F Statistics.

26. ANALYSIS OF VARIANCE

Variance

The Square of standard deviation is known as variance and it is denoted by (σ^2).

Variance = $(S.D)^2 = (\sigma^2)$

Coefficient of variance

- It is the product of the coefficient of standard deviation × 100
- **Coefficient of variance = S.D/AM × 100 it means Sigma/A.M × 100**

Covariance

- The measure of the statistical between two variables (X and Y) where the average product of the simultaneous deviation of the variables from their respective mean.

- $(\sigma^2) = \sum (X - X)^2/ n\text{-}1$ **or**
- $(\sigma^2) = \sum (X - X)^2/n$
- $COV_{XY} = \sum (X\text{-}X)(Y\text{-}Y)/n$
- Analysis of variance. Correlation and regression.

27. INTRODUCTION TO COMPUTERS, HARDWARE, STORAGE AND MEMORY DEVISES, INPUT AND OUTPUT DEVISES, FILE AND FOLDERS CONCEPTS. FILE MANAGEMENT, NETWORKS, PRINTERS, FLOPPIES, MOUSE AND KEYBOARD.

Computers are electronic devices that can follow instructions to accept the input, process the input and then produce information. Computers are made up of hardware and software.

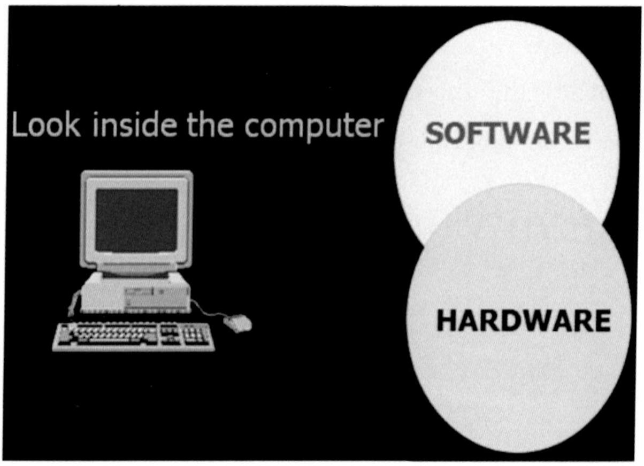

Hardware

- The parts of computer itself (tangible objects) including.

CPU (or Processor) and Primary memory (or Main Memory).

Input devices i.e. the keyboard and mouse.

- Output devices
- Storage devices
- Central Processing Unit (CPU)
- Input units

- Output units
- Memory (Main or Primary Memory & Secondary or Auxiliary Memory)

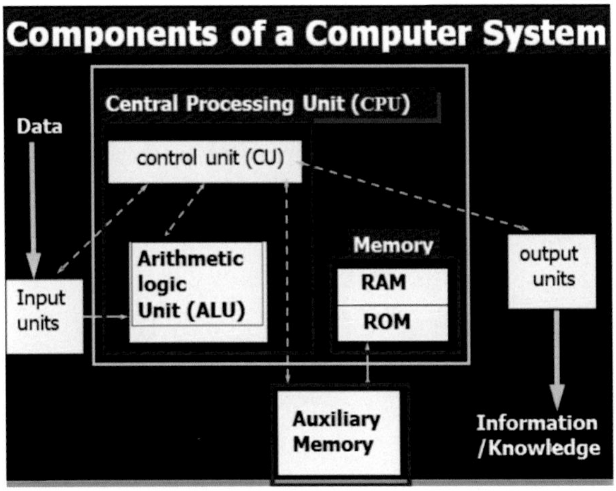

Input Devices

- Translate data from **form** that humans understand to one that the computer can work with
- Most common are the keyboard and the mouse

Examples of Input Devices

- Keyboard (QWERTY keyboard, ATMs keyboard)

1. ATM: automatic teller machine
2. Mouse
3. Scanner
4. Pre-storage Devise (Disk, CD's, … etc.)
5. Optical mark recognition (Light Pin, Bar code scanners)
6. Microphone
7. Joystick

Point and Draw devices

9. Trackball
10. Touchpad

11. Touch screen
12. Magnetic stripes and smart cars.
13. Digital Cameras

Central Processing Unit (CPU)

- A specific chip or the processor

  a CPU's performance is determined by the rest of the computer's circuitry and chips.
- The Central Processing Unit (CPU) performs the actual processing of data

The speed (clock speed) of CPU measured by Hertz (MHz)

The CPU consists of:

- Control Unit (CU)
- Arithmetic and Logical Unit (ALU)
- Some Registers

The Control Unit (CU):

- Coordinates all activities of the computer by:
- Determining which operations to perform and in what order to carry them out.
- The CU transmits coordinating control signals to other computer components.

The ALU:

It consists of electronic circuitry to perform:

- Arithmetic operations (addition, subtraction, multiplication and division)
- Logical operations (and, or, not, …) and to make some comparisons (less-than, equal, … etc.)

Primary Memory

- Memory (fast, expensive, short-term memory): Enables a computer to store, at least temporarily, data, programs, and intermediate results.

Two general parts:

1. RAM
2. ROM

RAM (Main Memory)

- It is the primary storage or random access memory (RAM).
- It temporarily holds data and programs for use during processing (volatile).
- Any information stored in RAM is lost when the computer is turned off.
- RAM is the memory that the computer uses to temporarily store the information as it is being processed. The more information being processed the more RAM the computer needs.
- RAM consists of locations or cells. Each cell has a unique address that distinguishes it from other cells.

ROM: Read Only Memory

- ROM is part of memory
- Programmed at manufacturing time
- Its contents cannot be changed by users
- It is a permanent store

Secondary storage

- Stores data and programs permanently: it is retained after the power is turned off.

Examples

- Hard Drive (Hard Disk)
- Located outside the CPU, but most often contained in the system cabinet
- Floppy Disk
- Optical Laser Discs
- CD-ROM, CD-RW, and DVD
- Common Secondary Media

Diskettes

- Data represented as magnetic spots on removable flexible plastic disks
- The most common size is 3 1/2 inches, in a rigid plastic case
- Disk drive holds the diskette, reads or retrieves the data and writes or stores data

Common secondary media

Hard drive

- Data is represented magnetically as with diskettes
- Normally more than one rigid platter in a sealed unit
- These disks are not removable
- Significantly more capacity and faster operating than diskettes

Common secondary media

Optical Laser Discs

- CD ROM & DVD's
- Data is represented as pits and lands
- Some kinds are read only (CD-ROM) and some Kinds are rewritable (CD-RW)
- Significantly more capacity and faster operating than diskettes

Output devices

- Pieces of equipment that translate the processed information from the CPU into a form that humans can understand.

Output devices

- Monitors
- Printers
- Dot matrix printers

- Ink-jet printers
- Laser printers
- Sound Blasters (Sound Card By Creative Lab)
- Controlling other devices

Software

The instructions that tell the computer what to do

- Application Software - helps end-users perform general-purpose tasks
- System Software - enables application software to interact with the computer

System software

- The software that controls everything that happens in a computer.
- Background software, manages the computer's internal resources
- Application Software - Basic Tools
- Word processors– **example: Microsoft word**
- Spreadsheets-- **example: Microsoft Excel**
- Database managers-- **example: Microsoft Access**

Graphics-example: photoshop

Units of measurements

- Bit (Binary Digit)(takes two values: 1 or 0)
- Byte = 8 bits
- KB (Kilo-byte) = 1024 bytes
- MB (mega-byte) = 1024 KB
- GB (Giga-byte) = 1024 MB
- TB (Tera-byte) = 1024 GB

Remark: $1024=2^{10}$

Four kinds of computers

- Microcomputer =>Personal Computer => PC
- There are **3** types of the Microcomputers:
  1. Laptop
  2. Desktop
  3. Workstation

Personal computer: A small, single-user computer based on a microprocessor.

Workstation: A powerful, single-user computer. A workstation is like a personal computer, but it has:

- A more powerful microprocessor.
- In general, a higher-quality monitor.

Minicomputer, Mainframe, and Supercomputer

Minicomputer: A multi-user computer capable of supporting up to hundreds of users simultaneously. Mainframe: A powerful multi-user computer capable of supporting many hundreds or thousands of users simultaneously. Supercomputer: An extremely fast computer that can perform hundreds of millions of instructions per second.

Characteristics of Computers

- Store a large amount of data and information for a long period of time.
- The process data and information in high accuracy level.
- Speed in processing data information.
- Sharing of information / network.

Understanding the difference between data, Information And Knowledge:

- **Data:** is the name given to basic facts such as names and numbers.
- **Information:** is data that has been converted into a more useful or intelligible form.
- **Knowledge:** arrangement of information and classified information of the same type or the same topic.
- Processing **data** produces **information**, and processing information produces **knowledge**.

Computer viruses

- A computer virus is an application program designed and written to destroy other programs.
- It has the ability to:
  - Link itself to other programs
  - Copy itself (it looks as if it repeats itself)

Examples of viruses

- Monkes
- ABC
- Crabs
- CIH

Viruses and virus protection

- A virus program
- Infects programs, documents, databases and more …
- It is man-made
- It can hide and reproduce
- It can lay dormant (inactive) and then activate

Sources of computer viruses

Three primary sources

- The Internet
- Via downloads and exchanges
- Diskettes
- Exchanging disks
- Computer networks
- Can spread from one network to another

How do you know if you have a virus

- Lack of storage capability
- Decrease in the speed of executing programs
- Unexpected error messages
- Halting the system

Virus protection

The software package distributed with new PCs always includes an **antiviral program.** The best way to cope with viruses is to recognize their existence and use an antiviral, or antivirus program.

28. DIFFERENT TYPES OF BOOTING, OPERATING SYSTEMS-SINGLE USER, MULTI-USER AND MULTI TASKING OPERATING SYSTEMS WITH EXAMPLES

Booting

Hardware doesn't know where the operating system resides and how to load it. Need a special program to do this job – Bootstrap loader.

E.g., BIOS – Boot Input Output System.

Bootstrap loader locates the kernel, loads it into main memory and starts its execution. In some systems, a simple bootstrap loader fetches a more complex boot program from disk, which in turn loads the kernel.

How Boot process occurs

Reset event on the CPU (power up, reboot) causes instruction register to be loaded with a predefined memory location. It contains a jump instruction that transfers execution to the location of Bootstrap program. This program is a form of ROM, since RAM is in an unknown state at system startup. ROM is convenient as it needs no initialization and can't be affected by a virus.

Tasks performed at boot up

Run diagnostics to determine the state of the machine. If diagnostics pass, booting continues. Runs a Power-On Self Test (*POST*) to check the devices that the computer will rely on, are functioning. BIOS goes through a preconfigured list of devices until it finds one that is bootable. If it finds no such device, an error is given and the boot process stops. Initializes CPU registers, device controllers and contents of the main memory. After this, it loads the OS. On finding a bootable device, the BIOS loads and executes its boot sector. In the case of a hard drive, this is referred to as the master boot record (*MBR*) and is often not OS specific. The MBR code checks the partition table for an active partition. If one is found, the MBR code loads that partition's boot sector and executes it. The boot sector is often operating system specific, however in most operating systems, its main function is to load and execute a kernel, which continues startup.

Secondary boot loaders

If there is no active partition or the active partition's boot sector is invalid, the MBR may load a secondary boot loader and pass control to it and this secondary boot loader will select a partition (often via user input) and load its boot sector examples of secondary boot loaders

GRUB – GRand Unified Bootloader

LILO – LInux LOader

NTLDR – NT Loader

Booting and ROM

A system such as cellular phones, PDAs and game consoles stores entire OS on ROM. Done only for small OS, simple supporting hardware, and rugged operation. Changing bootstrap code would require changing ROM chips.

EPROM – Erasable Programmable ROM.

Code execution in ROM is slower. Copied to RAM for faster execution.

29. INTERNET AND E-MAIL. IMPORTANT SERVICES PROVIDED BY INTERNET. USE OF INTERNET IN BIOTECHNOLOGY STUDIES AND RESEARCH

The Worldwide collection of networks connects millions of computers.

Access the internet

- Communications
- Information
- Shopping
- Banking and Investing
- Classes
- Entertainment

Billions of documents, called Web pages, available to anyone connected to the Internet

- A program is a series of instructions that tells the computer what to do
- Allows you to interact with the software using graphics and icons
- Controls how you enter data and how the screen displays information
- The Web is a collection of files that reside on computers, called Web servers.
- Web servers are connected to each other through the Internet
- The software you use to connect your computer to the Web server is your Web browser.
- Common Web browsers are Netscape Navigator and Internet Explorer

Email

- E-mail, or electronic mail, is an efficient way to exchange messages with others on a network.
- This network may be small and self-contained, such as within an office, or as large as the Internet.
- E-mail messages can go to only one recipient or to dozens of recipients at once.
- When you send a message, it travels through the network to an e-mail server.
- The e-mail server stores messages until the recipient(s) request them.
- You may have an e-mail account through a business network or you may create an e-mail account with a service that provides Internet access.
- An e-mail address consists of:
  - A user name or login ID
  - The "at" symbol (@)
  - The name of the e-mail server
- To access your e-mail account you also need an e-mail program, which is also called e-mail client software.
- You use the e-mail program to open, print, delete, reply to, forward, and save mail from your e-mail server.
- One such program is Microsoft's Outlook Express, which installs as part of Internet Explorer.
- Outlook Express can be accessed through the Start menu on the All Programs submenu.
- To send an e-mail message:
- Press the New Mail button on the Outlook Express toolbar.
- Type in the e-mail addresses of the recipient.
- Type your subject matter.
- Type your message and then click the Send button.
- To retrieve mail that has been sent to you
- Click the Send/Recv button on the toolbar.

- Outlook Express will contact your e-mail server and download your e-mail messages
- To reply to a message
- Click the Reply button.
- The recipient's address and the subject matter are automatically filled.
- When you have completed typing your reply, click the Send button.

Internet use for Biotechnology

- In the last few decades, advances in molecular biology and the equipment available for research in this field have increased the rapid sequencing of large portions of the genomes of several species.
- In fact to date several bacterial genomes more than 568, four animal genomes such as Drosophila, human being and at least three plant genomes (*Arabidopsis thaliana* from dicotyledons, Rice from monocotyledons and poplar from trees) have been sequenced. Several other genomes from microbial, animal and plant kingdom are actively being sequenced.
- The human genome project was designed to sequence all human chromosomes.
- Bioinformatics is a new discipline emerging from a combination of Biology, Mathematics and information technology.
- Bioinformatics is a science associated with database like activities involving persistent sets of data that are maintained in a consistent state over essentially indefinite periods of time.
- Bioinformatics is a science of developing computer databases and algorithms for the purpose of speeding up and enhancing biological research.
- It is an integration of mathematical, statistical and computational methods to analyze biological, biochemical and biophysical data.
- "Biology easily has 500 years of exciting problems to work on." *Donald Knuth (Stanford Professor & famous computer scientist).*

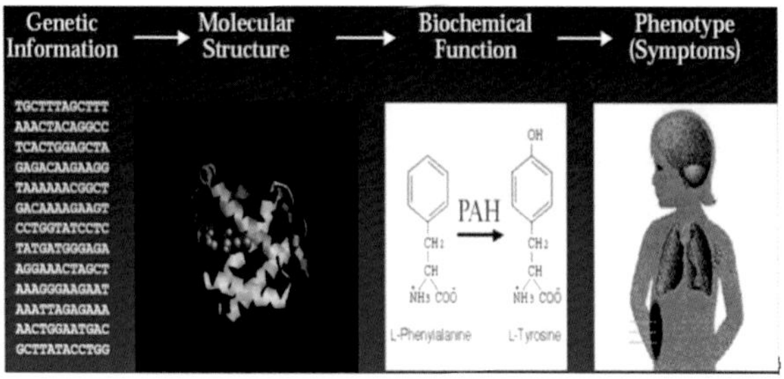

By developing techniques for analyzing sequence data and related structures, we can attempt to understand molecular basis of life.

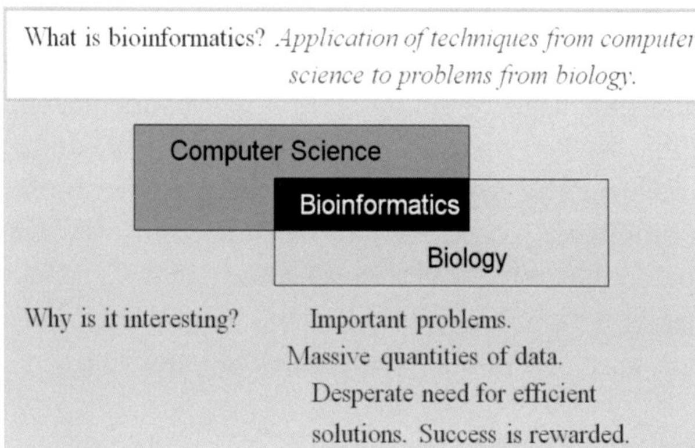

Different variety of Available Bioinformatics Tasks

- Creation of biological database
- Search similarity of sequences
- Discovery of coded genes and also their structure
- DNA and protein function and annotation
- Elucidation of protein structure
- Drug design
- Biochemical analysis (metabolic pathway).
- Tools for analysis of high-throughput molecular biology experimentation data (Microarrays, Mass spectrometry, etc.).
- Comparative genome analysis
- Building gene and protein networks
- Scientific literature search

Computation Biology

It refers to algorithmic tools to facilitate biological analysis, whereas bio information structure comprises the entire collection of information management system, analysis tools and communication networks supporting biology. Computational biology involves the following points which described below. Finding the genes in the DNA sequences of various organisms. Developing methods to predict the structure and or function of newly discovered proteins and structural RNA sequences. Clustering protein sequences into families of related sequences and the development of protein models. Aligning similar proteins and generating phylogenetic trees to examine evolutionary relationships (www.scribd.com., www. slideshare.com., www.coursehero.com., www. biocyclopedia.com., www. molecular-plant-biotechnology.info., www.slideplayer.com., www. documents.mx., www.docplayer.net., www.bmb.psu.edu.,www.archive. org.,www.mafiadoc.com., www.powershow.com., www.bioscience.jbpub. com., www.outhorstream.com., www.docslide.com., www.msluay.com., www.scienceprofonlue.com., www.csus.edu., www.krishikosh.egrouth. ac.in., www.tankonyvtor.hu., www.silo.pub., www.studyblue.com., www. Preservearticles.com., www.biotechuniverse.blogspot.com).

PART V

MOLECULAR BIOLOGY AND GENETIC ENGINEERING

30. NUCLEIC ACIDS AS GENETIC MATERIAL, STRUCTURE OF A-, B-AND Z-DNA, PALINDROMIC SEQUENCES, STRUCTURE OF RNA (T-RNA, M-RNA AND R-RNA), DNA DENATURATION AND RENATURATION

DNA is found in all plants, animals, prokaryotes and some viruses. DNA is a double helical structure that consists of genetic material. It is joined by a phosphodiester bond. It is acidic in nature. The entire genetic message that controls the chemistry of every cell of the body acting in a specific way is actually written in the language of four nitrogenous bases of DNA i.e., purine and pyrimidines. It is a polymeric compound containing four monomers known as deoxyribonucleotide monomers or deoxyribotides. Each DNA consists of a pentose sugar, a phosphate group and nitrogenous bases Figure **30.1.** Purines bases (adenine and guanine) are heterocyclic and two ringed bases and the pyrimidines (thymine and cytosine) are one ringed bases.

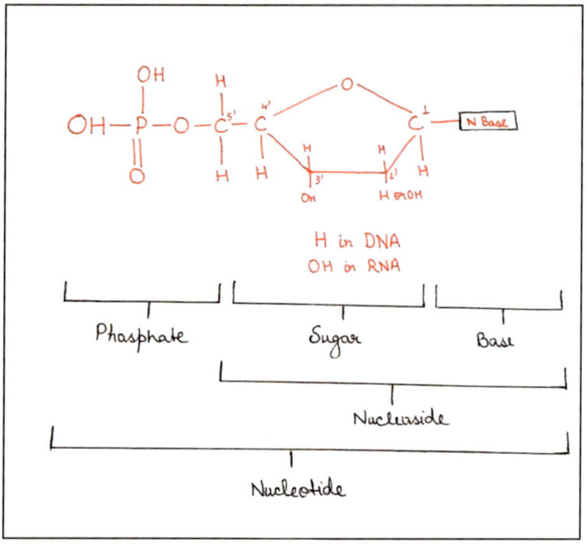

Figure 30. 1 Nucleotides

Nucleosides

- The nitrogenous bases combined with a pentose sugar are called nucleosides.

Nucleotides

A nucleoside linked with phosphate forms a nucleotide. Nucleoside = pentose sugar + nitrogenous base nucleoside + phosphate Nucleotide. On the basis of different nitrogenous bases the deoxynucleotides are of following types: (i) Adenine (A) deoxyadenosine-3'/5'-monophosphate (3'75'-d AMP) deoxyguanosine -5'-monophosphate (5'-d GMP) deoxythymidine -5-monophosphate (5'-d TMP) (ii) Guanine (G) (iii) Thymine (T) (iv) Cytosine (C) deoxycytidine -5-monophosphate (5'-d CMP). In addition to the presence of nucleosides in DNA helix, these are also present in the nucleoplasm and cytoplasm in the form of deoxyribonucleotide phosphates e.g. deoxyadenosine triphosphate (dATP), deoxyguanosine triphosphate (dGTP), deoxycytidine triphosphate (dCTP), deoxythymidine triphosphate (dTTP); Figure 30.2.

The advantage of these four deoxyribonucleotides in triphosphate forms is that the DNA polymerase acts only on triphosphates of nucleotides during DNA replication. Similarly, the ribonucleotides contain ribose sugar, nitrogenous bases and phosphate. Except sugar, the other components are similar. However, uracil (U) is found in RNA instead of thymine. Generally, RNA molecule is single-stranded besides some exceptions.

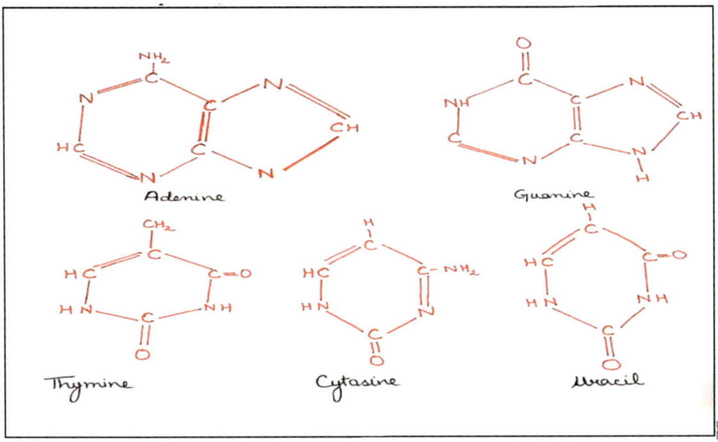

Figure 30.2 Nitrogenous base.

Polynucleotide

The nucleotides undergo the process of polymerization to form a long chain of polynucleotides. The nucleotides are designated by prefixing 'poly' to each repeating unit such as poly A (polyadenylic acid), poly T (polythymidilic acid), poly G (polyguanidylic acid), poly C (polycytidilic acid) and poly U (poly uridylic acid). The polynucleotides that consist of the same are called homopolynucleotides such as poly A, poly T, poly G, poly C and poly U.

Chargaff-Equivalence Rule

By 1948, a chemist Erwin Chargaff started using paper chromatography repeating unit to analyze the base composition of DNA from a number of studies. In 1950, Chargaff discovered that in the DNA of different types of organisms the total amount of purines is equal to the total amount of pyrimidines i.e. The total number of A is equal to the total number of T (A-T), and the total number of G is equal to the total number of C (G-C). It means that A/T = G/C i.e. A+T/G+C 1. In the DNA molecules isolated from several organisms, regularity exists in the base composition.

Watson and crick's model of DNA

J.D.Watson and F.H.C. Crick (1953) combined the physical and chemical data generated by earlier workers, and proposed a double helix model for DNA molecule. This model is widely accepted. According to this model, the DNA molecule consists of two strands that are connected together by hydrogen bonds and helically twisted. Each step on one strand consists of a nucleotide of purine base which alternates with that of pyrimidine base. Thus, a strand of DNA molecule is a polymer of four nucleotides i.e. A. G. T, C. The two strands join together to form a double helix. Bases of two nucleotides form hydrogen bonds i.e. a combines with T by two hydrogen bonds (A T) and G combines with C by three hydrogen bonds (G C). However, the sequence of bonding is such that for every A.T.G.C. on one strand there would be T, A, C, G on the other strand. Therefore, the two chains are nucleotides on one chain are the photocopy of the sequence of nucleotides on the other chain. The two strands of double helix run in an anti-parallel direction i.e. they have opposite polarity.

The left-hand strand has 5'-3' polarity, whereas the right hand has 3'-5' polarity to the first one. The polarity is due to the direction of phosphodiester linkage complementary to each other i.e. sequence of as compared to the first one. The DNA model also suggested a copying mechanism of the genetic material. DNA replication is the fundamental and unique event underlying growth and reproduction in all living organisms ranging from the smallest viruses to the most complex of all creatures including man. DNA replicates by a semiconservative mechanism which was experimentally proved by Mathew, Meselson and Frank W. Stahlin in 1958. If changes occur in the sequence or composition of base pairs of DNA, mutation takes place. Though the presence of adenine, guanine, thymine and cytosine is a universal phenomenon, yet unusual bases in DNA molecule also occur. In some bacteriophages, 5- hydroxymethyl- cytosine (HMC) replaces cytosine of the DNA molecule when the methylation of adenine, guanine and cytosine occurs. This results in changes of these bases; see **Figure 30.3.**

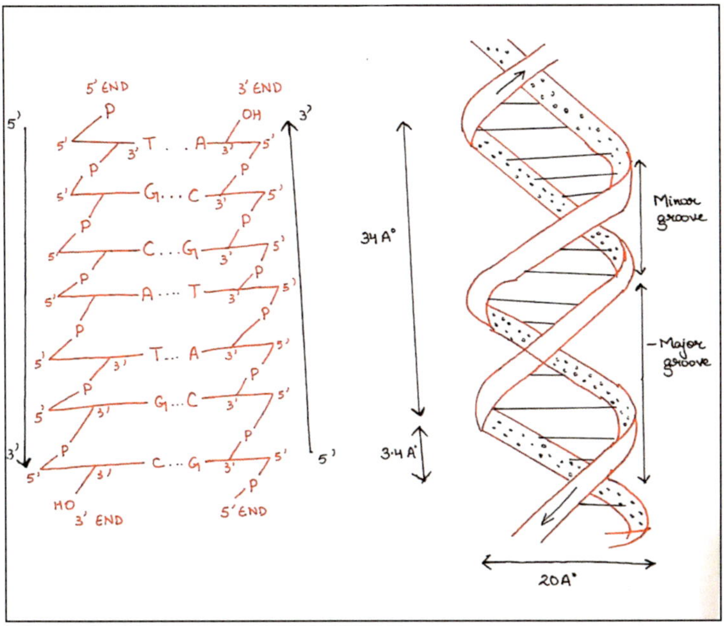

Figure 30.3 Watson Crick's Model of DNA

Three major forms

- A-DNA
- B-DNA
- Z-DNA

B-DNA is biologically the most common

- Right-handed (20 angstrom (a) diameter)
- Complementary base-pairing (Watson-crick)

A-T = G-C

Each base pair has the same width

- 10.85 a from c1' to c1'
- A-T and G-C pairs are interchangeable.

"Pseudo-dyad" axis of symmetry

- Ideal **B-DNA** has 10 base pairs per turn.
- Base thickness.
  - Aromatic rings with 3.4 A thickness to rings.

- Pitch = 10 x 3.4 = 34 A per complete turn.
- Axis passes through middle of each bp
- Minor groove is narrow.
- Major groove is wide in class exercise: explore the structure of B-DNA. Pay special attention to the major, minor grooves.
- Minor: exposes edge from which c1' atoms extend.
- Major: exposes the opposite edge of base pair.
- The pattern of H-bond possibilities is more specific and more discriminating in the major groove.

A-DNA

- Right-handed helix
- Wider and flatter than B-DNA.
- 11.6 bp per turn
- Pitch of 34A

An axial hole

- Base planes are tilted 20 degrees with respect to the helical axis

Helix axis passes "above" major groove

Deep major and shallow minor groove

- Observed under dehydrating conditions
- When the relative humidity is ~ 75%
- B-DNA → A-DNA (reversible)
- Most self-complementary oligonucleotides of < 10 bp crystallize in A-DNA conf.
- A-DNA has been observed in 2 contexts:
- At active site of DNA polymerase (~ 3 bp)
- Gram (+) bacteria undergoing sporulation
- Sasps induce B-DNA to → A-DNA
- Resistant to UV-induced damage
- Cross-linking of pyrimidine bases

Z-DNA

- A left-handed helix.

- It is seen in conditions of high salt concentrations.
- Reduces repulsions between closest phosphate groups on opposite strands (B-DNA) in complementary polynucleotides with alternating purines and pyrimidines.
- poly d(GC) · poly d(GC)
- poly d(AC) $\rightarrow$ poly d(GT)
- It might also be seen in DNA segments with the above characteristics.
- 12 base pairs per turn.
- A pitch of 44 degrees.
- A deep minor groove.
- No discernible major groove.
- Reversible change from B-DNA to Z-DNA in localized regions may act as a "switch" to regulate gene expression.
- Transient formation behind actively transcribing RNA polymerase.

Palindromes in DNA sequences

Genetic palindromes are similar to verbal palindromes. A palindromic sequence in DNA is one in which the 5' to 3' base pair sequence is identical on both strands (the 5' and 3' ends refers to the chemical structure of the DNA).

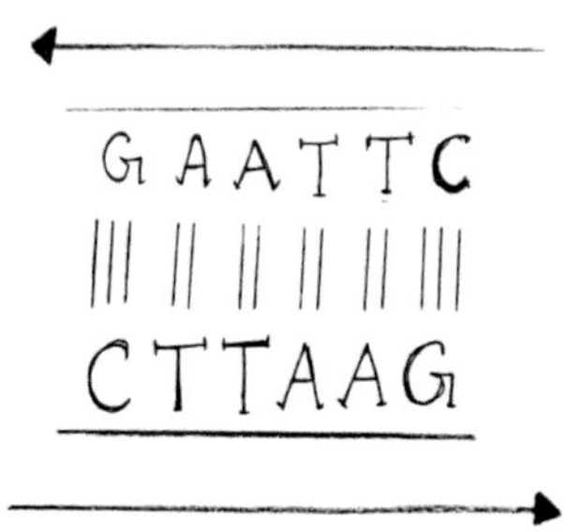

RNA is single-stranded and does not have to stay in the nucleus. RNA is not found in chromosomes because it does not carry the genetic code; however it can read the DNA code and take the information out of the nucleus. The main work is to build proteins. The building blocks of RNA are Nucleotides, just like DNA.

A Nucleotide in RNA is still made of 3 important things:

1. 5-Carbon Sugar - Ribose (instead of Deoxyribose)
2. Phosphate
3. Nitrogen base

- There are 4 nitrogen bases in RNA, A,G,C, and **U** that pair together)

  A → **U** C →G

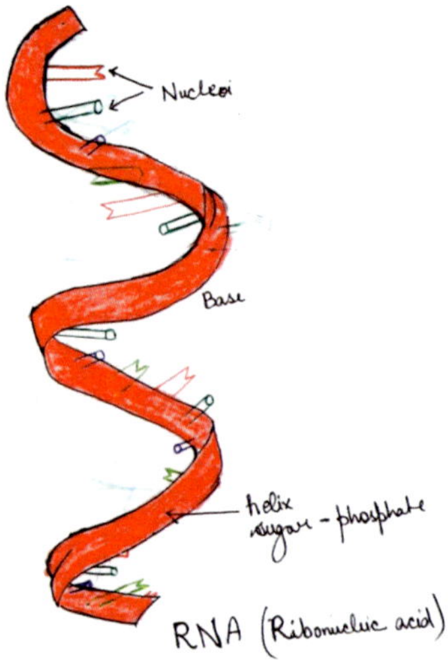

Types of RNA

- *Messenger RNA (mRNA):* Carries copies of instructions for the assembly of amino acids into proteins from DNA to the rest of the cell (serve as "messenger").
- *Ribosomal RNA (rRNA):* Makes up the major part of ribosomes, which is where proteins are made.
- *Transfer RNA (tRNA):* Transfers amino acids to ribosomes during protein synthesis.

DNA denaturation (melting)

- Strands are held together by weak, noncovalent bonds.
- Strands start to separate at a specific temperature.
- Within a few degrees, the process is complete.
- Solution contains single-stranded molecules.

- Higher single strand absorbance @ 260 nm.
- Hydrophobic base interactions reduced.
- Bases absorb photons more efficiently.
- Melting temperature (T_m) is half denatured.
- T_m increases with G-C content (%G + %C).
- A-T-rich sections melt before G-C-rich segments.

DNA renaturation

J. Marmur (Harvard, 1960) – first described

- Slowly cool heat-denatured DNA
- The temperature quickly to ~25°C below T_m & incubate awhile
- Complementary single-stranded DNAs can reassociate or reanneal

Renaturation very useful

- Genome complexity: variety & copy number
- Hybridization: molecular identification

Renaturation governed by

- Ionic strength of the solution
- Temperature
- Time
- DNA concentration
- Size of the interacting molecules

SV40:5.4 x 10^3 bp T4:1.8 x 10^5 *E. coli*: 4.5 x 10^6

- Force all DNAs through tiny orifice under high
- Random shear ~1000 bp
- Reanneal at the same DNA concentration (mg/ml)

Smaller genome - faster renaturation

- More copies of small genomes
- Increases chance of collision between complementary fragments
- Viral/bacterial genomes - symmetrical curve

Circular and Super Helical DNA

Almost in all the prokaryotes and a few viruses, the DNA is organized in the form of a closed circle. The two ends of the double helix get covalently sealed to form a closed circle. Thus, a closed circle contains two unbroken complementary strands. Sometimes one or more nicks or breaks may be present on one or both strands, for example, DNA of phage PM, Besides some exceptions, the covalently closed circles are twisted into super helix or super coils and is associated with basic proteins but not with histones found complexed with all eukaryotic DNA: these histone-like proteins appear to help the organization of bacterial chromatin with the result of a nucleosome-like structure, folding and super coiling of DNA and association of DNA polymerase with nucleoids.

These nucleoid-associated proteins include HU proteins, IHF, proteins HI, Fir A, H-NS and Fis. In archaeobacteria (e.g. Archaea) the chromosomal DNA exists in a protein associated form. Histone like proteins have been isolated from nucleoprotein complexes of *Thermoplasma acidophilum* and *Halobacterium salinarum*. Thus, the protein associated DNA and nucleosome like structures are in helix coils clockwise from the axis, the coiling is termed positive contrast, if the path of coiling is anticlockwise, the coil is called left-handed or negative coil. The two ends of a linear DNA helix can be joined to make each strand continuous. However, if one end rotates at 360° with respect to the other to produce.

These histone-like proteins appear to help the organization of bacterial DNA into a coiled variety. The unwinding of the double joints results in the formation of a twisted circle in the opposite sense i.e. opposite helix; the ends are in the unwinding direction. Such a twisted circle appears as an 8 i.e. it has one node or crossing over point. If it is twisted at 720° before joining, the resulting superhelix will contain two nodes. The enzymes topoisomerases alter the topological form i.e. super-coiling of a circular DNA molecule. Type I topoisomerases (e.g. *E.coli* top A) relax the negatively super-coiled DNA by breaking the 5'-phosphoryl end, and then resealing the nicked phosphodiester backbone (Moat and Foster, 1995).

Type II Topoisomerases need energy to unwind the DNA molecules resulting in the introduction of super-coils. One of type II isomerases, DNA gyrase is apparently responsible for the negatively super-coiled state of the bacterial chromosome. Super coiling is essential for the efficient replication and transcription of prokaryotic DNA. The bacterial chromosome is believed to contain about 50 negatively super-coiled loops topological unit, the boundaries of which may be defined by the sites on DNA that limit its rotation (Wang, 1982; Mout and Foster, 1995). One of the phosphodiester bonds in dsDNA allowing the 3'-OH end to swivel around or domains.

Organization of DNA

In addition to the organization of DNA in prokaryotes and lower eukaryotes as discussed earlier in eukaryotes, the DNA helix is highly organized into the well-defined DNA-protein complex termed as nucleosomes. Among the proteins, the most prominent are the histones. The histones are small and basic proteins rich in amino acids such as cells there are five types of histones e.g. H,, H,A, H,B, H, and H. Eight histone molecules (two each of H, A, H, B, H, and H) form an octameric ellipsoidal structure of about 1 nm to 7 nm in diameter. DNA coils around the surface of the ellipsoidal structure of histones 166 base pairs (about 7/4 turns) before proceeding onto the next and form. Thus a nucleosome is an octamer or four histone proteins complexed with DNA lysine and/or arginine. Almost in all eukaryotes is the nucleosome: long and 6.5- complex structure, histones play important role in determining the conformation of chromatin. The nucleosomes are the repeating units of DNA organization that are often termed as beads. The DNA isolated from chromatin looks like string: 146 base pairs of DNA lie in the helical path and the histone DNA assembly is known as the nucleosome core. The stretch of DNA between the nucleosome is known as the 'linker' which varies in length from 14 to over 100 base pairs. The H_1, is associated with the linker region and helps the folding of DNA into a complex structure called chromatin or beads. As a result of m chromatin becomes visible as chromosomes during cell division.

Structure of Ribonucleic Acid (RNA)

The RNA TMV, yellow mosaic virus, influenza virus, foot and mouth disease is usually single-stranded except one virus such as virus, etc. which have dsDNA. The virus, retrovirus, wound tumor entirely single strand of the RNA is folded either at certain regions or to form hairpin shaped structure. In the hairpin shaped structures, the complementary bases are linked by hydrogen bonds which give stability to the molecules. However, no complementary bases are found in the unfolded region. The RNA does not possess an equal purine-pyrimidine ratio, as it is found in the DNA. Like DNA, the RNA is also the polymer of four nucleotides each one contains D-ribose, phosphoric acid and a nitrogenous base. The bases are two purines (A, G) and two are not found in RNA. The pairing between bases occurs as A-U and G-C pyrimidines (C, U). Thiamine is not found in RNA. The pairing between bases occurs as A-U and G-C. The nucleotides formed by the four bases are adenosine monophosphate (AMP), guanosine monophosphate (GMP), cytosine monophosphate (CMP), and uridine monophosphate (UMP).

These are found freely in nucleoplasm but in the form of triphosphates e.g. ATP, GTP, UTP and CTP. As a result of polymerization, the ribonucleotides form a polynucleotide chain of RNA. If the RNA is involved in genetic mechanism, it is called genetic RNA as found in plant, animal and bacterial viruses. The DNA acts as a genetic material and RNA follows the order of DNA. In such cells, the RNA does not have a genetic role. Therefore, it is called non-genetic RNA. The non-genetic RNA is of three types: (i) ribosomal RNA (rRNA), (ii) transfer RNA (1RNA) or soluble RNA (sRNA), and (iii) messenger RNA (mRNA) or template RNA. These three types of RNA differ from each other in structure, site of synthesis in eukaryotic cell and function (www.scribd.com., www.slideshare.com., www.coursehero.com., www. biocyclopedia.com., www. molecular-plant-biotechnology.info., www.slideplayer.com., www.documents.mx., www. docplayer.net., www.bmb.psu.edu.,www.archive.org.,www.mafiadoc.com., www.powershow.com., www.bioscience.jbpub.com., www.outhorstream. com., www.docslide.com., www.msluay.com., www.scienceprofonlue. com., www.csus.edu., www.krishikosh.egrouth.ac.in., www.tankonyvtor. hu., www.silo.pub., www.studyblue.com., www.Preservearticles.com., www.biotechuniverse.blogspot.com).

31. DNA REPLICATION IN PROKARYOTES, DNA POLYMERASE I, II AND III, MODES AND MECHANISM OF DNA REPLICATION, DNA REPAIR MECHANISMS

DNA Replication

The transmission of chromosomal DNA from generation to generation is crucial to cell propagation. This can only be achieved when chromosomal DNA is accurately replicated, providing two copies of the entire genome for the faithful distribution into each daughter cell. Cells need to make a copy of

DNA before dividing so each daughter cell has a complete copy of genetic information.

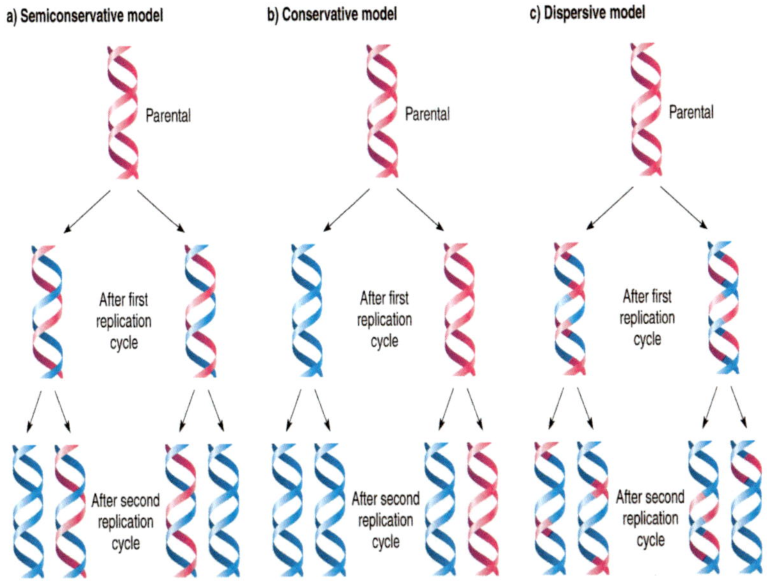

Semi-conservative model

Base pairing allows each strand to serve as a template for a new strand. The new strand is 1/2 parent template & 1/2 new DNA. Anti-parallel strands. Nucleotides in the DNA backbone are bonded from phosphate to sugar between the 3′ & 5′ carbons.

- A DNA molecule has "direction"
- The complementary strand runs in the opposite direction

Meselson and Stahl's Experiment

Two isotopes of nitrogen:

- ^{14}N common form; ^{15}N rare heavy form.
- *E. coli* were grown in a ^{15}N media first, and then transferred to ^{14}N media.
- Cultured *E. coli* were subjected to equilibrium density gradient centrifugation.

Modes of Replication

Replicons: units of replication. Replication origin. Theta replication: circular DNA, *E. coli*; single origin of replication forming a replication fork, usually a bidirectional replication. Rolling-circle replication: virus, F factor of *E. coli* and a single origin of replication

Eukaryotic cells; thousands of origins; a typical replicon: 200,000 ~ 300,000 bp in length.

Requirements of replication

- A template strand
- Raw material: nucleotides
- Enzymes and other proteins

Direction of replication

- DNA polymerase adds nucleotides only to the 3′ end of a growing strand.
- The replication can only go 5′→3′.

Bonding in DNA

- Hydrogen bonds
- Covalent phosphodiester bonds

Replication: 1st step

Initiation: 245 bp in the *oriC* (single origin replicon); an initiation protein. The unwinding of DNA is performed by Helicase. Gyrase removes supercoiling ahead of the replication fork. Single-stranded DNA is prevented from annealing by single-stranded binding proteins. Primers: an existing group of RNA nucleotides with a 3′-OH group to which a new nucleotide can be added; usually 10 ~ 12 nucleotides long.

Primase: RNA polymerase

Unwind DNA

- Helicase enzyme
- Unwinds part of DNA helix

- Stabilized by single-stranded binding proteins

Prevents DNA molecule from closing

- DNA gyrase
- The enzyme that prevents tangling upstream from the replication fork

Replication: 2nd step

RNA Primase adds asmall section of RNA (RNA primer) to the 3' end of template DNA polymerase 3 (an enzyme that builds new DNA strand) can only add nucleotides to existing strands of DNA.

- Elongation: carried out by DNA polymerase III
- Removing RNA primer: DNA polymerase I
- DNA ligase: connecting nicks after RNA primers are removed
- Termination: when a replication fork meets or by a termination protein

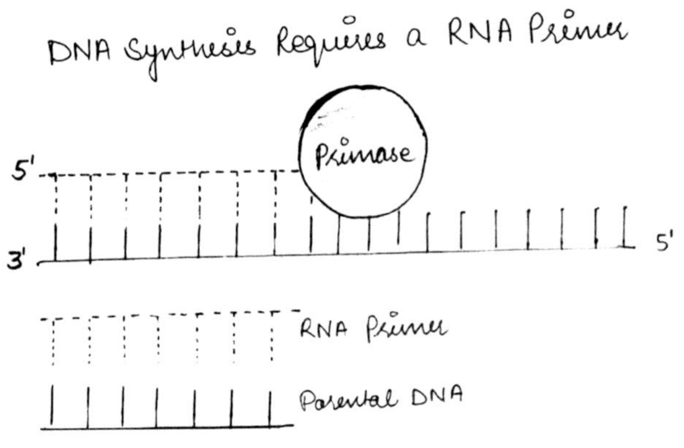

Replication: 3rd step

It builds daughter DNA strands, add new complementary bases and with the help of the enzyme DNA polymerase III.

Replication: 4th step

Replacement of RNA primer by DNA and it is done by DNA polymerase I.

Direction of replication

- Leading strand: undergoes continuous replication.
- Lagging strand: undergoes discontinuous replication.
- Okazaki fragment: the discontinuously synthesized short DNA fragments forming the lagging strand.

Leading & Lagging strands

- Limits of DNA polymerase III.
- It can only build onto 3′ end of an existing DNA strand.

Lagging strand

- Okazaki fragments.
- It is joined by ligase and "spot welder" enzyme.

Leading strand

It continuous synthesis.

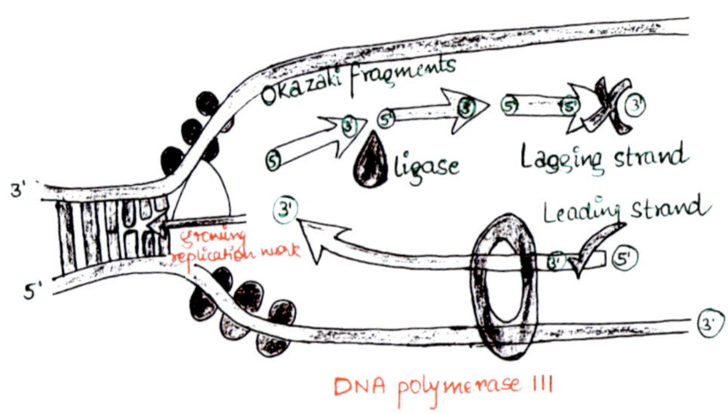

DNA replication on the lagging strand

- RNA primer is added
- It built by primase
- It serves as a starter sequence for DNA polymerase III
- However short segments called Okazaki fragments are made because it can only go in a 5 →3 direction
- Replacing RNA primers with DNA

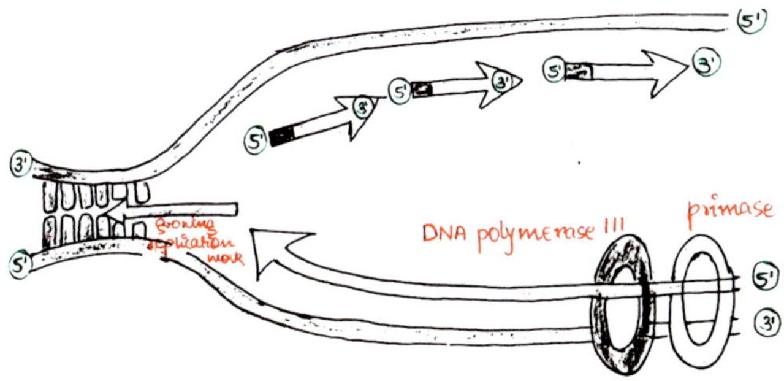

Next DNA polymerase I

- It removes sections of the RNA primer and replaces them with DNA nucleotides.

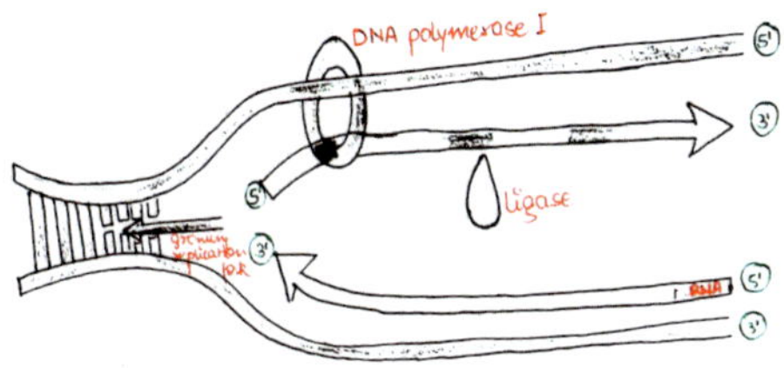

Replication fork

- SSB = single-stranded binding proteins.

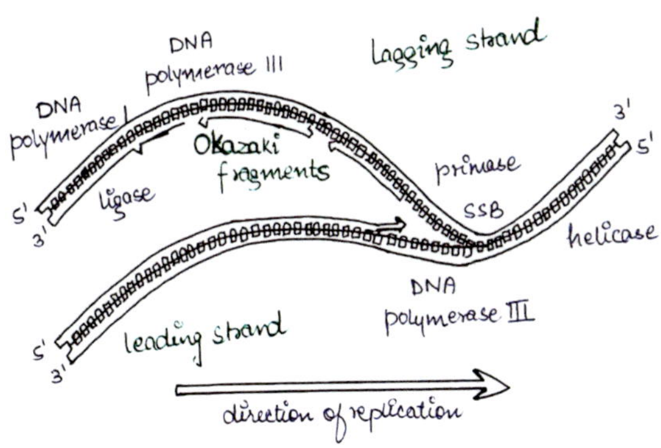

Stabilization of the Replication Machine

DNA polymerase delta and epsilon are attached to the DNA and held in place by other proteins. Sliding-clamp proteins (proliferating cell nuclear antigen

- PCNA), allow the stable binding of DNA polymerase and strand synthesis. Clamp-loading proteins (replication factor C - RFC), aid in attaching the sliding-clamp proteins. The DNA is copied by a replication machine that travels along each replication fork. One of the most important members of this complex is DNA polymerase. It can add DNA subunits and making new DNA.

DNA Polymerase

It was first discovered in 1956 by Kornberg and isolated from *E.coli* bacteria.

Bacteria have 3 types

- DNA Pol I, II, and III.
- DNA Pol III is involved in the replication of DNA.
- DNA Pol I involved in repair.
- Humans have 4 types (you need to know)
- DNA Pol alpha, beta, delta - nuclear DNA
- DNA Pol gamma - mitochondrial DNA
- All DNA Pol's have 2 properties
- Only synthesize DNA in one direction 5' to 3'.
- Only add nucleotides to the end of existing double-stranded DNA.
- Therefore, they cannot start the synthesis of DNA from scratch.
- RNA polymerases can, but not DNA polymerases.

DNA polymerase more facts:

- It has a proofreading mechanism built.
- Checks for base matching.
- Removes mismatched bases by going backward.
- The reason why it is not able to build DNA in the 3' to 5' direction. No energy from ATP hydrolysis.
- Makes just one error in 10E8 or 10E9 (billion) bases added.
- Unwinding of DNA strands by helicases.
  - Require ATP
- Single-stranded DNA-binding proteins.
  - Bind to single-stranded DNA to stabilize the structure.
  - RPA (replication protein A - in eukaryotes).

- Topoisomerase - helps with prevention of DNA strand twisting - 'swivels'
  - Two types
    - Type I - Break one strand only and then rejoin.
    - Type II - Break both strands and then rejoin.

Telomerase

- Enzyme.
- Reverse transcriptase.
- Acts without a DNA template, as it has its own RNA template stored inside.

The fidelity of DNA replication

Proofreading: DNA polymerase I: 3′→5′ exonuclease activity removes the incorrectly paired nucleotide.

Mismatch repair: correcting errors after replication is complete.

Eukaryotic DNA Replication

- Eukaryotic DNA polymerase.
- DNA polymerase α- acts like Primase to initiate.
- DNA polymerase - replicates lagging strand.
- DNA polymerase e- replicates the leading strand.

Replication at the ends of chromosomes

- Telomeres and telomerase.
- DNA repair mechanism.
- Damaged DNA must be repaired.
- If the damage is passed on to subsequent generations, then we use the evolutionary term - mutation. It must take place in the germ cells - the gametes - eggs and sperm.
- If the damage is to somatic cells (all other cells of the body bar germ cells) then just that one individual is affected.

Damage from where

- Consequences of DNA replication errors.
- Chemical agents acting on the DNA.
- UV light imparting energy into a DNA molecule.
- Spontaneous changes in the DNA.

Why DNA repair

DNA pol does a great job, but not good enough. Introduces errors in about 1 in 10E7 nucleotides added, which it does not correct. Other mechanisms exist (as we will see) to correct many of the errors left by the replication system. Most mistakes and damage corrected (99% -leaving just a few - only 1 in 10E9 errors are left). Mutations are permanent changes left in the DNA. Repair of non-replication related damage to the DNA must also be a priority for the cell. These defects also will prevent translation and duplication of the DNA. Cells will die. Again, any errors or changes to the DNA become Mutations - which are permanent changes left in the DNA.

Repair mechanisms

1. Reversal of damage
2. Excision repair
3. Mismatch repair
4. Recombination repair
5. Error-prone repair
6. Restriction-modification systems

1. Reversal of damage

Enzymatically un-do the damage

a) Photoreactivation

b) Removal of methyl groups

Photolyase breaks apart pyrimidine dimers

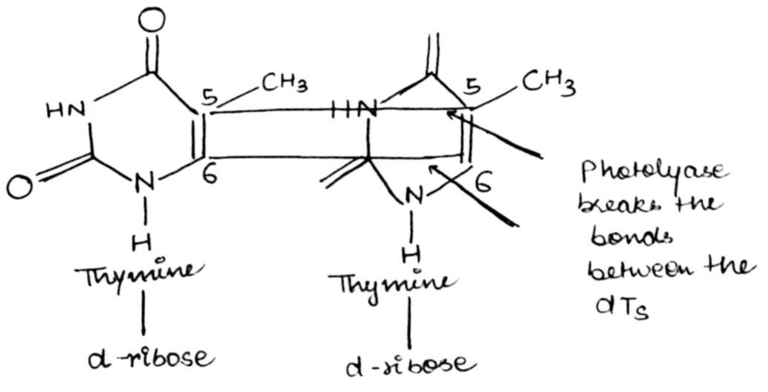

1a. Photoreactivation

Photolyase: binds pyrimidine dimers and catalyzes a photochemical reaction. Breaks the cyclobutane ring and reforms two adjacent T's. It contains 2 subunits, encoded by *phrA* and *phrB*.

Conversion of 6-O-methyl-dG back to dG

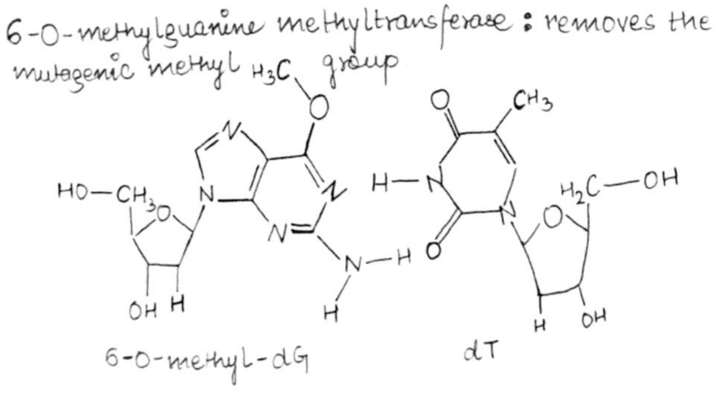

1b. Removal of methyl groups

- 6-O-methylguanine methyltransferase.
- Recognizes 6-O-methylguanine in DNA, removes the methyl group.
- Transfers the methyl group to an amino acid of the enzyme.
- "Suicide" mechanism.
- Encoded by the *ada* gene in *E. coli*.

Excision repair

General Process

- Remove damage (base or DNA backbone).
- ss nick/gap provides 3'OH for DNA Pol I initiation.
- DNA ligase seals nick.

Nucleotide excision repair

- Cut out a segment of DNA around a damaged base.

Base excision repair

- Cut out the base, then cut next to the apurinic/apyrimidinic site, and let DNA Pol I repair.

Mismatch repair

The action of DNA polymerase III (including proofreading exonuclease) results in 1 misincorporation per 10^8 bases synthesized. Mismatch repair reduces this rate to 1 change in every 10^{10} or 10^{11} bases. Recognize mispaired bases in DNA, e.g. G-T or A-C base pairs. These do not cause large distortions in the helix: the mismatch repair system apparently reads the sequence of bases in the DNA.

Role of Methylation in Discriminating Parental and Progeny Strands

The *dam* methylase acts on the A of GATC (note that this sequence is symmetrical or pseudopalindromic). Methylation is delayed for several minutes after replication. Mismatch repair works on the un-methylated strand (which is newly replicated) so that replication errors are removed preferentially.

Action of MutS, MutL, MutH

- **MutS**: recognizes the mismatch (heteroduplex).
- **MutL**: a dimer; in presence of ATP, binds to MutS-heteroduplex complex to activate MutH.

- **MutH**: endonuclease that cleaves 5' to the G in an unmethylated GATC, leaves a nick.

Eukaryotic homologs in mismatch repair

Human homologs to *mutL* (hMLH1) and *mutS* (hMSH2, hMSH1) have been discovered. Mutations in them can cause one of the most common hereditary cancers, hereditary nonpolyposis colon cancer (HNPCC).

Error-prone repair

Last resort for DNA repairs, e.g when the repair has not occurred prior to replication. The polymerase copy across a non-pairing, mutated base, or an apyrimidinic/apurinic site.

- DNA polymerase III usually dissociates at a nick or a lesion.
- But replication can occur past these lesions, especially during the SOS response ("Save Our Ship").

This translesion synthesis incorporates random nucleotides, so they are almost always mutations (3/4 times).

Role of *umuC* and *umuD* genes in error-prone repair

It named for the UV **non**-mutable phenotype of mutants with defects in these genes. It is needed for bypass synthesis; the mechanism is under investigation e.g. these proteins may reduce the template required for the polymerase. UmuD protein is proteolytically activated by LexA.

Evolution acts on mutations

If we did not have mutations then we would all be the same so any changes in the environment would be deleterious to all members of the population equally. There would be no evolution. But mutations do exist and it is supported by the comparison of related organisms.

32. TRANSCRIPTION IN PROKARYOTES, RNA POLYMERASE, TYPES AND FUNCTIONS OF RNA POLYMERASES IN EUKARYOTES, GENETIC CODE, TRANSLATION IN PROKARYOTES, POST-TRANSLATIONAL MODIFICATIONS. GENE ORGANIZATION, OPERON CONCEPT AND INTRODUCTION TO GENE REGULATION MECHANISMS

Transcription

It is the process of formation of the transcript (RNA). It takes place through the usual process of complementary base pairing, catalyzed and scrutinized by the enzyme RNA polymerase. It occurs unidirectionally in which RNA chain (transcript) is synthesized from 5' to 3 directions. It is chemically and enzymatically very similar to DNA replication. However, it differs from that of replication. The new strand is made from ribonucleotides (NTPs) rather than deoxyribonucleotides (NTPs). It is primer independent. RNA polymerase can initiate a new RNA chain on a DNA template without the need for a primer. It is a selective process i.e. transcription selectively copies only certain parts of the genome called a transcription unit. In contrast, during replication entire genome replicates. Only one of the DNA strands acts as a template on which the RNA strand is built. It is unidirectional.

Prokaryotes

Promoter

- Pribnow Box (also called the -10 element) – TATAAT.
- 35 element - TTGACA.

Eukaryotes

- Promoter – (asymmetrical sequence).
- Basic core promoter –TATA box (TATAAA(A)); within 50bp

upstream of start site; found in unicellular eukaryotes.

- Core promoter plus
  - Downstream promoters
  - Proximal promoter elements

Initiator sequences

Response elements are the recognition sites of certain transcription factors. Most of them are located within 1 kb from the transcriptional start site. Enhancer elements - binding with transcription factors (activators), it can enhance transcription, either located upstream or downstream of the transcriptional initiation site.

- Upstream enhancer elements
- Downstream enhancers
- Distal enhancer elements

Silencers - Binding with transcription factors (repressors) and it can repress transcription.

- Insulators

Proteins involved in transcription

- RNA polymerase
- Transcription factors are TFIIA, TFIIB, TFIID, TFIIE, TFIIF, TFIIH
- Transcription factors that bind to regulatory elements are holoenzyme or initiation complex
- Transcription factors have common DNA binding motifs such as
  - Zinc finger
  - Helix-turn-helix
  - Leucine zipper

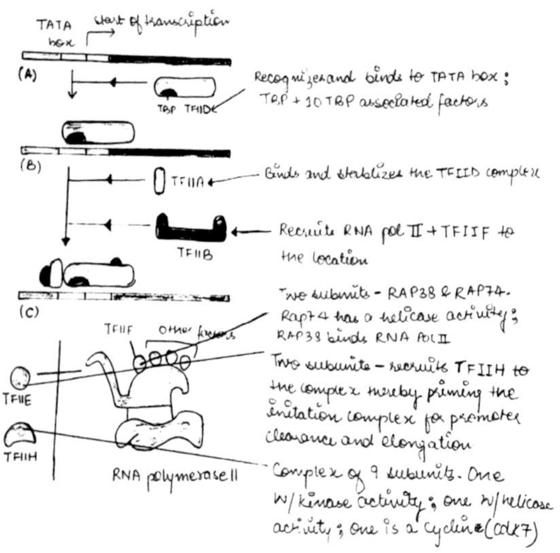

Synthesis of enzymes comes into the control of genetic material i.e., DNA in living organisms. Enzyme synthesis occurs by two processes such as translation and transcription. Translation refers to the synthesis of mRNA. It is controlled by the promoter region of the gene.

Pathway for Gene Expression

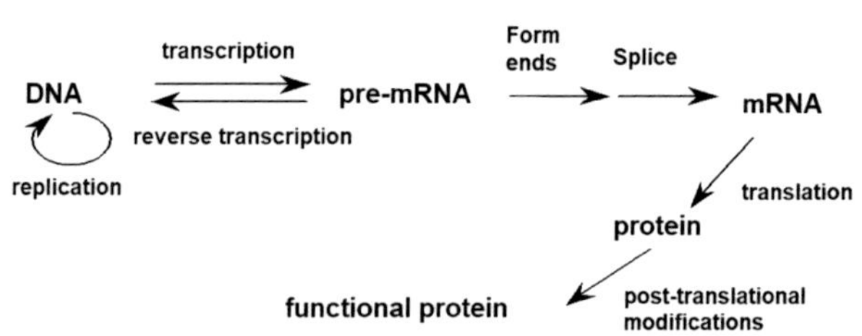

Reaction catalyzed by RNA polymerase

It catalyzes the synthesis of RNA directed by DNA as a template and in transcription it makes an RNA chain with a sequence complementary to the template strand of DNA. It does not require a primer and it easily can start RNA synthesis at a site on the DNA template *E. coli* RNA polymerase. It also synthesizes all classes of RNA.

- mRNA
- rRNA
- tRNA

It contain core= a_2bb' catalyzes elongation of an RNA chain and Holoenzyme = $a_2bb's$ catalyzes initiation or RNA synthesis specifically at a promoter.

Promoter

The sequence of DNA is required for accurate, specific initiation of transcription. The sequence of DNA binds with RNA polymerase accurately initiate transcription. In most cases, RNA polymerase binds to a DNA sequence including the initiation site, but it can be directed there by sequences flanking the initiation site.

RNA polymerase at a promoter

- **α: assembly and binds to UP.**
- **ß+ß**: form the catalytic center.
- **6: binds -10 and -35 of the** promoter to confer specificity during initiation.

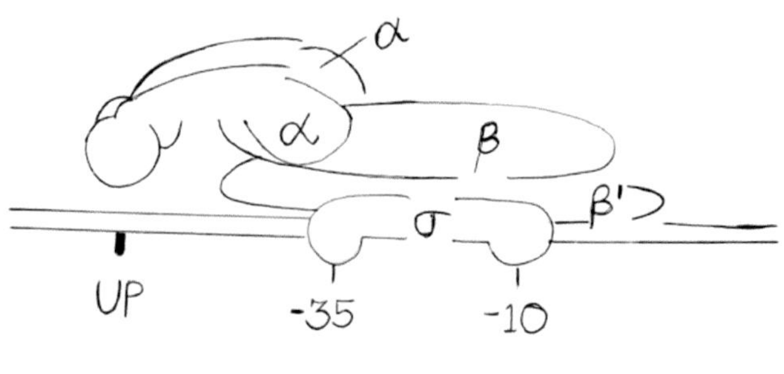

Mode of action of 6 factors

The 6 factor causes RNA polymerase to be selective in its choice of initiation sites by affecting the dissociation rate of polymerase from DNA. The core dissociates from general DNA with a half-time of 60 min; use in elongation. Holoenzyme dissociates from general DNA with a half-time of 1 sec. The holoenzyme dissociates from promoter DNA with a half-time of hours (in the absence of rNTPs).

Transcription cycle

- Initiation: The holoenzyme binds to the promoter, unwinds the DNA, and forms phosphodiester bonds between 7 to 12 nucleotides.
- Elongation: The core elongates RNA with high processivity and it may use NusA.
- Termination: Polymerase dissociates from the template DNA and releases new RNA.

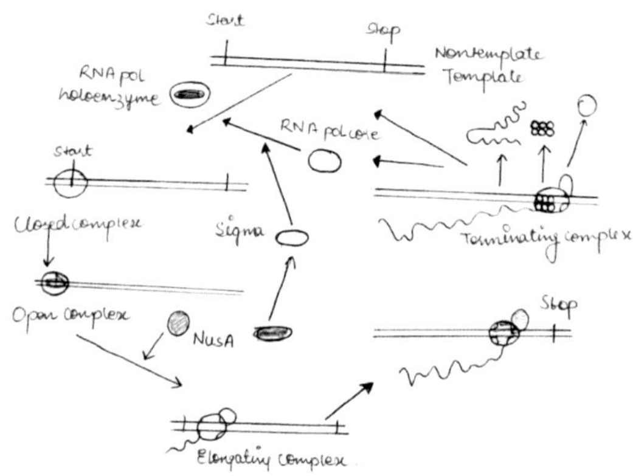

Sites on RNA polymerase core

The enzyme covers about 60 bp of DNA, with about 17 bp unwound is called a transcription bubble. The bubble must contact the active site for polymerization. At the beginning of the bubble, the DNA is unwound, implicating a helicase activity. At the end of the bubble, the DNA is rewound. The effects of transcription on the supercoiling of the template are discussed here. The unwinding (untwisting) and rewinding of the DNA template introduce positive supercoils ahead of the polymerase and negative supercoils behind it. The unwinding will decrease T by 1 for every 10 bp unwound..

RNA polymerases in eukaryotes

Name	Makes	α-amanitin
RNA Polymerase I	pre-rRNA	insensitive
RNA Polymerase II	pre-mRNA some snRNAs	very sensitive
RNA Polymerase III	pre-tRNA other small RNAs some snRNAs	less sensitive

Subunit structure of eukaryotic RNA polymerases

All three have multiple subunits (8 to 14). The molecular weight for each polymerase is about 500,000. Some subunits are common to all of 3 RNA polymerases. All 3 RNA polymerases have subunits that are homologous to the bacterial ß, ß' and α subunits.

Translation

The transcription and translation take place in different cellular compartments and transcription takes place in the nucleus and translation take place in the cytosol. In prokaryotes "transcription of mRNA and translation occur simultaneously. Thus, mRNA molecules undergo little or no modification after synthesis by RNA polymerase in prokaryotes. In contrast, pre-tRNA and pre-rRNA undergo processing like cleavage, the addition of nucleotides and chemical modification after synthesis. Although both prokaryotes and eukaryotes modify pre-tRNA and pre-rrna, eukaryotes very extensively process pre-mRNA destined to become mRNA. The primary transcript of an RNA polymerase is referred to as pre-RNA. Processing of eukaryotic pre-mRNA involves 5' capping, 3' polyadenylation, splicing and RNA editing before being transported to the cytosol, where they are translated by ribosomes". The final product of gene expression is a polypeptide chain of amino acids whose sequence is prescribed by the genetic code.

- mRNA transcribed from genomic DNA.
- Transfer RNA (tRNA) to transport amino acids.
- Ribosome to "read" mRNA, align amino acids attached to tRNA and create the peptide bonds between adjacent amino acids.

Ribosome

- Two subunits are composed of protein and ribosomal RNA (rRNA).
- The rRNA is a structural component of the ribosome subunits.

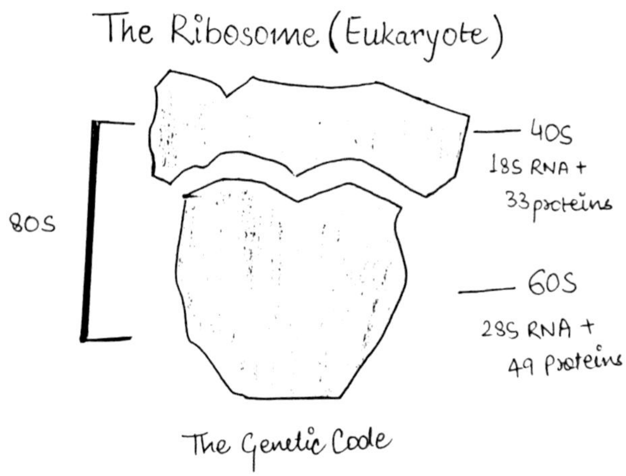

tRNA

tRNAs are small, extremely stable RNA structures shaped like a cloverleaf due to internal base pairing. They are almost identical in both prokaryotes and eukaryotes. They contain a number of post-transcriptional modifications and it can including non-traditional bases (other than (A, U, G and C). The amino acid attaches to the 3' end of the tRNA. The tRNA can base pair with the codon in mRNA at the anticodon on the tRNA.

Charging tRNA with an amino acid

An individual tRNA is specific for a single amino acid (There are at least 20 different tRNAs). Charging is the process of covalently attaching an amino acid to the tRNA. Charging is accomplished by aminoacyl tRNA synthetases and each of which is specific for an amino acid.

- ***Initiation***: The two subunits of the ribosome come together and the start codon on the mRNA in the ribosome is aligned to set the reading frame.
- ***Elongation***: It can charge tRNAs attach and peptide bonds form between the amino acids.

- ***Termination*:** It can start when ribosomes come to a stop codon (UAG, UAA and UGA), which indicate the signals of genetic information end.

Prokaryotes

Translation takes place directly after transcription. In prokaryotic mRNA is not modified. Transcription and translation take place in the same area.

Eukaryotes

A transcript is modified before leaving the nucleus (5' cap and 3' poly-A tail). The modifications increase the translation efficiency and lifespan of the mRNA. Translation takes place on ribosomes located in the rough endoplasmic reticulum (translation is physically separated from transcription). Influence on gene expression forms the basis of function in many antibiotics. The purpose is to impair function in the prokaryote without disrupting function in eukaryotes.

Antibiotics

- Tetracycline: It blocks the attachment of the AA-tRNA to the A site in the ribosome.
- Erythromycin: It blocks the movement of the ribosome down the mRNA.
- Streptomycin: It blocks the formation of the initiation complex and causes miscoding.
- Rifamycin**:** It blocks transcription by blocking the binding of RNA polymerase.

Genetic code

It consists of 64 triplet codons (A, G, C, U)$4^3 = 64$. All codons are used in protein synthesis. It has 20 amino acids. It has 3 termination (stop) codons: **UAA, UAG, and UGA**. So, **AUG** (methionine) is the start codon (also used internally). It has multiple codons for a single amino acid is called degeneracy. It has 5 amino acids are specified by the first two nucleotides only.

The Genetic code

UUU	Phe	UCU	Ser	UAU	Tyr	UGU	Cys
UUC		UCC		UAC		UGC	
UUA	Leu	UCA		UAA	STOP	UGA	STOP
UUG		UCG		UAG	STOP	UGG	Trp
CUU	Leu	CCU	Pro	CAU	His	CGU	Arg
CUC		CCC		CAC		CGC	
CUA		CCA		CAA	Gln	CGA	
CUG		CCG		CAG		CGG	
AUU	Ile	ACU	Thr	AAU	Asn	AGU	Ser
AUC		ACC		AAC		AGC	
AUA		ACA		AAA	Lys	AGA	Arg
AUG	Met	ACG		AAG		AGG	
GUU	Val	GCU	Ala	GAU	Asp	GGU	Gly
GUC		GCC		GAC		GGC	
GUA		GCA		GAA	Glu	GGA	
GUG		GCG		GAG		GGG	

Codon-Anticodon Interactions

The codon-anticodon base-pairing is antiparallel. The third position in the codon is frequently degenerate. One tRNA can interact with more than one codon (therefore 50 tRNAs). The Wobble rules are.

- C with G or I (inosine).
- A with U or I.
- G with C or U.
- U with A, G, or I.
- I with C, U, or A.

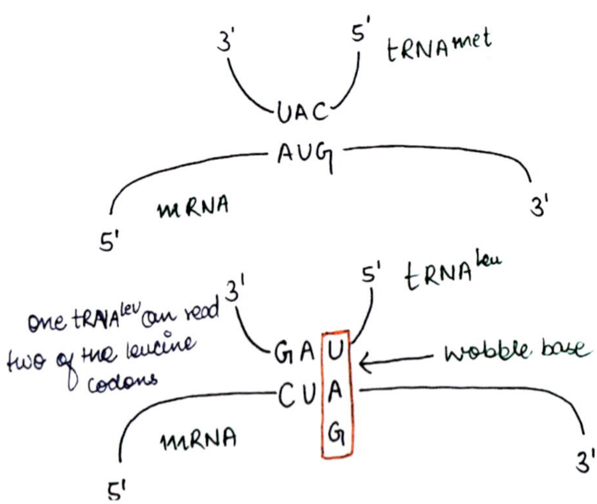

Termination Point

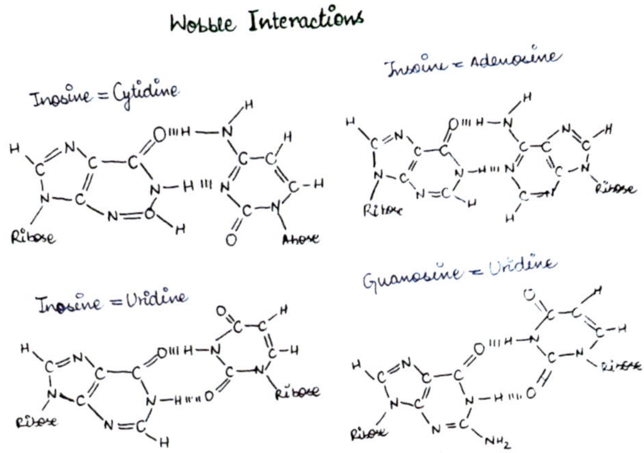

Initiation in Prokaryotes and Eukaryotes

Initiation can occur at internal AUG codons in prokaryotic mRNA. Initiation in eukaryotes occurs only at the first AUG codon. lac operon in *E. coli* is transcribed as a polycistronic mRNA with multiple AUG codons.

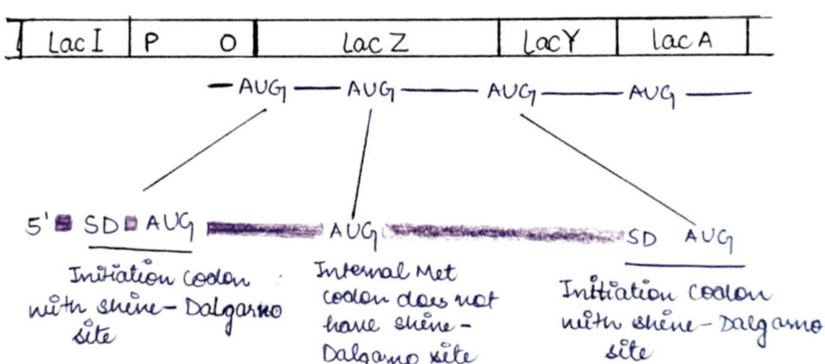

Reading frame

The reading frame is determined by the AUG initiation codon. Every subsequent triplet is read as a codon until reaching a stop codon.

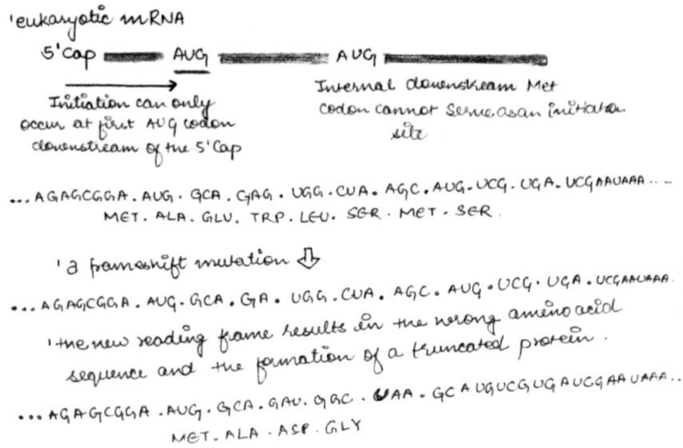

Mutations Affecting Translation

- Hemoglobin Wayne (3' terminal frameshift mutation).
- Missense mutations (e.g., AGC, Ser to AGA Arg).
- Nonsense mutations (e.g., UGG, Trp to UGA Stop).
- Read through, reverse terminator, or sense mutations (e.g., UAA Stop to CAA Gln) as in hemoglobin constant spring.
- Silent mutations (e.g., CUA Leu to CUG Leu) do not affect translation.

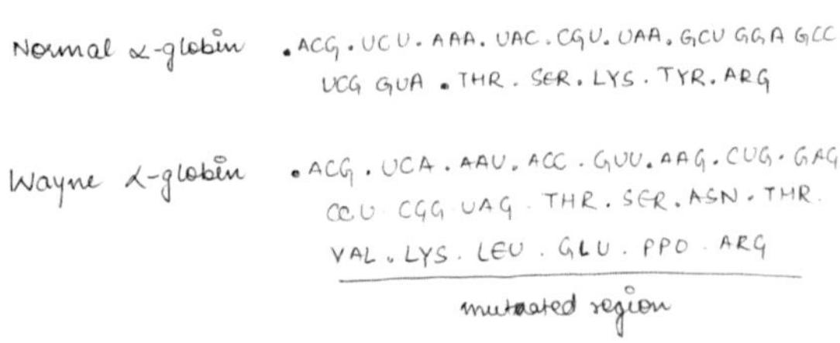

Post-translational modifications

- Cleavage of signal peptides
- Phosphorylation
- Amidation
- Glycosylation
- Hydroxylation
- Ubiquitination

- Addition of prosthetic groups
- Iodination
- Adenylation
- Sulfonation
- Prenylation
- Myristoylation
- Acylation
- Acetylation
- Methylation
- Oxidative crosslinking
- N-Glutamyl cyclization
- Carboxylation

N-terminal modifications

- N-formyl Met (by-product of translation process and incomplete cleavage by deformylase).
- Aminopeptidase cleavage of Met.
- N-terminal acetylation.
- N-terminal myristoylation.
- N-terminal glutamyl cyclization.

C-terminal modification

- C-terminal prenylation
- farnesylation
- geranylation/geranyl-geranylation
- C-terminal amidation

Phosphorylation

- Hydroxyl groups - Ser, Thr, Tyr.
- Amide groups of His, Lys & Carboxy group Asp.
- It often causes changes in protein conformation and/or protein-protein interactions.

C-terminal amidation

- Neuropeptides and hormones
- C-terminal glycine is hydroxylated alpha-hydroxy-glycine.
- *Glyoxylate*: The C-terminal amidation is essential to the biological activity of many neuropeptides and hormones.

Hydroxylation and carboxylation

- Hydroxylation (previous lecture).
- Lys and Pro (e.g. collagen).
- Requires vitamin C.
- Carboxylation
- Requires vitamin K.

Iodination

- Special case: Tyr → thyroxin.

Protein Acylation

- Methylation and acetylation.
- Acetylation of histone Lys.
- S-acylation (Cys).
- Prenylation (C-terminus).
- Myristoylation (N-terminus).

Acetylation

N-terminal acetylation

It does not have a simple sequence consensus and is usually accompanied by cleavage of the N-terminal Met.

Methylation

It can occur at carboxyls, amides, sulfhydryl, etc and S-Adenosyl Methionine (S-AM) is the activated carbon source.

Histones

- It is DNA binding proteins and chromatin structure and also affect the gene expression.

C-terminal Prenylation

Prenylation

- CAAX motif.
- AAX cleaved and Cys methylated then prenyl group added by thioether linkage (C-S-C).

Adenylation

The toxins are frequently adenylate of host target proteins to inactivate them. The glutamine synthetase is inactivated by adenylylation and activated by deadenylylation. The adenylation of a DNA ligase Lys residue is a key step in DNA replication and repair.

Proteolytic cleavage

All non-cytoplasmic proteins must be translocated. The leader peptide retards the folding of the protein so that molecular chaperone proteins can interact with it and direct its folding. The leader peptide also provides recognition signals for the translocation machinery. A leader peptidase removes the leader sequence when folding and targeting are assured.

Proteolytic processing

The proteolytic cleavage of the hydrophobic N-terminal signal peptide sequence is called proteolytic processing. It is done at a site defined by pairs of basic amino acid residues. It has designated by single Arg residues. The signal prediction of the algorithm is (e.g. Signal P).

Hormones

All secreted polypeptide hormones are synthesized with a signal sequence (that directs them to secretory granules, then outside). It is usually synthesized as inactive preprohormones ("pre-pro" implies at least two processing steps). The proteolytic processing produce the prohormone and the hormone.

Gastrin

The product of preprogastrin, having 101-104 residues, the signal peptide cleavage, prepropeptide is propeptide at 80-83 residues, Cleavage at Lys and Arg residues and C-terminal amidation leaves gastrin.

Propeptide → peptide

The N-terminal residue of gastrin is pyroglutamate; C-terminal amidation involves the destruction of Gly. Heptadecapeptide secreted by the antral mucosa of the stomach stimulates acid secretion.

***Protein glycosylation*:** It is usually extracellular or at the cell surface. It occurs along the secretory pathway. It often stabilizes the structure. It is difficult to get the crystal structure for more than one or two carbohydrate residues.

O-linked glycosylation

It has no consensus sequence for Ser and Thr, the consensus for Hyp (Gly – X – Hyl – Y – Arg). It begins with GalNac transferase (N-acetylgalactosamine) and mannose is a common addition to the core.

The ABO (H) blood group determinant is an O-linked polysaccharide

In 1901, Karl Landsteiner discovered blood group antigens. Since that time nearly 200 antigens have been identified. Carbohydrate is dependant on blood group antigens are carried by both glycoproteins and glycolipids

N-linked glycosylation

Most proteins contain several potential sites but usually only a few (in any) are actually glycosylated, it has many different patterns and the consensus is (-N-X-T/S-).

N-linked glycosylation begins in the cytoplasm

The glycosyl-dolichol-phosphate (lipid) core is transferred to protein.

Operon

The bacterial gene expression and regulation including structural genes and control elements are present in DNA recognized by regulator gene product. Trans-acting: A product that can function on any copy of its target DNA.

This implies that it is a diffusible protein or RNA. Cis-acting: A site that affects the activity only of sequences on its own molecule of DNA (or RNA) and this property usually implies that the site does not code for protein. Regulator gene: A gene that codes for a product (typically a protein) that controls the expression of other genes (usually at the level of transcription). Structural gene: A gene that codes for any RNA or protein product other than a regulator. In negative regulation, a repressor protein binds to an operator to prevent a gene from being expressed. In positive regulation, a transcription factor is required to bind at the promoter in order to enable RNA polymerase to initiate transcription.

- In inducible regulation, the gene is regulated by the presence of its substrate (the inducer).
- In repressible regulation, the gene is regulated by the product of its enzyme pathway (the corepressor).
- Gene regulation *in vivo* can utilize any of these mechanisms, resulting in all four combinations: negative inducible, negative repressible, positive inducible and positive repressible.

The lac Operon is Negative Inducible

Transcription of the "*lacZYA* operon is controlled by a repressor protein (the *lac* repressor) that binds to an operator that overlaps the promoter at the start of the cluster. It shows the constitutive expression-A state in which a gene is expressed continuously. In the absence of β-galactosides, the *lac* operon is expressed only at a very low (basal) level. The repressor protein is a tetramer of identical subunits coded by the *lacI* gene. β-galactoside sugars are the substrates of the *lac* operon, are its inducer. The addition of specific β-galactosides induces transcription of all three genes of the *lac* operon. The *lac* mRNA is extremely unstable" and induction can be rapidly reversed.

lac repressor is controlled by a small-molecule inducer

An inducer functions by converting the repressor protein into a form with lower operator affinity. Repressor has two binding sites, one for the operator DNA and another for the inducer. The repressor is inactivated by an allosteric interaction in which binding of inducer at its site changes the properties of the DNA-binding site (allosteric control). The true inducer is allolactose, not the actual substrate of β-galactosidase.

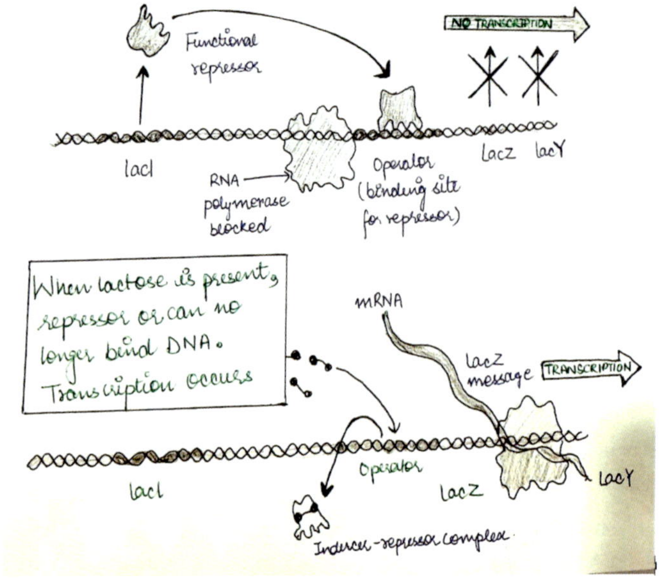

Lac Operon

cis-acting constitutive mutations identify the operator

Mutations in the operator cause constitutive expression of all three *lac* structural genes. These mutations are *cis*-acting and affect only those genes on the contiguous stretch of DNA. Mutations in the promoter prevent the expression of *lacZYA* and are uninducible and *cis*-acting. **Cis-dominant:** A site or mutation that affects the properties only of its own molecule of DNA, often indicating that a site does not code for a diffusible product.

trans-acting mutations identify the regulator gene

Mutations in the *lacI* gene are Trans-acting and affect the expression of all "*lacZYA* clusters in the bacterium. Mutations are that eliminate *lacI* functioning cause constitutive expression and are recessive (*lacI*$^{-}$). Mutations in the DNA-binding site of the repressor are constitutive because the repressor cannot bind the operator. Mutations in the inducer-binding site of the repressor prevent it from being inactivated and cause uninducibility. When mutant and wild-type subunits are present, a single *lacI*$^{-d}$ mutant subunit can inactivate a tetramer whose other subunits are wild-type. It is dominant negative. Interallelic complementation: The change in the properties of a heteromultimeric protein brought about by the interaction of subunits coded by two different mutant alleles. The mixed protein may be more or less active than the protein consisting of subunits of only one or the other type. Negative complementation**:** This occurs when interallelic complementation allows a mutant subunit to suppress the activity of a wild-type subunit in a multimeric protein. *lacI*$^{-d}$"mutations occur in the DNA-binding site. Their effect is explained by the fact that repressor activity requires all DNA-binding sites in the tetramer to be active.

lac repressor is a tetramer made of two dimers

A single repressor subunit can be divided into the "N-terminal DNA-binding domain, a hinge, and the core of the protein. The DNA-binding domain contains two short α-helical regions that bind the major groove of DNA. The inducer-binding site and the regions responsible for multimerization are located in the core. Monomers form a dimer by making contacts between core subdomains 1 and 2. Dimers form a tetramer by interactions between the tetramerization helices. Different types of mutations occur in different domains of the repressor protein.

lac repressor binding to the operator is regulated by an allosteric change in conformation

lac repressor protein binds to the double-stranded DNA sequence of the operator. The operator is a palindromic sequence of 26 bp. Each inverted repeat of the operator binds to the DNA-binding site of one repressor subunit. Inducer binding causes a change in repressor conformation that reduces its affinity for DNA and releases it from the operator.

lac repressor binds to three operators and interacts with RNA polymerase

Each dimer in a repressor tetramer can bind an operator so that the tetramer can bind two operators simultaneously. The full repression requires the repressor to bind to an additional operator downstream or upstream as well as to the primary operator at the *lacZ* promoter. The binding of the repressor at the operator stimulates the binding of RNA polymerase at the promoter but precludes transcription.

Tryptophan operon

It is present in bacteria and was first described in *E.coli.* A group of genes is transcribed together that codes for the production of tryptophan.

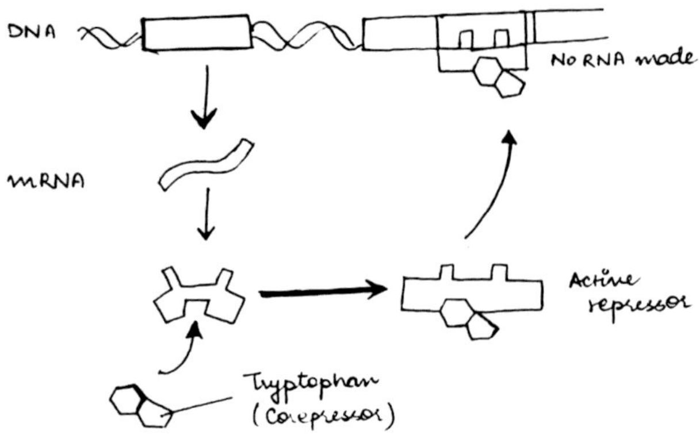

(b) Tryptophan present, repressor active, operon off

Tryptophan is the allosteric regulator of the repressor protein

Lactose operon

It is needed for the metabolism and transport of lactose in *E.coli* and other many different bacteria. The product of gene related to lacZ is β-galactosidase which splits lactose into glucose and galactose.

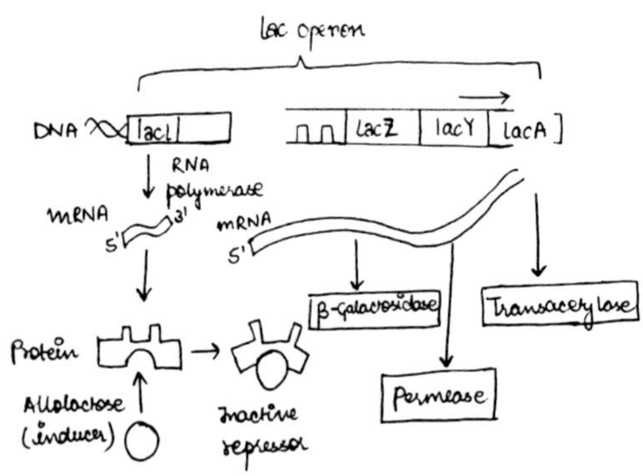

(b) lactose present, repressor inactive, operon on

Lactose is the allosteric regulator of the repressor protein

Positive gene regulation

If glucose levels are low (along with overall energy levels), then cyclic AMP (cAMP)binds to cAMP receptor protein (CRP) which activates transcription. If glucose levels are sufficient and cAMP levels are low (lots of ATP), then the CRP protein has an inactive shape and cannot bind upstream of the *lac* promoter.

Control of gene regulation

The control of gene expression can occur at any step in the pathway from gene to a functional protein.

1. Packing/unpacking DNA
2. Transcription
3. mRNA processing
4. mRNA transport
5. Translation
6. Protein processing
7. Protein degradation

33. TRANSPOSONS: AN ELEMENTARY IDEA

Transposon

A DNA sequence can move in the genome. It is also called a mobile DNA element or transposable element. It has "selfish DNA" that exist only to maintain themselves. Transposition**:** The process by which these sequences are copied and inserted into a new site in the genome. Probably, it had a significant influence on evolution. In the 1940s, Barbara McClintock discovered the first transposable element in maize, earned the Nobel prize in 1983. The late 1960s, transposition was also found in bacteria.

Retrotransposons

It is mostly mobile elements in bacteria are DNA transposons. In contrast, most mobile elements in eukaryotes are retrotransposons, but eukaryotic DNA transposons also occur. The bacterial insertion sequence element is the P element in *Drosophila.*

- LTR retrotransposons.
- Non-LTR retrotransposons: the most common type of transposons in mammals.

Non-LTR retrotransposons

- Long interspersed elements (LINEs): It is 6 kb in size present in the human account for 21% of the genome.
- Short interspersed elements (SINEs): It has 300 bp in a human account for **13%** of the genome.

Transposable Elements

It is mobile genetic element and comprises 45% of human chromosomal DNA is called "middle repetitive DNA". It contributes to spontaneous mutation, genetic rearrangements, horizontal transfer of genetic material and aid speciation and genomic changes (in bacteria transposons are often associated with antibiotic resistance genes). The cells must depress transposition to insure genetic stability.

Composite bacterial transposons

It has repeated ends, usually inverted, sometimes direct. It has IS elements and can independently transpose and mobilize all intervening DNA. It is often antibiotic resistance genes (examples Tn3 (ampicillin), Tn5 (kanamycin), Tn10 (tetracycline). It frequently resides on plasmids.

DNA transposons

- Excisive mechanism; examples: Tn5, Tn10, P elements.
- Replicative mechanism; examples: Tn3, bacteriophage Mu.

Excisive transposons

It has a cut-and-paste mechanism and cut them out of the original site and producing double-strand break. It cut the target site and ligate to element ends, thereby inserting at a new site. It has original site break repaired usually with sister chromosome, restoring the transposon at the original site and sometimes end healed without the transposon, can also be associated with deletion at the excision site.

Replicative transposons

The original cut of the transposon is only a nick and only one strand is initially ligated. Element replicates through itself. It has produces as intermediate a "co-integrate" structure. It has co-integrate is resolved by resolvase (as TnpR of Tn3) and at a specific site (as *res* of Tn3).

"Degenerate" transposons

Many naturally occurring transposable elements have suffered mutation and are no longer active. Some of these may have cis-acting end mutations and cannot be mobilized. Others may have intact ends but no transposase: these can be mobilized by an element that is tnp+ ("autonomous" element). Ac-Ds system is an example of the latter: Ac (activator) can mobilize Ds (dissociator).

- MITEs (minature inverted repeat transposable elements) are nonautonomous DNA elements.
- SINEs are retrotransposon version (LINEs).

Classification of retroelements

It has an obligate RNA intermediate, uses reverse transcriptase (RT, RNA-dependent DNA polymerase).

- LTR-retroelements: long terminal repeatsTy1/Copia, Ty3/gypsy and retroviruses.
- Non-LTR-retroelements "retroposons"LINES.

34. BASIC CONCEPT OF RECOMBINANT DNA TECHNOLOGY, PRINCIPLES OF GENE CLONING. RESTRICTION-MODIFICATION SYSTEMS, USE OF RESTRICTION ENZYMES IN BIOTECHNOLOGY, CLONING VECTORS

Chimeric DNA

The recombinant DNA is called chimeric DNA because it are made up of two different organisms.

Restriction Enzymes for Cloning

For the cloning of DNA, often we need to cut the DNA at specific sites, recognized and cleaved by specific enzymes. Also by locating the positions of the cleavage site of a number of restriction enzymes, restriction maps can be prepared. These restriction enzymes recognize short sequences of double-stranded DNA as targets for cleavage. Different enzymes recognize different but specific sequences, each ranging in length from 4 to 8 base pairs. The enzymes are named by a threeletter code that identifies their origin. Roman numerals are added to distinguish several enzymes with the same origin. For instance, *EcoRI* is derived from *E.coli* and *HpaI* is derived from *Haemophilus parainfluenzae.* Besides cleavage, modification in the form of methylation is also brought about by some enzymes called modification enzymes. This methylation distinguishes genes in different states of functioning. There are also some enzymes that perform the functions of both restriction and modification. Based on the attributes, restriction enzymes have been grouped into two classes: (i) Type II restriction enzymes systems which have separate enzymes for modification and restriction; and (ii) type I and type III enzymes systems, in which the same enzyme possesses both activities, although the restriction and modification sites differ in position.

- The above two classes of restriction enzymes, type II enzymes are most important for cloning purposes.
- The target sites for these enzymes are 4-8 bp with symmetry. Most type II restriction enzymes cleave DNA at the unmethylated target site.
- Some enzymes introduce staggered cuts and others generate blunt ends.
- Enzymes with 4 bp target sites are used when frequent cuts are desired and those with 8 bp are used when rare cuts are desired to get long DNA segments.
- Otherwise majority of enzymes used have 6 bp target sites.
- Some of them can cleave both methylated as well as unmethylated targets, but the majority of them cleave only unmethylated targets.

Construction of Chimeric DNA

DNA molecules are produced by inserting a foreign DNA segment into the DNA molecule of a vector. These chimeric DNA molecules are the means of cloning genes or other small DNA sequences. The construction of the chimeric DNA, as we know now, involves the joining of the broken ends of the vector with the two ends of the sequence to be cloned. There at least three methods available for the construction of this chimeric DNA

by palindromes and staggered cleavage. This is the most common meth palindromic sequence; a restriction enzyme causes staggered cuts producing short complementary single-stranded sticky ends (a palindromic sequence is one, which has complementary sequences at the two ends of a single strand, e.g. AT.GAT).

This can be illustrated using the example of enzyme *EcoRI*, which cuts the sequences at specific positions. When another DNA molecule is similarly cut by the same order enzyme, similar sticky ends having the same sequences in the single-stranded ends will be produced, so that when it is mixed with the previous molecule similarly treated, the two will anneal producing a chimeric DNA or chi plasmid if plasmid DNA is involved. Enzyme DNA ligase helps in joining the bonds at the cut ends of two molecules. The above technique has the advantage of regenerating two restriction sites (*EcoRI* site) in the chimeric DNA, so that the foreign DNA segment can be retrieved rather easily from the cloned copies of chimeric DNA by cleavage again with the same enzyme. There are also the following disadvantages with this technique. The two cleaved ends of a vector or of a foreign DNA may join end to end before getting inserted. Therefore while isolating chimeric DNA, one will have to select chimeric molecules each having only a single insert. This can be achieved by separating molecules of different sizes by gel electrophoresis. The recognition site, particularly in the sequence to be cloned may not lie at a convenient position so that sometimes only a part of the desired segment will be inserted.

DNA cloning in bacteria and eukaryotes

Cloning in bacteria

Bacteria are often used as hosts for the multiplication of chimeric vectors or recombinant DNA molecules. Therefore, a bacterial strain should be available, which should be the host for the vector and should be modified or selected for high transformation efficiency. Bacteria can be classified into.

- Gram-negative bacteria e.g. *Escherichia coli. Pseudomonas, Rhizobium* and
- Gram-positive bacteria e.g. *Bacillus subtilis* and *Streptomycetes sp.*

Ordinarily the same plasmid can not maintain itself in a stable form in both classes of bacteria and, therefore, plasmid vectors for two classes usually differ. While gram-negative bacteria can be used for plasmids, phages and cosmids vectors, gram-positive bacteria are used as hosts for only plasmids. Chimeric vector is inserted bacterial cells and transformed bacterial colonies are selected and for multiplication in suspension cultures. The chimeric DNA is retrieved whenever required.

Cloning in eukaryotes

Since chromosomes found in the nucleus of eukaryotes are separated from the rest of cell through nuclear membrane, and since many of the genes are split genes with exons and introns, genetic engineering with eukaryotes requires new methods and tools. When eukaryotic genes are cloned prokaryotes, the split genes cannot be correctly expressed, because prokaryotes do not have the equipment necessary for splicing out the RNA transcribed from the introns of a gene. In view of this, eukaryotic cells may sometimes be needed for cloning and particularly for the expression of cloned eukaryotic genes.

Among eukaryotes, DNA cloning has been done in yeast, mouse and to some extent even in some higher plant species. In yeast, a plasmid called 2u DNA (63 bp) is found, which is an appropriate cloning vehicle. An efficient transformation method is also available, which involves protoplast production followed by PEG directed introduction of DNA into protoplast. In animal cells like mouse cells, special animal viruses were used as cloning vehicles. Simian virus 40 (SV40) is one such virus, in which the globin gene could be integrated. This gene integrated into SV40 could be transcribed and translated in mouse kidney cells. A variety of human cell lines in culture is also used for cloning of foreign DNA, using some animal viruses as vectors.

Exonuclease

These enzymes act upon the genome and digest the base pairs on 5' or 3' ends of a single-stranded DNA or at single-strand nicks or gaps in double-stranded DNA **Figure 34.1**.

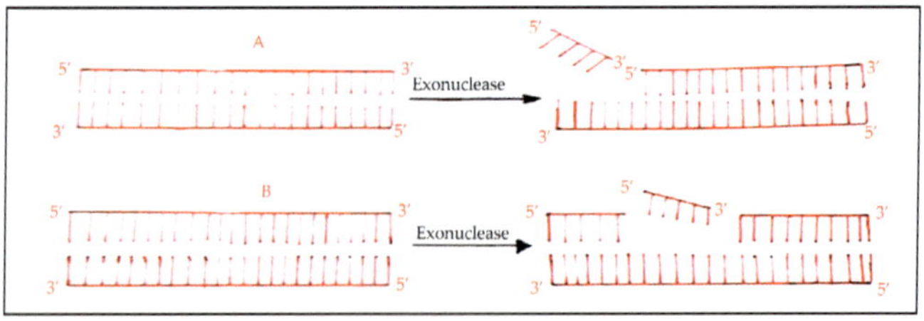

Figure 34.1 Action of exonuclease (a) endonuclease (b).

Endonuclease

They act upon genetic material and cleave the double-stranded DNA at any point except the ends, but their action involves only one strand of the duplex **Figure 34.2**.

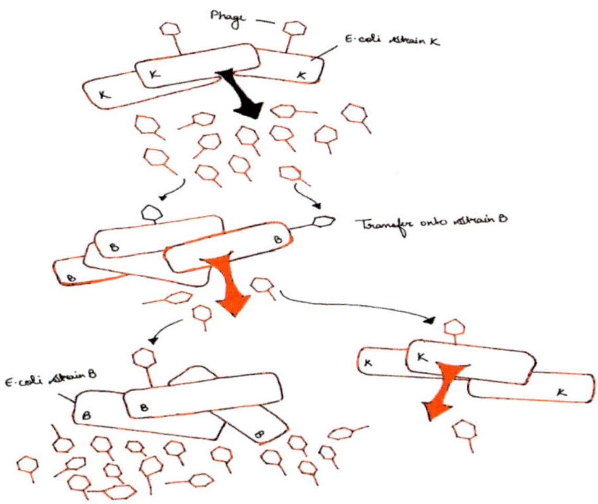

Figure 34.2 Host controlled restriction and modification of phage I in *E.coli* strain K and strain B.

Restriction Endonucleases

The restriction enzymes are called as 'molecular scissors'. These act as a foundation of recombinant DNA technology. These enzymes are present in bacteria and provide a type of defense mechanism called the 'restriction-modification system. The molecular basis of these systems was elucidated first by **Werner Arber** in 1965. However, there was no understanding about cutting and joining of DNA molecules using DNA enzymes, before 1970. During the early 1950s, it was discovered when phage A infects *E. coli*, phage DNA is delivered inside the cell. Host-control restriction-modification system provides defense to *E.coli.* A bacterium protects itself through (i) restriction mechanism (i.e. identifies the introduced foreign DNA and cuts into pieces called restriction endonucleases, and (ii) modification system (i.e. methylation of certain bases of its own DNA by methylase so that phage A-expressed restriction endonuclease could not cleave its DNA).

Nomenclature

The Restriction Enzymes are named based on some of the following principles:

- Name of the organism is identified by the first letter of the genus name and the first two letters of the species name to form a three-letter abbreviation in the italic, for example, *E. coli* = *Eco* and *H. influenzae* = *Hin*, etc.
- A strain or type identified is written as subscript e.g. *EcoK* for *E. coli* strain K. Hind for *H. influenzae* strain Rd.
- In such cases where the restriction and modification systems are genetically specified by a plasmid, the extra-chromosomal element is identified by a subscript e.g. *Eco RI, EcoPI*, etc.
- When a strain has several restriction and modification systems, these are identified by Roman numerals, for example, HindI, HindII, HindIII for *H. influenzae* strain Rd. etc These Roman numerals should not be confused with those in the classification of restriction enzymes into Type I.

Type III Restriction Enzymes: The Type III enzyme is made up of two sub-units, one specifies for site recognition and modification and the other for cleavage. In a reaction, it moves along the DNA and requires ATP as a source of energy and Mg as a co-factor. ATPase activity is lacking in these enzymes.

Examples of some enzymes

There are some other restriction enzymes that do not fall under Type I-III. Hence there is a need for reclassifying restriction enzymes under the other categories from Type I to Type IV. Examples of the other restriction enzymes are *Eco* 571, MCR (modified cytosine reaction), etc. *Eco*571 is made up of single polypeptide. It shows both nuclease modification and nuclease activity. S-adenosyl methionine stimulates the nuclease activity of this enzyme. It bears some other properties that resemble Type II enzymes. Therefore, *Eco* 571 should be put under Type II to Type IV

S1 Nuclease

It degrades the single-stranded DNA or single strand protrusion of double-stranded DNA Cohesive ends. As a result of the action of the nuclease, cohesive ends are converted into blunt ends. Therefore, if nuclease is used to remove the incompatible ends so that overlapping ends may be developed for annealing of two fragments of DNA molecules, mode of action of S1 nuclease.

DNA ligases

Mertz Davis (1972) for the first time demonstrated that cleaved DNA molecules could be covalently sealed with *E. coli* DNA ligase and were able to produce recombinant DNA molecules which have 5-33 '-OH (hydroxyl) termini. There are two enzymes that are extensively used for covalently joining restriction fragments the ligase from that encoded by T4 phage. The main one of DNA phage, hence, the enzyme is known as T4 DNA ligase.

Alkaline phosphatase

When a plasmid vector, for joining a foreign DNA fragment, is treated with the restriction enzyme, the greatest difficulty arises at the same time. Because the cohesive ends of broken plasmids, instead of joining with foreign DNA join the cohesive end of the same DNA molecules and get recircularized. To overcome this problem, the restricted plasmid (i.e. plasmid treated with restriction enzymes) is treated with an enzyme, alkaline phosphatase that digests the terminal 5 'phosphoryl group. The restriction fragments of the foreign DNA are not treated with alkaline phosphatase. Therefore, the 5 'end of the foreign DNA fragment of the plasmid. The hybrid ligase will only join 3 'and 5' ends of recombinant DNA together if the 5 'end is phosphorylated. Thus, alkaline phosphatase and ligase prevent recircularization of the vector and increase the frequency of production of recombinant DNA molecules. The nicks between the two 3 'ends of DNA fragment and vector DNA are repaired inside the bacterial host cells during the transformation can covalently join to 3 'end or recombinant DNA obtained has a nick with 3' and 5' hydroxy ends Figure 34.3.

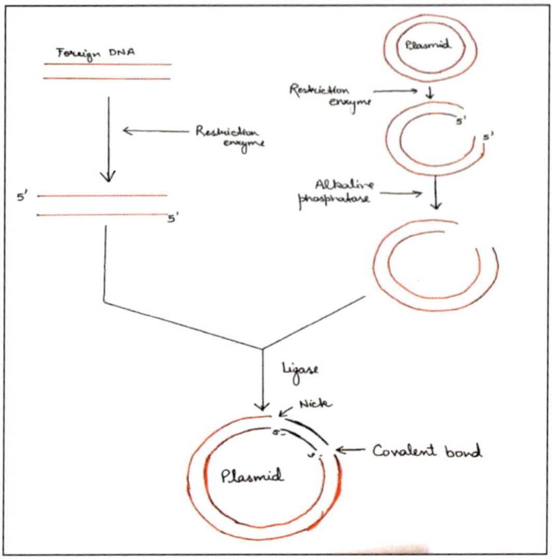

Figure 34.3 Alkaline Phosphatase

Reverse Transcriptase

In addition to these enzymes, reverse transcriptase is used to synthesize the copy DNA complementary DNA (cDNA) by using an mRNA template. Reverse transcriptase is very useful in the synthesis of cDNA and construction of or as a cDNA clone bank. Until recently, it was known that the genetic information of DNA pass to protein through mRNA. During the 1960s, Temin and co-workers postulated that in certain cancer causing animal viruses that contain RNA as genetic material, transcription of cancerous genes (on RNA into DNA) takes places probably by DNA polymerase directed by virus RNA. Then DNA is a template for the synthesis of many copies of viral RNA in a cell. Baltimore discovered that information can also pass back from RNA to DNA. They found that retroviruses (possessing RNA) contain RNA-dependent DNA polymerase which is also called as reverse transcriptase. This produces single-stranded DNA, which in turn functions as a template for the complementary long chain of DNA.

DNA polymerase

This enzyme polymerizes the DNA synthesis on DNA template (or cDNA template) and also catalyzes a 5'- 3' and 3'-5' exonucleolytic degradation of DNA (**Kornberg, 1974**).

T4 polynucleotide kinase

The T4 polynucleotide kinase catalyzes the transfer of yPO, of ATP to 5' terminus of a dephosphorylated DNA (i.e. DNA lacking -PO, residue) or RNA as below: 5'GGOH 3 3'CCTTAAp 5'-GGOH Polynucleotide kinase -3' 3'-CCTTAA5', S-7P-ATP. This enzyme shows both phosphorylation as well as phosphatase activities. In phosphorylation , the y-PO is transferred to the 5'end of dephosphorylated DNA. When ADP is in excess, it causes the T4 polynucleotide kinase to transfer the terminal phosphate (attached to a nucleotide) from phosphorylated DNA to ADP.

Terminal transferase

This enzyme is also known as from calf thymus. It has a property to add 'oligodeoxynucleotide tail' to the 3'-OH end of double-stranded DNA fragments. Thus it extends homopolymer tails and this phenomenon is called homopolymer tailing deoxynucleotidyl transferase.

Use of linkers and adaptors

As discussed earlier the treatment of DNA with Type II restriction enzymes results in the formation of fragments having sticky ends or blunt ends. Such DNA fragments cannot be ligated with cloning vectors. Therefore, artificially cleavage sites at the blunt end can be added as linker molecules. Chemically synthesized double-stranded oligonucleotides are called linkers. They are added double-stranded DNA molecules that act as adaptor mol- are added recognition site for restriction enzymes. For example, the following oligonucleotide possess a recognition sequence for *EcoRI* (asterisk nucleotides). It can be ligated to the blunt end of any double-stranded DNA and cut by *EcoRI* to generate sticky ends to DNA **Figure 34.4**.

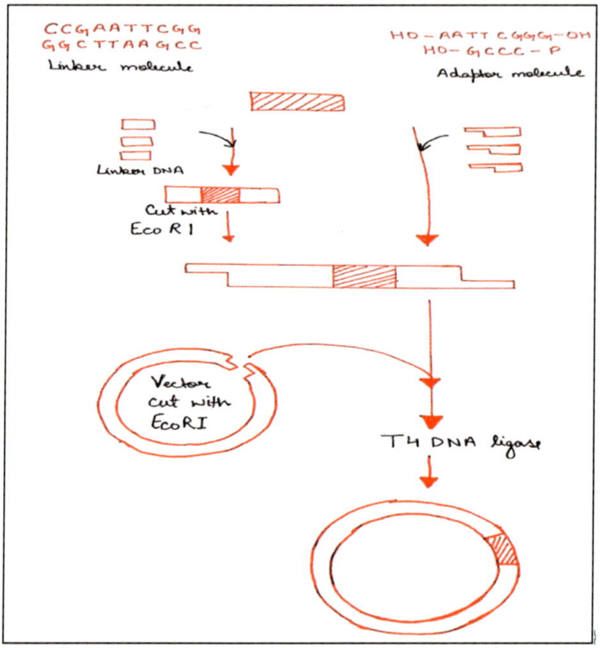

Figure 34.4 Linker and Adaptor molecules

Cloning and Expression Vectors

This has allowed breakage of a DNA molecule at two desired places to isolate a specific DNA segment and then insert it in another DNA molecule at a desired position. The product thus obtained is called **recombinant DNA** and the techniques often called **genetic engineering.** Using, this technique we can isolate and clone single copy of a gene or a DNA segment in to an indefinite number of copies, all identical. This became possible because vectors like plasmids and phages reproduce in a host in their usual manner ever after insertion of foreign DNA, so that the insertion DNA will also replicate faithfully with the parent DNA. This technique is called **gene cloning.** A variety of vectors have been developed which not only allow multiplication, but may also be manipulation in such a way that the inserted gene may express in the host.

Cloning vector for Recombinant DNA

One of the most important uses of recombinant DNA technology is in the cloning of (i) random DNA or cDNA segments, often used as a probe or (ii) specific genes, which may be either isolated from the genome or synthesized organochemically or in the form of cDNA from mRNA. This other DNA molecule is often used in the form of a vector, which could be a plasmid, a bacteriophage, a derived cosmid or phagemid, a transposon or even a virus. Another critical desired feature of any cloning vector is that it should possess a site at which foreign DNA can be inserted without disrupting any essential function. Therefore, in each case, an enzyme will also have to be selected which will cause a single break. Sometimes vectors are modified by inserting a DNA segment to create a unique site for one or more enzymes to facilitate its use in gene cloning. This inserted DNA with restriction sites for several enzymes is sometimes called a polylinker.

Plasmids as vectors

Plasmids are defined as autonomous elements, whose genomes exist in the cell as extrachromosomal units. They are self-replicating circular duplex DNA molecules, which are maintained in a characteristic number of copies in a bacterial cell, yeast cell or even in organelles found in eukaryotic cells. These plasmids can be single-copy plasmids that are maintained as one plasmid DNA per cell or multicopy plasmids, which are maintained as 10-20 genomes per cell. These are also plasmids, which are under relaxed replication control, thus permitting their accumulation in very large numbers. These are the plasmids that are used as cloning vectors, due to their increased yield potential. Circular plasmid DNA which is used as a vector can be cleaved at one site with the help of a restriction enzyme to give a linear DNA molecule. A foreign DNA segment can now be inserted, by joining the ends of broken circular DNA to the ends of foreign DNA, thus regenerating a bigger circular DNA molecule that can now be separated by electrophoresis on the basis of its size. The selection of chimeric DNA is also facilitated by the resistance genes, which the plasmid may carry against one or more antibiotics. If a plasmid has two such genes conferring resistance against two antibiotics and if the foreign DNA insertion site lies within one of these two genes, then the chimeric vector loses resistance against one antibiotic, the gene for which has foreign DNA inserted within its structure. In such a situation, the parent vector in bacterial cells can be selected by resistance against two antibiotics and the chimeric DNA can be selected by retention of resistance against only one of the two antibiotics.

pBR322 and pBR327 vectors

The naturally occurring plasmids may not possess all the above and other essential properties of a suitable cloning vector. Therefore, one may have to restructure them by inserting genes of relaxed replication and genes for antibiotic resistance. This has actually been done and suitable plasmid vectors have been obtained. One is pBR322, which is 4,362 bp DNA and was derived by several alterations in earlier cloning vectors. It has genes for resistance against two antibiotics an origin of replication and a variety of restriction sites for cloning of restriction fragments obtained through cleavage with a specific restriction enzyme. Another vector pBR327 was derived from pBR322, by deletion of nucleotides between 1427 to 2516. These nucleotides are deleted to reduce the size of the vector and to eliminate sequences that were known to interfere with the expression of the cloned DNA in eukaryotic cells. pBR327 still contains genes for resistance against two antibiotics. Both pBR322 and pBR327 are very common plasmid vectors Figure 34.5.

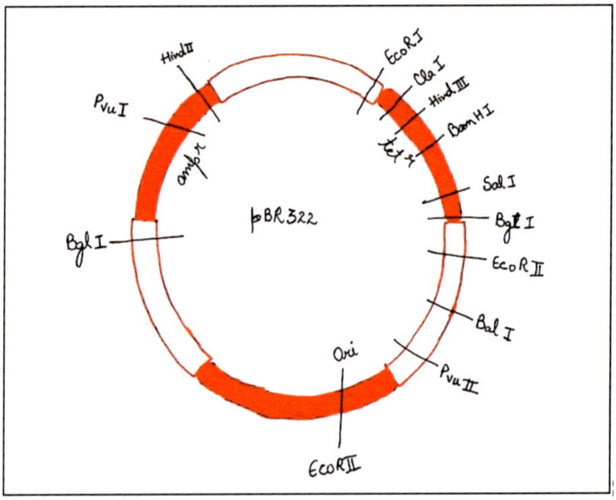

Figure 34.5 pBR322 vector

pUC vectors

Another series of plasmids that are used as cloning vectors belong to the pUC series. These plasmids are 2700 bp long and possess(i) ampicillin resistance gene, (ii) the origin of replication derived from pBR322 and (iii) the lac Z gene derived from *E.coli*. Within the lac region is also found a polylinker sequence having unique restriction sites. When DNA fragments are cloned in the region of pUC, the lac gene is inactivated. These plasmids when transformed into an appropriate *E.coli* strain, which is lac, and grown in the presence of IPTG and X-gal, will give rise to white or clear colonies. On the other hand, a pUC having no inserts and transformed into bacterial will have an active lac Z gene and therefore will produce blue colonies, thus permitting identification of colonies having pUC vector with cloned DNA segments. The cloning vectors belonging to the pUC vector with cloned DNA segments. The cloning vectors belonging to the pUC family are available in pairs with reversed orders of restriction sites relative to the lac Z promoter. pUC8 and pUC9 make one such pair. Other similar pairs include pUC12 and pUC13 or pUC18 and pUC19 Figure 34.6.

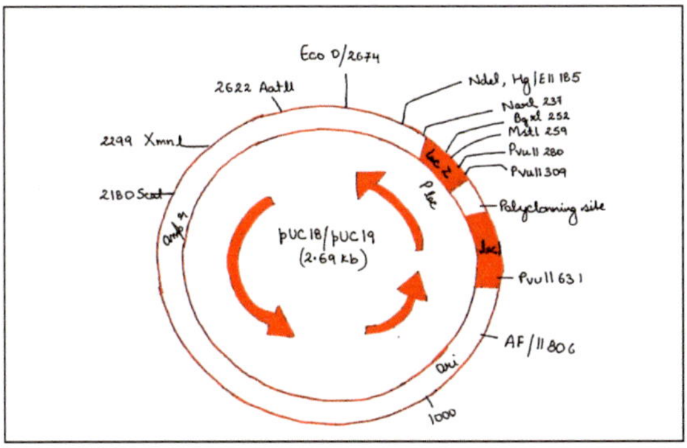

Figure 34.6 pUC 19 cloning vector

As discussed above, in pBR322and Pbr327, the DNA is inserted at a site located in one of the two genes for resistance against antibiotics, so that it will inactive one of the two resistance genes. The insert bearing plasmid can be selected by their ability to grow in a medium containing only one of the two antibiotics and by their failure to grow in a medium containing both the antibiotics. The plasmids carrying no insert, on the other hand, will be able to grow in media containing one or both the antibiotics. In this manner, the presence of the lac Z gene in pUC and resistance genes against ampicillin and tetracycline in pBR322 and pBR327 allows the selection of *E.coli* colonies transformed with plasmids carrying the desired foreign cloned DNA segment.

Yeast plasmid vectors

Special vectors are also known for introducing DNA segments in yeast cells, a eukaryotic system that has been used extensively for developing genetically engineered yeast cells. Although *E.coli* plasmids or phages can be are used for the transfer of genes to yeast cells, the frequency of transformation is rather low. For instance, yeast cells that are LEU could be transformed using chimeric plasmid carrying LEU + gene yeast. This involves integrative transformation having low frequency. To overcome this difficulty of low-frequency transformation, efforts were made to prepare a variety of vectors for yeast The possible examples of yeast plasmid vector are : (i) Yep or yeast integrative plasmids, which allow transformation by crossing over and have no replication origin; (ii) Yep or yeast episomal plasmids, which carry 2 micron DNA sequence including the origin of replication and rep genes; (iii) YRp or yeast replication plasmids, which carry autonomously replicating sequence; this sequence is very common with many yeast genes so that stable transformation can be achieved by crossing over; (iv) YCp or yeast centromere (CEN) containing plasmids, which function as true chromosomes and segregate during mitosis and meiosis; (v) pYAC or yeast artificial yeast chromosome vector which carries both centromere and telomere sequences and, therefore, can be used to obtain artificial chromosomes.

The above vectors have been extensively utilized for a study of yeast genome and are also used as shuttle vector, which allows genic sequences to be routinely transferred back and forth between yeast and *E.coli* cells, provided they contain origins of replication active in both yeast and *E.coli*. These vectors also have markers enabling the selection of *E. coli* cells or colonies transformed with these vectors. The vectors can therefore be amplified in *E.coli* and then used to transform yeast cells. Retriever vectors are other class vectors, which are used to retrieve specific genes from the normal chromosome of an organism like yeast through recombination. They will multiply both in *E.coli* and yeast. In such a vector the gene in question is removed by enzymes to produce a gapped vector having flanking sequences. Through recombination, they acquire lost sequences from the host. Retriever vectors are very useful in the isolation of genes from yeast for further molecular studies including sequencing.

Ti and Ri Plasmids as Vectors for Higher Plants

Among higher plants, the Ti plasmid of *Agrobacterium tumefaciens* or Ri plasmid of *A.rhizogenes* are the best-known vectors. **T-DNA,** from Ti or Ri plasmid of *Agrobacterium*, is considered to be a very potential vector for cloning experiments with higher plants. This will involve the following steps: (i) foreign DNA has to be first cloned into T-DNA of Ti or Ri plasmids; (ii) modified hybrid T-DNA can be transferred to the genome of plant cells by *Agrobacterium* infection. Recombinant Ti plasmid can also be obtained by cloning a large fragment of Ti plasmid in pBR322. These recombinant Ti plasmids can then be used for the transformation of higher plants. This system can be widely used since *Agrobacterium* infects nearly all dicotyledonous plants. Such cloning in plants has proved to be of immense use to modify agricultural plants to increase their productivity. Monocotyledonous plants to which our cereal crops belong are ,however, difficult materials to be used for improvement through such techniques; some success, however, has been achieved in cereals also in recent years; see **Figure 34.7.**

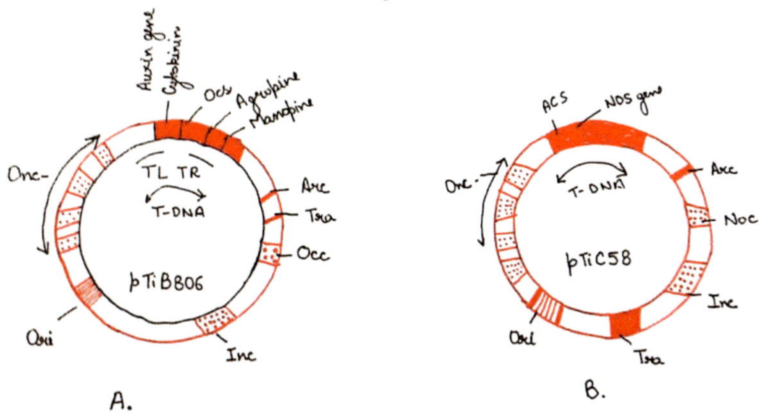

Figure 34.7 Octopine plasmid (a) Nopaline plasmid (b).

Bacteriophages as Vectors

Bacteriophages serve as another source of cloning vectors. Since usually a phage has a linear DNA molecule, a single break will generate two fragments, which are later joined together generating two fragments, which are later joined together with foreign DNA to generate a chimeric phage particle. The chimeric phage can be isolated after allowing it to infect bacteria and collecting progeny particles after a lytic cycle. The use of phage particles as vectors imposes a limitation on the size of foreign DNA is too long, size of phage DNA may not be accommodated in phage head. In order to overcome this problem, those segments of DNA, which do not contain essential genes, may be removed. Such a technique has been followed in phage **lambda** to create a smaller vector genome having a single restriction site for the enzyme *EcoRI.* Since the reduced size also fails to be adequately packed in the phage head, this automatically provides a selection method, in which only the chimeric particles will be obtained in the phage progeny and the vector particles lacking cloned segment will be eliminated due to its reduced size.

Lambda phage vectors

Plasmid vectors described in the previous section are often used for cloning DNA segments of small size. However, while preparing a genomic library in a eukaryote, the cloned fragments should be large enough to contain a whole gene. The above properties and other requirements of cloning the whole genome in eukaryotes are fulfilled by the phage lambda and cosmid vectors, the former permitting cloning of segments up to 20-25kb long and the latter accommodating segments up to 45kb long. Phage lambda ,however, is easier and more efficient for making genomic and cDNA libraries.

(a) lambda gt 10 and lambda 11

Lambda 10 and lambda gt 11 are modified lambda phages designed to clone cDNA fragments. The major difference between those two vectors is that gt11 is an expression vector, where the inserted DNA is expressed as a B-galactosidase fusion protein. Lambda gt 10 is a 43 kb double-stranded DNA for cloning fragments that are only 7 kb in length. The insertion of DNA causes the inactivation Cl^{+} gene generating a Cl^{-} bacteriophage. Non-recombinant lambda gt 10 forms cloudy plaques an appropriate *E.coli* host, while recombinant plaques. Further, in an *E.coli* strain carrying hf La150 mutation only cl- phage will form plaques, become cl+ will forms lysogens and will not undergo lysis to form any plaques. Recombinant Lamba gt 10 plaques can thus be easily selected. Lambda gt 11 is a 43.7kb double-stranded lambda phage for cloning DNA segments, which are less than 6 kb in length. Foreign DNA can be expressed as beta-galactosidase fusion proteins. Recombinant lambda gt 11can be screened using either nucleic acid or antibody probes. The recombinant lambda gt 11becomes gal-, while non –recombinant lambda gt 11 remains gal+ so that an appropriate *E.coli* host, and with recombinant phage will form white or clear colonies and that with recombinant phage (gal+) will form blue colonies permitting screening in the presence of IPTG (inducer) and X gal (substrate).

(b) EMBL3 AND EMBL4 (replacement vectors)

EMBL3 and EMBL4 are the two vectors that are designed so that a central non- essential part of the 44 kb-long phage can be replaced by a foreign

DNA. Cleavage of the phage with an appropriate enzyme generates three fragments. The central fragment representing 40% of the phage genome is nonessential for the propagation of the phage and can be replaced by foreign DNA that may be as long as 20-30kb. Non-recombinant DNA resulting from the fusion of the left arm and right arm will give a DNA that is too small for packaging, so that there is automatic selection against non- recombinant phages. EMBL3 and EMBL4 are two such replacement vectors used for preparing genomic libraries in eukaryotes with cloned fragments, 15-25 kb in size. The two vectors have polylinkers with reverse orders of restriction sites with respect to each other.

(c) Charon 34 and Charon 35

Charon 34 and Charon 35 differ from each other only in their central fragments and will accept fragments 9-20 kb long. They have a more extensive range of restriction targets within their polylinkers than do EMBL3 and EMBL4. In these vectors, physical separation of lambda arms from the central fragments is required before ligation to donor fragments. In EMBL3 and EMBL4, removal of central fragments is not required, because there is a mechanism that ensures that 98% of packageable DNA is recombinant.

M13 phage as cloning vector for DNA sequencing

M13 is a filamentous bacteriophage of *E.coli* and contains a 7.2 kb long single-stranded circular series of cloning vectors that can be used for cloning of a wide variety of DNA. M13 phage has been variously modified to give rise to a M13 mp series of cloning vectors which can be used for cloning a wide variety of DNA fragments particularly for the purpose of sequencing through Sanger's method of dideoxy chain termination. Cloning vectors of the M13 mp series have a lac Z gene that complements gal host giving blue colonies. But if a foreign DNA segment is inserted at one of the polylinker sites associated with the lac Z gene, it inactivates the lac Z gene and no complementation is possible. Thus on transformation, only white or clear plaques are obtained thus permitting selection of recombinant M13mp plaques. Once the foreign DNA is cloned in the M13 vector, commercially available oligonucleotide primers are used for copying the insert in presence of dideoxynucleotides, so that fragments of different sizes with known termini are produced, permitting the determination of the sequence. Reversed order of the restriction sites in the polylinker is present in a pair of vectors like M13mp8 and M13mp9 permitting sequencing from both the ends of double-stranded DNA molecule. The extension of the commercial primer or cloned fragments in the M13 vector in the presence of radioactively labeled nucleotides will also allow the generation of radioactivity single-stranded probes.

Cosmids as vectors

Cosmids are plasmids particles, into which certain specific DNA sequences, namely those for cos sites are inserted. Since these cos sites enable the DNA to gel packed in lambda particle, cosmids allow the packaging of DNA in phage particle in vitro, thus permitting their purification. Like plasmids, these cosmids perpetuate in bacteria and do not carry the genes for lytic development. The advantage of the use of cosmids for cloning is that its efficiency is high enough to produce a complete genomic library of 10/6 to 10/7 clones from a mere 1 micro g of insert DNA. The disadvantage, however, is its inability to accept more than 40-50 kbp of DNA. Bacteriophage P1 system and F- factor based vectors can allow cloning of DNA segments, as large as 100 to 1000 kbp length.

Phagemids as vectors

Phagemids are prepared artificially by combining features of phage with plasmids, as the name suggests. One such phagemid, which is commonly used in molecular biology laboratories, is pBluescript 11 KS, which is derived from pUC19 and 2961 bp long. The KS designation indicates the orientation of polylinker, such that the transcription of lac Z gene proceeds from the restriction site for KpnI to that for SacI. The detailed structure of pBluescript II KS. It may be noted that it has the following features: (i) a multiple cloning site (MCS) flanked by T3 and T7 promoters to be read in opposite directions on the two strands, (ii) an inducible lac promoter (Lac I), upstream of lac Z region, which complements with *E. coli* and provides the facility for selection of chimeric vector DNA using the criterion of white colonies (iii) fl (+) and fl(-) origins of replication derived from a filamentous phaGE for recovery of sense (+) and antisense (-) strands of the lac z gene, when the host is coinfected with a helper phage; (iv) an origin of replication derived from plasmid, and used in the absence of helper phage; (iv) a gene for ampicillin resistance for antibiotic selection of the chimeric phagemid vector.

P1 cloning vectors for cloning large DNA segments

The bacteriophage P1 cloning vectors can allow cloning of 100 kbp long DNA segments with an efficiency of 10 raised to power 5 clones of insert DNA. Therefore, in their capability, they fall between YACs and cosmids. The vector with insert DNA is amplified in *E.coli* and several micrograms of cloned DNA can be recovered from 5-10 ml of the exponential phase of *E.coli* cells.

F- factor based vector (which behave like a phage)

F-factor based vectors have recently been developed for cloning large DNA segments in *E.coli.* The cloning of large DNA segments is achieved by a method called chromosomal building, in which through repeated recombination size of cloned segments can be increased. These bacterial vectors will complement the YAC vectors for cloning segments larger than 100 kbp in length and offer some advantages over the YAC system. These vectors have already been used for cloning large DNA segments from the bithorax gene of *Drosophila*

35. METHODS OF GENE TRANSFER

Gene transfer in dicots by using Agrobacterium TI-DNA as a Vector

It is not possible to transfer Ti-plasmid into plant cells because of its large size, most likely up to 235 Kb. Therefore, T-DNA is excised and used as foreign DNA for expression in plant cells. From a bacteriological point of view opine production and development of gall is not required for the successful integration and expression of DNA. Hence, replacement of **nos** and **ops** genes with a foreign DNA fragment retaining the opine synthase promoter for expression of DNA DNA fragment has no harm. Schell and Van Montague (1983) undertook some experiments where genes for resistance to kanamycin and methotrexate expressed in cultured callus cells. Generally speaking, promoters for the expression of foreign DNA. It has become possible that specific tissues of plants when they are with specific regulated promoters.

Gene transfer in monocots

The limitation of Ti-plasmid is that it is specific only to dicot plants. *A. tumifaciens* does not induce tumor in monocot plants. However, the major crops including cereals are monocots. **Hooykaas Van et al.** (1984) discovered that *A. tumifaciens* could transfer T-DNA into certain monocots with the consequences of expression of opine gene within the plant cell without inducing the tumors. This discovery made it possible that T-DNA can be expressed into cells of monocot plants.

Plant Cell Transformation

Plant cell suspension culture coupled with DNA transfer techniques has become a new field of gene manipulation in plants. Cell suspension cultures have been the traditional source of plant material for biochemical selection of plant cell mutants, such as (d) numerous cell lines resistant to various amino analogs (Maliga, 1980). Suspension of protoplasts constitutes an identical plant material from one laboratory to another. The approaches made for plant protoplast transformation are (i) Introduction of DNA sequence via integration into T-DNA of Ti-plasmid of *A. tumifaciens*. (ii) Transfer of naked DNA into protoplast including bacterial spheroplasts and microinjection and (iii) Transformation with plant viruses carrying inserts e.g. *Gemini and Caulimoviruses.*

Plant cell transformation by ultrasonication:

Ultrasonication is done for various biological experiments. But in some plants, this technique has been done for successful gene delivery The Biotechnology Research Centre, Beijing (China) has used this technique for gene transfer in plants like wheat, tobacco and sugarbeet. When the cultured explants were sonicated with plasmid DNA carrying marker genes such as cat, nptll, and gus, and sonicated calli transferred of a selective medium, shoots were grown successfully. The calli in control experiments that were not sonicated with the plasmid did not grow on selective medium but they died. Transgenic tobacco plants were obtained at a frequency of 22 percent.

Liposome-mediated gene transfer

Liposomes are microscopically small-sized particles. They contain phospholipid bilayers which are concentric in nature. Liposomes enclose an aqueous chamber and can entrap water soluble molecules. Therefore, they are called lipid bags. Many plasmids are enclosed in them. By using polyethylene glycol (PEG) they may be stimulated to fuse with protoplast. In several plants like carrot, tobacco, petunia etc. this technique has been used for the successful transfer of genes. Due to endocytosis of liposomes, the DNA enters the protoplast. It gets adhered first to the protoplast surface and gets fused with the protoplast t the site of the surface. Consequently, plasmids are released inside the cell. There are several advantages in using this technique such as (E) low toxicity, (i) protection of DNA/RNA from nucleases that lyse them, (iii) long stable storage of DNA fragments in the liposome, (iv) applicability in various cell types, and (v) high level of reproductivity.

Animal Cells

Culture of animal cells on a large scale is more difficult and expensive than plant cells, bacteria, yeast, etc. This is why the commercialization of natural products of animal origin is limited. However, most of the animal products have been produced by using bacteria as a biological tool. In India, Dr. K. Chandrashekharan, Center for Cellular and Molecular Biology (CCMB) Hyderabad emphasized the use of recombinant DNA technology for the introduction of foreign DNA into mammalian cells. This would help in curing certain genetic defects. Some of the examples of gene cloning in animals (a) Animal A variety of animal viruses are known, for example, simian virus 40 (SV40), adenovirus, retrovirus (ssRNA), vaccinia virus, etc. They can increase the efficiency of animal cell transformation if they are used as a vector. A special feature of viruses is that they contain the strong promoters which can of the viruses. The most commonly used virus is SV40 which contains a circular DNA of about 5.2 Kb. In addition to the region, the DNA contains early genes and late genes. Early genes are required for DNA replication, whereas late genes encode coat proteins of the virus. DNA fragment and phage, the amount of DNA to be packed into virus capsid is limited.

The genome has 3 identified non-essential regions; hence the use of non-defective viral vectors is limited. The defective virus vectors have missing genes whose infection process has been helped by viral DNA constituting functional copies of missing genes. If the cos cells (the host monkey cells) are used as a host, they defective replication origin growth of SV40. Recombinants containing foreign DNA inserted into the early region can be propagated without the need for helper virus (Glover, 1984). In this way, the recombinant and helper can be separately recovered from the host.

Electroporation

Electroporation (using an electric field) is now used to transfer the foreign DNA into the fragile cells. Brief pulses of high voltage electricity (about 350 V) are applied to protoplasts suspension containing naked or recombinant plasmids. The electric pulses induce the formation of large pores in the cell membrane. These pores give a passage through which the foreign DNA can enter into the protoplasts and thus, increase the transformation frequency. The transformed protoplasts are cultured for about one month. These develop microcaaalli which are plated on solid medium containing selective marker (e.g. kanamycin). After 37-45 days, the calli are analyzed for the presence of differences in transformed cells by using the electroporation method, genes of choice have been successfully transferred in the protoplasts of wheat, rice, petunia, sorghum, maize and tobacco. Moreover, the transformation frequency can be improved by using linear DNA instead of a circular one, by giving heat sock to protoplast at 45°C for 5 minutes, by adding PEG and using 1.25 kV/em voltages.

Particle Bombardment Gun

It was developed by Prof. Stanford and coworkers of the Cornell University (USA) in 1987. As the term denotes, it shoots foreign DNA into plant cells or tissue at a very high speed. This technique is also known as particle bombardment, particle gun method, biolistic process, microprojectile bombardment or particle acceleration. This technique is most suitable for those plants that hardly regenerate and do not show a sufficient response to gene transfer through *Agrobacterium* for example, rice, wheal corn, sorghum Chickpea and pigeon-pea. The apparatus consists of a chamber connected to an outlet to create the vacuum. At the top, a cylinder is temporarily sealed off from the rest of the chamber with a plastic rupture disk. Helium gas flows into the cylinder. A plastic microcarrier is placed close to the rupture disk.

Pollen transformation through particle bombardment

Genetic improvement in alfalfa has been made by plant tissue culture methods. This method is time-consuming, requires special techniques and efficiency of getting stable regenerate is low. Ramaiah and Skinner (1997) produced transgenic alfalfa by direct delivery of DNA into pollen grains by particle bombardment method. Plasmid PBI121 bearing the GUS reporter gene was transferred into pollen grains. The microprojectile bombarded liquid pollen suspension and tripped the recipient female flower. Pollinated flowers set seeds in about a month. Thirty percent of plants derived from fertile seeds showed the integration of GUS plasmid. It was confirmed by Southern analysis. After 10 vegetative generations, some were allowed to fertilize the tagged transgenic plants lost the integrated GUS plasmid, whereas, in few others, copies of the GUS gene decreased due to unknown reasons.

Microinjection

- Microinjection is a technique of delivering foreign DNA into a living cell (a cell, egg, oocyte, embryos of animals) by a glass micropipette.
- One end has a plasmid carrying the early region of SV 40 with cos cells express wild-type large T antigen and can support the micropipette is heated until the glass becomes some forms to very fine tip at the heated end.
- The tip of the pipette attains to about 0.5 mm diameter which resembles an injection needle.
- The process of delivering foreign DNA is done under a powerful microscope. Cells to be microinjected are placed in a container.
- A holding pipette is placed in the field of view of the microscope.
- The holding pipette holds a target cell at the tip when gently sucked.
- The tip of the micropipette is injected through the membrane of the cell.
- Contents of the needle are delivered into the cytoplasm and the empty needle is taken out.

Direct Transformation

Like plant cells, mammalian cells can be transformed by foreign DNA fragments. For transforming mammalian cells, it is necessary to precipitate the DNA with calcium phosphate and mix the cells to be transformed. The DNA molecule passes through the cell membrane and integrates randomly with the mammalian chromosome. Using this technique, a selective marker can be linked up with the DNA fragment to be cloned and can be introduced into mammalian cells. The transformed cells are, thereafter, separated from cell line after plating them on a selective medium. Many techniques transformed cell lines. Following these biochemical processes. **Perucho *et al.*** (1980) successfully isolated the chicken thymidine gene by ligating with pBR322 and transforming the thymidine -deficient mouse.

Gene DNA sequencing

The structure of DNA (e.g. gene insert, a recombinant plasmid or entire genome) can be analyzed by determining the nucleotide sequences. In molecular cloning, the information of nucleotide sequences is essential. In 1965, Robert Holley and his research group at the Cornell University completely sequenced nucleotides of tRNA (tRNA for yeast alanine). In 1977, the following two methods were ONA developed. Allan Maxam and Walter Gi**lbert** developed the chemical method of sequencing. In this method, end- labeled DNA is subjected to base specific cleavage reaction before gel separation. In the routine sequencing of DNA, this method is not commonly followed. In the same year (1977), Frederick Sanger and co-workers developed. It is also called the dideoxynucleotide chain termination method because dideoxynucleotides are used in this enzymatic method of DNA sequencing.

Maxam and Gilbert's Chemical Degradation Method

This method is not very popular because it is time-consuming and labor-intensive. In this method, the DNA molecule can be radiolabeled at either the 5 'end by using polynucleotide kinase, or the 3' end by terminal transferase. One end of radiolabeled double-stranded DNA is removed by using endonuclease. A base is modified chemically followed by cleavage of the sugar-phosphate backbone of DNA. No specific reaction for the four bases is carried out, except specific reactions to G only and purine-specific reactions that remove A or G. A difference in these reactions indicates the presence of A.

Cleavage of purine

The mixture is separated into four sets, each treated with a different reagent which degrades only G or C or A and G or C and T. In one set, DNA is treated with acid followed by dimethyl sulfate. This causes methylation of A, (at 3 'position and G (at 7' position). Subsequently, the addition of alkali (-OH) and pyrimidine results in DNA cleavage and removal of purines (A or G).

Cleavage of pyrimidine

Similar to the cleavage of purine, pyrimidine (C or T) is cleaved in the presence of 1-2 M NaCl solution. It works only with C. Differences between these two indicate the presence of T in the DNA sequence. Cleavage of pyrimidines (C or T) through hydrazine hydrolysis. Partial chemical cleavage of DNA fragments as done above generates the populations of radiolabeled molecules extending from radiolabeled terminus to the site of chemical cleavage. These fragments of different sizes that represent unique pairs of 5 and 3' cleavage products in the random collection. A complete set is formed by these products: the length of each number is short by a nucleotide. These can be separated by gel electrophoresis. The fragments containing the labeled terminus can be observed by autoradiography of the gel. Following the order of fragments obtained from different digestion,s the sequence of nucleotides is deduced and interpreted.

Sanger method (Dideoxynucleotide chain termination method)

Twice Nobel Prize winner Frederick Sanger's powerful method for developed DNA sequencing utilizes single-stranded DNA as a template. This method is also called the dideoxynucleotide chain termination method. The requirements are: a free 3' -OH ends for a primer to start DNA synthesis, DNA polymerase and dNTPS. Shows the presence of free 3'-OH group at 3' end in dATP and no 2',3' -OH in 2',3'ddATP In 2',3', ddATP, the a hydrogen atom is attached at 2' and 3 carbons instead of an -OH hydroxyl group. If any of the four ddNTPs binds, the chain elongation is terminated. Because ddNTPs do not have a free 3'-OH end that is required for chain elongation no phosphodiester bond will be formed.

Automatic DNA sequencers

Automatic DNA sequencing machines were Sanger's method. In this new method, a different fluorescent dye is tagged to the ddNTPS. Using this technique a DNA sequence containing thousands of nucleotides can be determined in a few hours. Each dideoxynucleotide is linked with a fluorescent dye that imparts different colors to all the fragments terminating in that nucleotide. All four labeled ddNTPs are added to a single capillary tube. It is a refinement of gel electrophoresis that separates quickly. DNA fragments of different colors are separated by their respective size in a single electrophoretic gel. A current is applied to the gel. The negatively charged DNA strands migrate through the pores of gel towards the positive end.

DNA fragments migrate faster and vice versa. All fragments of a given length migrate in a single peak. The DNA fragments are illuminated with a laser beam. Then the fluorescent dyes are excited and emit light of specific wavelengths which is recorded by sequences that are read by determining the sequence of the colors emitted from specific peaks as they pass the detector. This information is fed directly to a computer that determines the sequence. A tracing electrogram of emitted light of the four dyes is generated by the computer. The color of each dye represents the different nucleotides.

Site-directed mutagenesis

Now, techniques have been developed for a specific portion in the genome in a view to getting some novel products of enormous value. Mutations can be done by directing insertions or deletions to a site on the DNA but these minor changes may have some single amino acid is altered to improve its properties. For such modifications, point-mutations i.e. changing of a single nucleotide is done on a specific portion in the gene. Thus, the most suitable method that could bring about point-mutation in a gene to mutate specific are unlikely to be of use.

Importance

For a protein: oligonucleotide-directed mutagenesis. This is known as a technique that has potential for protein engineering, the engineered enzymes would be better than the wild type ones. Change in a single nucleotide base pair is called 'point mutation'. Some useful properties in proteins (like stability in subtilisin) that results in the substitution of selected amino acids. Using this technique, specific points of a gene can be mutated. Therefore, this method has been used to understand the function of many genes. Moreover, this technique can only be used when the nucleotide sequence of the gene is known.

Methods of mutagenesis

When it has been decided which base is to be subjected to site-directed mutagenesis using short synthetic DNA. The oligonucleotide sequence has a desired mutation at a specific site. The oligonucleotide corresponds to the mutated nucleotide and its neighboring regions, mainly between 15 and 20 nucleotides in length. A single-stranded clone of the wild-type gene is produced by using an M13 phage-based vector. The oligonucleotide is allowed to hybridize with the single-stranded clone. The hybridization is performed under the conditions. The oligonucleotide will base-pair with wild type complementary sequence (B) and acts as a primer to produce double-stranded DNA by using DNA polymerase (C).

36. INTRODUCTION TO DNA LIBRARIES PCR, DNA SEQUENCING, BLOTTING TECHNIQUES.

It is a collection of genes that can be made and stored in gene libraries. There are two types

- Genomic libraries
- cDNA Libraries

***Genomic Libraries*:** It is made from DNA and contains entire genes (exons and introns).

***cDNA Libraries*:** It is made from mRNAs that are converted into DNA (only exons).

A genomic library is a population of host bacteria, each of which carries a DNA molecule that was inserted into a cloning vector, such that the collection of cloned DNA molecules represents the entire genome of the source organism. Genomic libraries are made of DNA and contain entire genes (exons and introns).

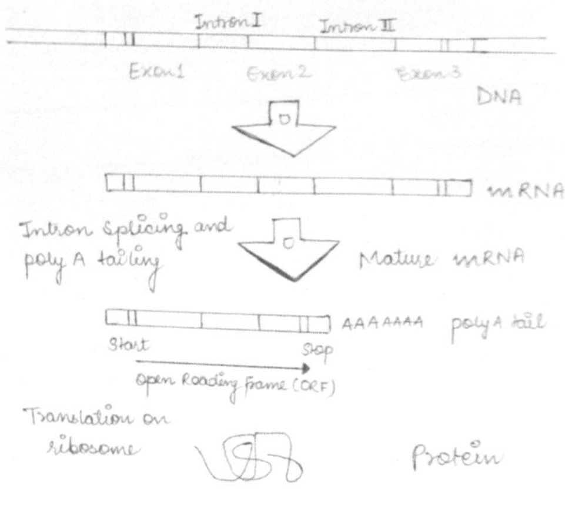

DNA Libraries

Much of the genomic DNA is not expressed as mRNA. Many issues about gene function are best addressed by examining the product that they encode. The cDNA copies of mRNA contain primarily sequences that encode a protein. Therefore, cDNA clones are useful for many studies of gene function. Use the enzyme *reverse transcriptase* to copy mRNA into complementary DNA, called cDNA. This is equivalent to the template strand of the duplex DNA. Use a DNA polymerase to copy that cDNA into the non-template (message synonymous) strand. Insert the duplex cDNA product into a cloning vector and propagate in a host, e.g. *E. coli.* First-strand synthesis often does not go to completion. Individual cDNA clones will frequently have the reverse complement of only part of the mRNA. Multiple cDNA clones from a single mRNA will be present in the library. Priming second strand synthesis is inefficient. Some methods necessarily result in the loss of sequences at the 5' end of the nontemplate strand. A cDNA library has >100,000 individual clones. It contains copies of as many as 50,000 different mRNAs. The frequency of occurrence of a cDNA from a given gene reflects the abundance of the mRNA for that gene and hybridization with a gene-specific probe. Express the cDNA in the host cell (i.e. make a functional protein product).

- Specific antisera.
- Labeled ligand to a receptor.
- Assay for a function (complementation).
- Differential analysis

Screening by Hybridization

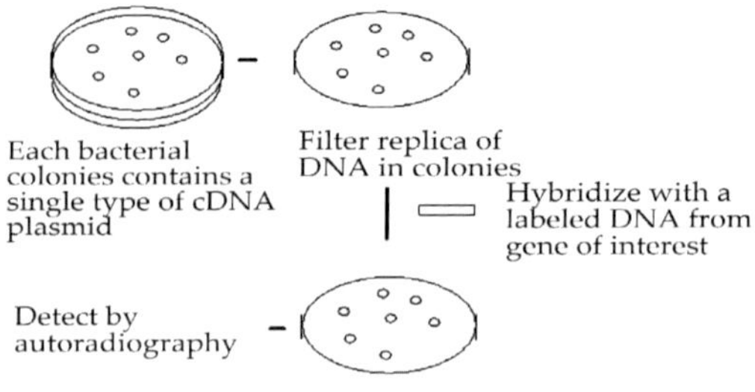

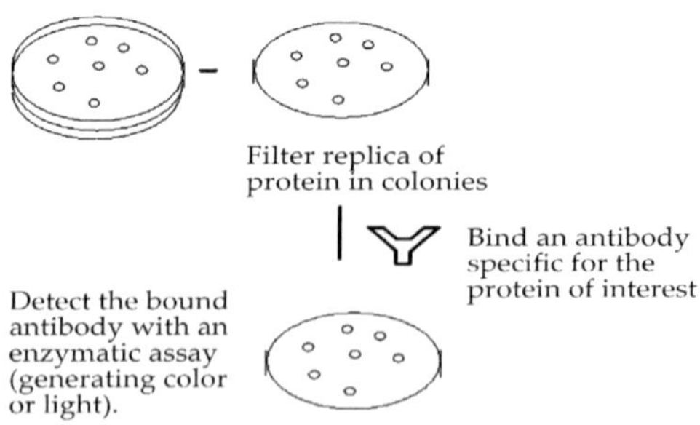

Screening for an expressed product

Genomic DNA clones

Clones of genomic DNA contain fragments of chromosomal DNA. They are used to: It obtains detailed structures of genes. It is identify regulatory regions. Map and analyze alterations to the genome, e.g. isolate genes that when mutated cause a hereditary disease. The direct alterations are occured in the genome. Sequence the genome. Instead of screening for one gene at a time, an entire genome can be sequenced, and one can use experimental and bioinformatic approaches to find many (all) genes of interest.

Made possible by

- Substantial increases in the speed of sequencing.
- Larger insert libraries for larger genomes.
- Combination of hierarchical sequencing (based on maps) and whole-genome shotgun sequencing.

Polymerase chain reaction (PCR)

Gene Amplification: Basic PCR and its modifications

We discussed the techniques of recombinant DNA and gene cloning, which permit us to obtain an unlimited supply of identical copies of a gene sequence or DNA segment, which is cloned in a prokaryotic a eukaryotic cell with the help of a vector. This technique, discovered around 1975, proved extremely useful in experiments of molecular biology and genetic engineering and thus became an essential tool in all molecular biology laboratories. In 1985, yet another remarkable tool in molecular biology was discovered, which is known as polymerase chain reaction (PCR) nicknamed now as people's choice reaction. The methodology is so important that the prestigious journal Science published from the USA considered PR as the major scientific development of the year 1989, and had chosen Taq DNA polymerase, the enzyme used in PCR, as the mol ee of the year 1989. Kary Mullis who discovered this Scientific American, how he got the basic idea this reaction while driving through the redwood mountains of California, USA. The polymerase chain reaction is such a powerful technique that may replace completely the gene cloning with vectors in due course of time.

The basic Polymerase Chain Reaction (PCR)

- The DNA replication involves the polymerization of nucleotides using a template DNA strand with the help of an enzyme DNA polymerase, but this reaction invariably requires a primer strand to which further nucleotides can be added using the DNA polymerase enzyme. If the primer strand is not available, the reaction cannot proceed.
- In living cells, this primer strand is not a DNA strand but is a small single-stranded RNA molecule synthesized with the help of an RNA polymerase enzyme.
- In PCR, a similar reaction takes place in an 'Eppendrof tube', where the primer strand is added from outside in the form of a deoxyoligonucleotide, and DNA polymerase enzyme is added to help in polymerization.
- Unlimited supply of amplified NA is obtained by repeating the reaction, which is made possible by regular denaturation of freshly synthesized double-stranded DNA molecules by heating it to 90-98°C.
- At this high temperature the two strands separate. Once the double-stranded DNA is made single-stranded by heating upto 90-98°C, the mixture with two primers, recognizing the two strands and bordering the sequence to be amplified is cooled to 40-60 C.
- This allows the primers (which are in excess) to bind to their complementary strands through renaturation.
- The presence of the **Taq** DNA polymerase enzyme and all the four essential nucleoside triphosphates allows the synthesis of complementary strands in the usual manner.
- In a thermal cycler, this process is automatically repeated 20 and 30 times, so that in a single afternoon, a billion copies of the sequence flanked by the left and right primers can be produced.
- In order to continue the synthesis, the temperature of the mixture is alternately increased (for denaturation) and decreased (for renaturation) once every 1-3 minutes (as fixed by the computer device).

- Therefore during temperature rise, the enzymatic activity of DNA polymerase should not be destroyed, otherwise one may have to add a fresh aliquot of the enzyme in each cycle of amplification.
- This became possible only by the discovery of the thermostable enzyme Taq DNA polymerase, isolated from *Thermus aquaticus* growing in hot springs.
- This enzyme acts best at 72°C and the needed denaturation temperature of 90°C does not destroy its enzymatic activity.
- Later, other thermostable enzymes like *Pflu* DNA polymerase isolated from *Pyrococcus fiuriosus* and Vent polymerase isolated from *Thermococcus litoralis,* were discovered and were found more efficient.
- These enzymes allowed automation of the entire process and automatic PCR thermal cyclers are now available, which can amplify DNA sequences at a fast speed unattended Figure 36.1.
- As pointed out earlier, PCR may eventually replace the gene cloning technique. This technology is much easier and requires a much smaller quantity nanogram or ng 10g of DNA. For cloning, DNA is needed, as the starting material, in microgram (ug) quantities.

Different schemes of PCR

In the previous section, we discussed the basic outline of PCR technique, outlining how billions of copies of a DNA sequence can be obtained within a few hours, without the use of a vector and a host as done in gene cloning. However, there are several variations to this basic PCR technique, which allow varied applications of this technique.

Inverse PCR

In this technique, the amplification of those DNA sequences takes place, which is away from the primers and not those flanked by the primers. For instance, if the border sequences of a DNA segment are not known and those of a vector arc known, there the sequence to be amplified may be cloned in the vector and sequences of the vector may be used as polymerization sequence flanked by the primers and towards the DNA inserted segment.

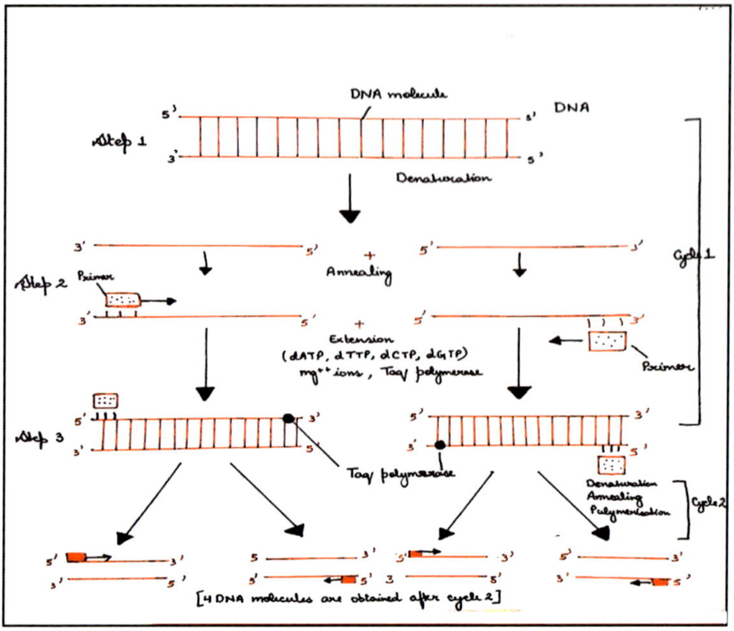

Figure 36.1 Polymerase chain reactions

Anchored PCR

In the basic PCR technique and the inverse PCR, one has to use two primers representing the sequences lying at the ends of sequences to be amplified. But sometimes, we may have knowledge about the sequence at only one of the two ends of the DNA sequence to be amplified. In such cases, anchored PCR may be used, which will utilize only one primer instead of two primers. In this technique, due to the use of one primer, only one strand will be copied first, after which a poly G tail will be attached at the end of the newly synthesized strand. This newly synthesized strand with a poly-G tail at its 3' end will then become a template for the daughter strand synthesis utilizing an anchor primer with which a poly C sequence is linked to complement with poly G of the template. In the next cycle, both the original primer and anchored primer will be used for gene amplification.

PCR for site-directed mutagenesis

This technique is used for introducing mutations at the desired place in a DNA sequence by sequences of primers. A variation of this technique allows mutation to be introduced mutations are limited to the ends of the gene sequence to be introduced at any place of interest in the gene the method is described as overlap extension which works as follows: In two separate PCR reactions, a particular gene is amplified into two separate segments. In each of these two reactions, there is one primer at the end of the gene and the other (with the desired alteration for mutation) internal to the sequence. The internal modified primers in two reactions are complementary to one another so that the amplified products will have their ends internal to the original sequence actions will overlap. The internal ends of the two PCR products, when mixed, function as primers. Extension of these primers by Taq polymerase results in the formation of a complete gene, with mutations incorporated at internal sites.

Cloning and expression of the PCR product

The amplified PCR product can be cloned if restriction sites for specific enzymes are introduced at the 5' end of each primer. These restriction sites do not interfere with amplification and can be used for cloning with a variety of vectors in the usual manner. Similarly, incorporation of a T7 promoter at the 5' end of one primer will allow copies of the PCR product

Asymmetric PCR for DNA sequencing

PCR products can also be used as templates for direct DNA sequencing (ligation-mediated PCR - LMPCR). Single-stranded DNA can be produced for this purpose using asymmetric PCR, in which the two primers are used in a 100:1 ratio, so that after 20-25 cycles of amplification, one primer is exhausted so that single-stranded DNA is produced in the next 5-10 cycles.

Polymerase chain reaction application

During the last few years, with the improvement of PCR protocols and due to the availability of automatic thermal cyclers commercially the applications of PCR have increased manifold. However, for the application of PCR, we need a pair of primers that should be based on the knowledge of the nucleotide sequence of DNA to be amplified. Therefore, the non-availability of this information about the DNA segment or gene to be amplified becomes a limitation in the applications of PCR, although this difficulty has been overcome for the study of DNA polymorphism, through the primers. Some of the applications of the PCR technology of random DNA primers are described in this section.

Study of DNA polymorphism using PCR

Using PCR, DNA polymorphism can be studied using the sequence at loci with (using primers based on known sequence) or at random site primers having unknown but random DNA sequences (these are called random primers). In either case, DNA may be amplified from one or more genomic DNAs and subjected to digestion with one or mo to electrophoresis and the restriction patterns may be photographed to reveal polymorphism. Sequences of the prolamin gene and phytochrome gene have been used for the study of polymorphism for these genes in different species of Oryza (wild and cultivated rice). Similarly, storage protein genes in wheat and many genes in other animal and plant species have used specific restriction endonucleases. The digested DNA may be subjected used, for the study of DNA polymorphism. The non-availability of sequences of specific genes for designing primers for DNA amplification has been overcome by a variety of approaches including the following (i) In the case of multigene families like those for RNA, tRNA and heat shock proteins, there are conserved consensus sequences known, which are used for designing universal primers. Similarly for the amplification of antibody genes also, universal primers have been used in PCR (ii) Gene bank Sequence Data have also been used for designing primers for amplification of microsatellites like poly (A), poly (G), pol (TC) and poly (TG), etc.

PCR and RAPD markers

The most important use of PCR for e study of DNA polymorphism, however, involves the use of random prime giving Random Amplified Polymorphic DNA. This technique is considered simpler and offers several advantages over more commonly used techniques known as RFLP and RAPDs have actually been used for a study of polymorphism chromosome mapping in maize, soybean, mouse, humans,.

PCR and VNTR loci

Alec Jeffreys and his colleagues in the U.K suggested the use of PCR for the study of DNA polymorphism at VNTR (variable number tandem repeats, also known as **'minisatellites** loci in humans. Primers were developed for the conserved flanking regions of VNTR loci so that all available VNTR loci could be amplified. The PCR products differ according to the number of repeat units present in the different VNTR loci and this variation is visualized in the positions occupied by different bands after electrophoresis.

PCR and SSR loci

The VNTR loci discussed above consist of repeat electrophoresis units in the range of 11-60 base pairs in length. A much simpler type repeat units exist in the form of dinucleotides, trinucleotides or tetra nucleotides, e.g. (CA)n, (GT)n (AAT)n or (AGAT)n. These repeat units were described as microsatellites, but later also called short tandem repeats or simple sequence repeats (SSR). Some of these SSR loci, such as (CA)n or (GT) n occur in the human with n varying from 10 to 60. The DNA sequences flanking SSRs are conserved, allowing the selection of PCR primers that will amplify the intervening SSR in all genotypes of the target species PCR reaction includes a small amount of 3P labeled nucleotide (only one of the four nucleotides is labeled) or it may include primers, one or both of them being end-labeled. This allows visualization of PCR products using autoradiography after electrophoresis on a standard sequencing gel. Variation in the length of products is a function of the number of SSR units. The principle underlying is the detection of the polymorphism using two parents and their F1 hybrid. Markers resulting from SSR length polymorphisms are placed on genetic maps such as RFLPs, RAPDs, VNTRs, other SSRs and even morphological markers.

A comparison of very limited sequence data from plants (including algae) those of vertebrates (including humans) indicated that SSRs plants are found with the same frequency as in vertebrates. Recently DNA sequence data for five tree species and Zea mays have been screened for the presence of (AC)n and (AG)n and the frequency in these six species was found to be close to one microsatellite (SSR). Other plant and animal species have also been examined for the presence of a number of SSRs and a wide occurrence of these markers in plants and animals preparing saturated was suggesting their utility in preparing saturated molecular maps.

Molecular mapping using PCR Genetic and physical chromosome maps have also been prepared, both in plants and animals, using PCR. Once RAPD's are detected, recombinant inbreeds (RI's) or F2/backcross segregating populations, or doubled haploids (DH) derived from haploids (in plants) can be used to detect linkage and recombination frequencies. The molecular marker which cosegregates RI's or F,'s or DH's suggests linkage and the frequency of cosegregation among RIs or Fa/backcross plants will be a measure of the strength of linkage. The molecular markers identified through PCR can also be assigned to specific chromosomes using aneuploids or alien addition lines e.g. wheat-barley additions -21 pairs of wheat chromosomes +1 pair of a specific barley chromosome). A number of molecular markers have been mapped using PCR in maize, soybean, barley, mouse, humans, etc.

Sequence tagged microsatellite sites (STMS) or SSR loci

It has also been examined and mapped using PCR. The gene sequence databases are screened for a microsatellite and the flanking sequences are used for designing primers for PCR. Several STMS or SSR loci using this technique have been mapped in mice, where as many as 200 microsatellites have been detected in the genome database. Besides SSR's or STMS's, other markers including sequence tagged sites or STS's (a short DNA sequence in a molecular genomic DNA marker) and expression sequence tags or ESTs (a short DNA sequence from a cDNA clone representing an expressed gene) have also been for genetic mapping. In both these cases, PCR can be profitably utilized for screening RI's (mouse), somatic hybrid cell lines (humans) or F2 individuals (in plants). This will allow the detection of linkage and the calculation of recombination frequencies leading to the preparation of genetic maps. Complete molecular maps of several human chromosomes could be prepared using STS's and PCR.

Gene tagging using PCR

PCR has also been utilized for developing molecular m to specific genes of economic importance. For instance in tomato 144 random primers were developed to produce 625 PCR products from a set at markers closely link instance in tomato, near isogenic lines (NIL's, which differ only for the presence and absent of a specific gene) for *Pseudomonas* resistance. At least three PCR products were present in one and not in the other of a pair NIR's showing linkage with the *Pseudomonas* resistance gene.

PCR for confirming the presence of transferred gene

When a gene is transferred in a vector to cultured cells or primers can be designed to conduct PCR for amplification of the sequence so transferred. This technique allows the confirmation of transfer and maintenance of the gene of interest. In gene therapy experiments, the transfer and presence of a marker gene for neomycin resistance (NeoR) could be detected in the blood of patients, even after 60 days, although the frequency of marked lymphocytes gradually decreased necessitating a second transfusion of such marked cells. Similarly, ADA (adenosine deaminase deficiency) gene was transferred for therapeutic purposes and was used to detect the presence of this gene up to 6-5 months, s that the patient (4 year old girl) was then placed on a program of maintenance infusions at 3-5 months intervals. Thus PCR proves to be of immense help in monitoring a gene in genetic engineering or gene therapy experiments.

Human genetics using PCR

PCR has also found extensive use in human genetics. Some of this use includes the following.

Prenatal diagnosis using PCR

Prenatal diagnosis of sickle cell anemia with enhanced sensitivity was perhaps the first application of PCR. Using PCR, this is done in less than one day in contrast to several weeks needed, when Southern blots are used for hybridization with a probe. The test has also been used for the diagnosis of phenylkeptonuria (PKU), β thalassemia, hemophilia, etc. The PCR product in all these cases is examined using a labeled probe, to suggest whether or not mutant sequence causing the disease times, RFLP pattern of PCR products in healthy and defective feti singular- fetus) differ, thus enabling prenatal diagnosis.

Recombination data using PCR

PCR can be used for DNA amplification utilizing individual sperms, which are the products of meiosis and bination. Therefore, by studying PCR products from a large number of individual sperms one can calculate the proportion of sperms that are recommbinant for linked markers. This allows the construction of genetic maps at a resolution that is not possible with pedigree analysis. Microdissected chromosome bands are also used for PCR to allow the physical location of recombination at genetic lesions.

Sexing of embryos using PCR

Since DNA sequences single cells can study using PCR, the sex of human or livestock embryos, fertilized *in vitro*, can be determined before implantation. For this purpose, PCR primers and probes specific for sex chromosomes (e.g. ZFY or ZFX sequences) can be used. Y-specific primers and probes humans are available for this purpose. This technique can also be used to detect sex-linked disorders in the fertilized embryos.

DNA Fingerprinting

DNA fingerprinting is more successfully used in forensic science to search out criminals, rapists, solving disputed parentage and uniting the lost children to their parents or relatives by confirming their identity. This is done through linking the DNA recovered from samples of blood, semen, hairs, etc. at the spot of crime and the DNA of the suspected individuals or between the child and his/ her parents/relatives. We will discuss briefly the use of DNA fingerprinting in forensic medicine for the identification of criminals and disputed parentage. Some well-characterized sequences of microsatellites being used for designing primers, so that DNA fingerprinting may be achieved through PCR. PCR allows amplification of DNA from individual hairs, stains of blood or seminal fluid having partially degraded DNA, which could not be used earlier for characterization of individuals.

Blotting techniques

DNA, RNA and proteins are separated by blotting techniques. These are described below:

Southern blotting techniques

A method, developed genes in a DNA restriction fragment is called as Southern blotting technique. Southern blots can easily provide a physical map of restriction sites within a gene located normally and reveal the number of copies of the gene in the genome, and the degree of similarity of the gene when compared with the other complementary genes. A molecular biologist E.M. Southern (1975) developed the system for analyzing the associations on a chromosome. The procedure starts with the digestion of the DNA population by one or many restriction enzymes. Consequently, DNA fragments of unequal length are produced. This preparation is passed through agarose gel electrophoresis which results in the separation of DNA molecules based on their size. DNA restriction fragments present in the gel are denatured by alkali treatment. The gel is then put on top of the buffer saturated filter paper. Upper surface of the gel is covered with nitrocellulose filter and overlaid with dry filter paper. The dry filter paper draws the buffer through the gel. The buffer contains single-stranded DNA. Nitrocellulose filter binds DNA fragments strongly. After baking at 80 ° C, DNA fragments are permanent and fixed to the when come in contact of the nitrocellulose filter. Then the filter is placed in a solution containing radio-labeled RNA or denatures DNA probe of known sequences. These are complementary in sequence to the blot-transferred DNA. The radiolabeled nucleic acid probe hybridizes the DNA on a nitrocellulose filter. The filter is thoroughly washed to remove the probe. The hybridized regions are autoradiographically detected by placing the nitrocellulose filter in contact with a photographic film.

Northern blotting technique

The southern blotting technique could not be directly applied to the blot transfer of mRNA separated by gel electrophoresis, because RNA was found not to bind with nitrocellulose filter Alwine's et al (1979) devised a technique in which RNA bands are blot transferred from the gel to chemical reactive paper. Aminobenzyloxymethyl cellulose papers prepared from Whatman filter paper No. 540 and diazotized and rendered into the reactive paper and, therefore, hecomes available hybridization with radiolabeled DNA probes. The hybridized bands are found by autoradiography. Thus, Alwine's method extends that of Southern method and tor this reason it has been grunt the jargon Northern blotting. "There is not this northern or western like Southern." These blot transfers are reusable because of the firm covalent bonding of RNA to the reactive paper. Effective in binding the DNA while small fragments of DNA can most effectively be transferred to the diazotized paper derivative to nitrocellulose. These techniques were more advanced and demonstrated that mRNA bands can also be blotted directly on nitrocellulose paper and appropriate conditions.

Western blotting

Towbin et al. (1979) developed the western blotting technique to find out the newly encoded proteins by a transformed cell. Its working principle lies on the antigen-antibody reaction: hence is an immuno detection technique. In this method, radiolabeled nucleic acid probes are not used.

This technique follows the following steps: (Extraction of protein from transformed cells. (I) Separation of protein by using SDS-PAGE (sodium dodecyl sulfate polyacrylamide electrophoresis) where SDS acts as a solvent for electrophoresis. Transfer of the electrophoresed gel in a buffer at low temperature (40 ° C) for half an hour (h) Blotting of proteins on nitrocellulose filter paper of nitrocellulose filter, Whatman filter and coarse filter in the transfer buffer. Placing of the Whatman filter paper on a cathode plate is followed by a stack of coarse filter Whatman filter, electrophoresed gel, nitrocellulose filter, Whatman filter paper, coarse filter stack, Whatman filter and anode plate. Putting the complete set up in the transfer tank containing sufficient transfer buffer. Application of an electric field (30 V overnight for 5 hours) to cause the migration of proteins from the gel to nitrocellulose filter and binding on its surface. (The nitrocellulose filter has the exact image of the pattern of proteins as present in the gel. This type of blotting is called western blotting). Hybridization of proteins by using radiolabeled antibodies (I^{123}- antibodies) of known structure, isolated from a rabbit. Washing of the nitrocellulose filter with a wash solution (Tris-buffered saline + tween 20) to facilitate the removal of unhybridized antibodies. Detection of the hybridized sequences by autoradiography.

PRACTICALS

1. PERFORMING DIFFERENT METHODS OF STERILIZATION USED IN BIOTECHNOLOGY

Principle

Maintaining the sterile environment during the transfer or culturing of cell/ tissues of microbes plants and animal cells is most important. The concept of sterilization, for making the materials free from any type of contamination was given by **L. Pasteur**. Thus sterilization is a process of making an article. Surface or medium free from any type of microorganisms that contaminate the object and provide unwanted results. However, sterilization is one of the most important steps for cultivation isolation and study of purified cells or tissues in the laboratory. The other important things to be sterilized are the surgical tools, culture vessels, nutrient media and plant materials some other methods used to make these sterile are disinfection and incineration. Disinfection is a similar process of killing the harmful microorganisms especially the objects but not the culture media. Disinfection of table tops, equipment and others surface are usually done by using glycolic acid compounds, carbolic acid and formaldehyde, ethanol, etc. It is a process of killing microorganisms by using a flame, therefore, it is called flame sterilization. It is done by keeping the inoculation needle over the flame of the Bunsen burner till it becomes red hot. Thus, the microorganisms present on the surface of the needle are destroyed. Besides, physical and chemical methods are used for sterilization.

Physical methods of sterilization

There are several physical methods of sterilization of materials and objects. These are briefly described below:

(a) Moist Heat: Culture media (liquid and agar), water, glassware, surgical blades and scalpels are sterilized by using moist heat i.e. steam under pressure. It is done by using an autoclave, and also by a pressure cooker (used for cooking purposes at home).

(b) Dry Heat: Dry heat sterilization is carried out by using a hot air oven. Glassware, glass syringe, forceps, scalpel, pipettes, flasks metallic instruments. Petri dishes, etc. are sterilized in an oven at 150 ° C for 1 hour, or 250 ° C for 30 minutes..

(c) Radiation: Normally ultraviolet (UV) radiation is used in the inoculation chamber or a laminar air flow. Expose the working area under UV radiation before 2 hours to start the work. The source of UV radiation is a UV lamp or UV tubes enclosed in quartz because glass will not transmit UV radiation. Care should be taken not to see the UV radiation with naked eyes. Otherwise, any abnormality may occur to the eyes.

(d) Membrane Filtration: Sterilization of heat-sensitive substances like enzymes, antibiotics, amino acids could not be done by autoclaving because these may be denatured and non-functional. Hence, these are sterilization through various types of filters which may retain bacteria. Millipore membrane filters are commonly used for this purpose. It is commercially available e.g. Nalgene, Millipore and Whatman (England) membrane filters. It is made up of cellulose acetate and / or cellulose nitrate that contains small-sized pores of varying diameter (0.22, 0.45 and 0.5 um), But a membrane filter of 0.22 um is usually the preferred Millipore membrane filter placed inside the filtration assembly that is made up of autoclavable plastic materials, stainless steel or glass.

The whole assembly with the Millipore filter paper is sterilized by autoclaving before use. In such cases, the solutions to be sterilized usually are passed through membrane filters by a negative pressure applied through suction or centrifugal force. The filtrate so obtained is collected in a sterile container; the filtrate becomes microbe-free.

II. Chemical sterilization

(a) **Alcohol:** Ethanol (70%) or propane (70%) is used to sterilization working table top, inoculation chamber, hands, materials to be used, glass apparatus, etc.

(b) **Aldehyde:** Generally a laboratory or chamber is fumigated with formaldehyde when the number of contaminants gets increased.

(c) **Inorganic chemicals:** There are certain chemicals toxic to any organism such as salts of copper, mercury, etc. HgC1, solution (0.1%) is most commonly used as disinfectants for seeds, explants satorals For the same purpose, other chemicals used are sodium hypochlorite (NaOCl) (10%). The materials to be disinfected in the solution are kept in Hg, solution for 5-10 minutes (or naked hypo-chlorite). Soon take out the materials, transfer into sterilized distilled water and washed properly. Again repeat the process of washing 5-8 times to remove the traces of chemicals.

Requirements

- Autoclave (for moist heat sterilization), laminar airflow (fitted with UV tube),
- Oven (tor dry heat sterilization.
- Seed / leaves / root pieces / plants of any plants.
- Sprit, Ethyl alcohol (70%), mercuric chloride (0.1% HgCI,). sodium hypochlorite / calcium

hypochlorite (10%) (for chemical sterilization.

- Vacuum pump, filtration unit and Whitman filter paper (0.22 m) membrane filtration sterilization.
- Sterile distilled water, glassware, cotton plugs, Bunsen burner, absorbent cotton, plasticware Petri dishes, etc.
- Surgical blades, holders, forceps, etc. Plant seedling (10 days old).

Procedures

- **Sterilization of Glassware and Object / Tools:** For complete sterilization, syringe, glasswares, forceps, surgical blades holder, needles, scalpel, and metal instruments are kept in an oven at 150C for 1 hour or 250 ° C for 30 minutes.

- **Sterilization of Nutrient Media:** medium media (broth or medium agar), water and other waste (millipore filter) are sterilized by an autoclave (high pressure steam using moist heat) at for 20-30 minutes which gives 121°C Glassware must be wrapped with almunum foil and flasks containing nutrient medium, must be plugged with cotton and then wrapped with aluminum foil. After sterilization, the materials should not be taken out immediately. You must wait for 15-20 minutes to lower down the temperature to about 60 after releasing the pressure. The suitable containers should be taken out.
- **Surface Sterilization:** For the surface sterilization of seeds and other parts of a plant (leaves, bark, roots), these are washed by tap water. Then it is dipped in disinfectant chemical solutions first in 70% ethanol (ethyl alcohol) for 30 seconds, then in 10% sodium or calcium hypochlorite for about 10-20 minutes. The time of sterilization varies with the materials used. The material should be thoroughly washed (6-10 times) with double distilled water (5-8 times repeat this process). Sterilization procedures are not to be performed under aseptic conditions (laminar airflow). Sometimes, seeds are dipped only in 10% Na / Ca hypochlorite for 5-10 minutes of sterilization without any other treatment.
- **Membrane Filtration Method:** The membrane filtration method is used only for those chemicals which are heat sensitive / unstable at high temperature (eg enzymes, antibiotics, amino acid, vitamins Commonly, Whatman filter (0.22 um) is used for sterilization. These substances are passed through ultra-filtration membrane filters using a vacuum pump.

Results

If a culture is not autoclaved properly, it will be contaminated by a microorganism. Due to excess autoclaving, the culture media turns to brown color. Surface sterilization of seeds with chemicals for a long time affects the viability of seeds and other plant parts.

2. ISOLATION OF MILK PROTEIN (CASEIN) FROM THE MILK SAMPLE.

Principle

Proteins are present inside as well as outside the cells of all living organisms. They may function inside the cells occurring as enzymes, hormones, etc. Most of them contain 20 amino acids. Proteins are the building block of all cells, cellular components and give structural topography. Caseins are the main group of milk protein. Further several proteins can be resolved from this group such as, b, g and k on the basis of electrophoretic mobility under an electric field. It is selectively precipitated at pH 4.6. At this pH, caseins move under an electric field. It explains that the number of positive charges is equal to the number of negative charges at this pH. This is called the isoelectric point. Besides, the minimum solubility of any zwitterionic molecule (those having both a negative and positive charge generating residues within a molecule) is found at its isoelectric point. Such molecules may be separated by filtration or net charge zero because they do not decantation. There are many methods for the estimation of proteins in the sample (milk) e.g. Lowry's methods and Bradford's methods. In this method, the isoelectric precipitation principle is used for the isolation of casein from milk.

Requirements

- Beaker (100 ml), pH meter, cheesecloth.
- Water bath, oven, refrigerator, centrifuge, thermometer, balance, dropper, glass rod.
- 1N HCI (100 ml); 0.4 M HCl (90.7 ml of stock solution and diluted to 2 liters) 0.4 M NaOH
- Skimmed milk sample (centrifuge 50 ml milk at 2,000 rpm for 10 minutes, remove fat layer and use the solution for the test).

Procedure

1. Take 50 ml of skimmed milk sample and dilute it by adding 50 ml water.
2. Keep the mixture in a water bath at 30°•C.
3. By using a dropper add 0.4 M HCl in the mixture drop by drop stirring continuously.

(IV) Cheek the pH continuously by using a pH meter until a pH of 4.6 is obtained.

- Filter the content through cheese cloths when attaining pH of 4.6
- Collect the precipitated casein curd and measure the weight of it.
- Take the known amount of wet sample and dry it in an oven at 100 ° C.
- Cheek the weight of the dried sample continuously till it attains a constant weight.
- Then store the remaining sample in the freezer.

Results

- **Record the results as given below**
- **Volume of the milk used =**
- **Temperature of the milk =**
- **Initial pH of the milk =**
- **(Amount of 0.4 M HCL was used (ml), till the pH rose to 4.6 (used volume)**
- **Wet weight of the sample = gram after drying at 100 C**
- **Dry weight of the sample = gram**

3. DETERMINATION PROTEINS IN AN AQUEOUS SAMPLE AND MEASURED QUANTITATIVELY BY BIURET METHOD OR LOWERY'S METHOD

A) Biuret test for quantitative estimation of proteins

The proteins are made of amino acids containing peptide bonds. In the aqueous phase, two or more peptide bonds react with copper sulphate (blue solution of biuret) and result in a purple-violet colored product. The number of peptide bonds presented in a sample governs the color intensity of the sample. The color intensity of this product is measured at 540 nm by a colorimeter. The absorbance value is directly proportional to protein concentration in the sample. Biuret reagent is a solution consisting of copper sulfate in an alkaline solution of sodium potassium tartrate. Cu^{++} reacts with proteins coordinates with the peptide bond and is reduced to Cu^{++} ions resulting in a purple-violet color to the solution.

Requirements

- Test tubes (10) kept in a test tube rack, glass pipettes (1 ml, 5 ml
- Colorimeter
- Unknown protein solution
- Reagents as below:
- Biuret reagent: Weigh 1.5 g and 4.25 g of sodium / potassium tartrate and 4g of NaOH and dissolve in I liter of distilled Store solution in a reagent bottle at room temperature It is stable for a week.
- Standard BSA (bovine serum albumin): Weigh 1 g of BSA and dissolve in a little amount of distilled water in a volumetric flask and make the volume to 100 ml. The frozen solution should be liquefied before use.
- Solution of an unknown protein: The BSA solution should be made in such a way that each student must get 3 ml. Each student should be given solutions of different concentrations calculate the amount. Prepare BSA solution of different concentrations e g. 4.6, 8, 10 mg/ml and give the concentration to different students and maintain the record accordingly.
- **Procedure**
- Take test tubes and mark them with marking pen 1, 2,3…..and 8
- Add in each tube BSA, solution of test protein and distilled water as below
- To each tube add 4 ml of Biuret reagent and mix the contents by shaking the tubes.
- Mix the contents and incubate in a water bath set at 37 C for 10 minute
- Cool the test tube by keep in the test tube stand for 20 minutes at room temperature to develop the color
- Measure the optical density at 540 nm using a colorimeter

Results

- The standard solution of BSA has 0.1 ml will contain 1 mg Tube I containing 0.2 ml of BSA will have 2, and so on.

- Calculate the protein content from a standard curve prepared using mg /ml of albumin, casein, etc.
- The protein standard is also prepared as described above.
- Serially dilute the protein solution by adding gradually increasing the amount of distilled water decreasing the concentration of protein solution.
- Prepare a ratio of protein:water in the order : 10:0, 9:1, 8:2. 7:3, 6:4, 5:5, 4:6, 3:7. 2:8, 1:9, 10:0. Measure the optical density in a colormeter.
- Plot a graph using concentration versus optical density.

B) Lowry's method for quantifying proteins

Principle

Lowry et al (1951) gave a simple and less time-consuming colorimetric method for quantitative estimation of proteins. The principle of this method is based on the fact that the Folin-Ciocalteau reagent reacts with aromatic residues of protein and yields a blue color that is read in a colorimeter. The different proteins contain different aromatic residues. Blue color develops because the alkaline copper reacts with protein. Tyrosine and tryptophan present in protein reduce phosphomolybdate (present in the folin ciocalteau reagent).

Requirements

- Protein sample, albumin or a standard solution.
- Pipettes (1 ml, 5 m).
- Test tubes, water bath, colorimeter.
- 1N NaOH solution, alkaline sodium carbonate solution (2 g Na, CO, dissolved in 100 md of rd due to lib 0.IN NaOH solution).
- Reagents as below.
- Copper sulphate---sodium-potassium tartrate solution in 1% of sodium-potassium tartrate solution. Prepare fresh Alkaline copper reagent: Mix 50 ml of alkaline sodium carbonate solution and I ml of copper sulphate sodium potassium tartrate solution only on the day of use.
- Folin-ciocalteau reagent: It is a solution of sodium tungstate and

sodium molybdate in phosphonic and hydrochloric acid available commercially. Dilute the commercial reagent with an equal volume of distilled water only on the day of use.

Procedure

- Take a test tube, transfer 1ml of 1 N NaOH solution and heat up to 100 °
- Suspend I ml of protein sample into the above solution for 5 minutes.
- Add 5 ml of alkaline copper reagent mix properly and leave this mixture at room temperature for 10 minutes.
- Add 0.5 ml of Folin-ciocalteau reagent reagent rapidly with immediate mixing.
- Leave it for 30 minutes; measure the absorbance of the solution at 750 nm.

Results

- **Prepare a standard curve using casein or bovine serum and measure absorbance at 750 nm taking 1-10 ml solution with appropriate serial dilution.**
- **Prepare a graph between the protein concentration and absorbance.**
- **Calculate the amount of protein present in the unknown sample following the standard curve of known protein.**
- **Report the result as mg protein/ml of solution.**

4. DETERMINATION OF CATALYTIC POWER OF ACID PHOSPHATASE

Principle

Enzymes are the major class of proteins, called biological catalysts or biocatalysts because they speed up the rate of reaction occurring inside the cells of all organisms. In our body, numerous biological reactions occur which 1878 by Friedrich W. Kuhne to designate the biological catalysts that had earlier been called ferments. Acid phosphatase is located in lysosomes having acidic cytoplasm is nearly neutral. When an enzyme is active only in acidic pH. It is called acid-phosphatase. The enzyme catalyzes the release of inorganic phosphate from organophosphate i.e. nucleotides, nucleic acid, phosphorylated proteins, etc. Enzyme activity is assayed by measuring the conversion of the substrate into products which is formed within a given time. For determining the activity of acid phosphatase an artificial substrate, p-nitrophenyl phosphate, is used which is broken without the amount of enzymes, The term enzyme was called is pH falling between 5 and 6., The pH of by acid phosphatase is as below: Enzyme p-nitrophenyl phosphate + water p-nitrophenol + inorganic phosphate (P) (yellow) p-nitrophenyl phosphate (PNPP) is a colorless substrate but after the reaction, it has yellow particles due to the liberation of p-nitrophenol which can be easily measured by a colorimeter. The color intensity can be correlated with enzyme activity by using a standard graph. Enzyme fraction is procured mm any source as an aqueous solution. It is mixed with PNPP in a suitable buffer at a pH of 5.7 and allowed to react. At the end of the reaction, an aliquot of potassium hydroxide is added that stops the reaction and inactivates the enzyme. Besides, this alkali intensifies the yellow color through favoring of PNP. This facilitates to detect the color change in spite of the presence of a small amount of PNP

Requirements

- Acetate buffer (1 M. pH 5.7).
- p-nitrophenyl phosphate (0.05 M) solution, p-nitrophenol (200 mM solution) solution.
- Potassium hydroxide (0.5 M solution, 500 ml).
- Sodium acetate, acetic acid, sodium chloride (0.9%, saline solution) Sprouted seed of mung dal (legume), beaker.
- Test tubes, mortar and pestle, glass pipettes (1ml, 5 ml), cheese clothes.

Procedure

- Take mung seeds and soak overnight to allow germination under dark condition.
- Weigh 10g of sprouted seeds; wash with distilled water (5 times) in a beaker.
- Place in the pestle and mortar, transfer a small amount of sprouted seeds and pour a small amount of saline solution.

- Care should be taken that the total amount of ground material should be about 50 ml.
- Take a cheesecloth, make 5 layers of it and squeeze the ground material.
- Collect the filtrate in a test tube. This filtrate consists of the required enzyme. Since the room temperature fluctuates during different seasons, the enzyme fraction should be placed in a freeze till further use.

Assay of an enzyme by pouring the solutions in two different tests tubes in the following order:

- 0.4 ml Acetate buffer
- PNPP solution = 0.2 ml
- Distilled water 0.2 ml
- enzyme fraction (0.2 ml),
- (Immediately after adding enzyme mix the contents and note the titer cover the test and place in the dark for 20-25 minute incubation at 25 C).
- After 20 minutes of incubation, add 3 ml of KOH solution with the help of sterile autopipettes or sucker on the mouth of a pipette (mouth pipetting of KOH is very dangerous).
- Mix the contents properly with a gentle shaking of the tube.
- Take 1 to 6 & tubes and mark them properly.

Results

Record the absorbance serially from 1 to 6. The solution of PNP is 200 mM means 1 ml of solution contains 0.2 mmol of PNP. Record the absorbance after recording the readings.

5. IDENTIFICATION OF N-TERMINAL AMINO ACID OF A PROTEIN USING DANSYL CHLORIDE METHOD

Principle

Crystals of amino acids are white and its solution appears colorless which depends on the amount of carbon properties, etc. Specifically, dansyl chloride reacts with the free amino group of the N-terminal acid residues of a protein. Upon acid hydrolysis, it results in the formation of dansylated amino acid and the other free amino acids are separated. They can be identified using ultraviolet light. Under UV light, dansyl amino acids appear greenish-yellow in color.

Requirements

- Dansyl chloride solution (2.5 mg/ml in acetone) (prepare fresh) and keep at 4°C in the refrigerator.
- 0.2 M aqueous solution of sodium bicarbonate (keep at 4°C, when not in use).
- Ignition tubes/test tubes referred to as 'dansyl tubes' (5 cm x 2 mm) plates [polyamide thin layer plates both sides)] on both sides. Number the plates with a pencil in the top corner.
- Mark the origin for loading with the lower hand corner of the numbered side of the plate (the origin for loading on the reverse side of the plate should be immediately behind the loading position for the front of the plate i.e. 0.3 0.5 mm from each edge in the lower right-hand corner).
- Prepare two chromatography solvents as below:
- Solvent 1: Formic acid: water (4.5: 100, (v/v)
- Solvent 2: Toluene: acetic acid (8: Ultraviolet source (254 Lum or 265 um), vacuum pump, desiccator

Procedure

- For Amino Acids
- Prepare the stock solution (1 mg/ml) of standard dansyl amino acid in 0.2 M NaHCC (standard dansyl amino acid are-Pro, Leu, Phe. Thr. Glu, Arg Store the stock solution at 4°C when not in

use).

- Transfer 10 ul of each amino acid solution in a dansyl tube and mix gently. Add 30 ul of dansyl chloride solution (5 ul for each amino acid).
- Add 10ul of unknown amino acid and 5 ul of dansyl chloride in a separate dansyl tube.
- Seal the tube with parafilm and incubate at 37°C for 1 hour at room temperature for 3 hour).
- Dry the samples in a vacuum because the small volume of the liquid present will take about 5 minutes.
- Dissolve the dried samples in 10 ul of acetone using a capillary tube.
- Load 1 µl aliquots of the standard mixture at the origin (the diameter of the spot should not be allowed to exceed 1-2 mm).
- Load 1 ul of the unknown amino acid on the reverse side.
- Place the plate in the first chromatography solvent when the loaded sample is fully dry Allow the plate to develop and the solvent to reaches the top. Dry both sides of the plate.
- Develop the dansyl plate in the first solvent. Dry the plate until the solvent front reaches the top.
- Observe the plate under UV light

2. For Proteins

- Transfer 10ul of protein in 0.2 M NaHCO, in a one-ml stoppered tube.
- Pour 10 µl of dansyl chloride solution into one ml tube.
- Stopper the tube and incubate at 37 °C for 1 hour (or at room temperature for 3 hours)
- Dry the samples in vacuum.
- Transfer 20 µl of 6 N HCI to ml tube and close the tube tightly with a stopper. Hydrolyze it keeping in an oven maintained at 105 °C for 24 hours.
- Dry the samples in a vacuum.
- Dissolve the dried mixture using a fine capillary tube. Diameter of the spot should not be allowed to exceed 1-2 mm

- Load 1 µl of protein hydrolysate containing the unknown dansyl amino acids on the reverse.
- Place the plate in the first chromatography solvent when the loaded completely dry. Allow the plate to develop and reach the solvent to the top. Dry both sides of the plate.
- Develop the dansyl plate in the second solvent at right angles to the direction of development as found in the first solvent. Wait to reach the solvent to the top, and dry the plate.
- Observe the plant under UV light.

Results

- **You should identify three major bands at the bottom of the plate.**
- **Dansyl chloride is hydrolyzed into dansyl hydroxide.**
- **The hydrolyzed product appears as blue color: Dansyl amide (produced by side reaction with NiS) shows blue-green fluorescence and about one-third way up the plate.**
- **These two spots are useful internal markers.**
- **The third spot fluoresces green and will correspond to the dansyl derivative of s amino acid.**
- **Solvent 2 separates the dansyl derivatives of hydrophobic amino acids and some natural amino acids.**
- **Derivatives of charged and other neutral amino acids remain at the lower end of the chromatogram.**
- **The chromatogram is compared with the standard provided in your laboratory.**

6. ANALYSIS OF PROTEIN BY GEL ELECTROPHORESIS UNDER DENATURING CONDITIONS (SDS-PAGE)

Principle

The proteins, DNA and RNA (biomolecules) have electric charges that depend on the molecule and the conditions of the medium (pH of the buffer used). Charged molecules can be separated by electrophoresis in gels. Polyacrylamide gel is used for protein isolation differences in the amino acid composition proteins have a unique mass and charge. Hence, the proteins have a negative charge and a net positive charge or isoelectric point (no clarification at a given buffer SDS- PAGE is a high resolution method used universally for analyzing the mixture of proteins according to their respective size. SDS solubilized in soluble proteins makes possible the analysis of the other insoluble mixtures Separation of the proteins does not occur due to similar charge: mass ratios. Therefore, such proteins are treated first with an ionic detergent called sodium dodecylsulfate(SDS) before the start and during the course of electrophoresis (PAGE). Therefore, such electrophoresis is called SDS PAGE identical proteins are denatured by SDS resulting in their sub-units. The polypeptide chains get opened and extended. On the basis of the mass but not the charge, The molecule is separated Electrophoretic separation is normally used for these reasons i.e. (i) the gel acts as molecular sieves hence separates the molecules on the basis of their size, and (ii) the gel suppresses conventional current produced by the small temperature gradient that improves the resolution. Polyacrylamide gel is better for the size fraction of proteins.

SDS-PAGE by Laemmli

Procedure

- Laemmli invented the gel disc electrophoresis which is a superior system to resolve proteins in a mixture.
- According to the procedure, proteins are electrophoresed by a discontinuous polyacrylamide gel. A stacking gel (1/ 3rd of the total gel) which includes sample loading wells consists of 4% acrylamide prepared in Tris HCl buffer (pH 6.8).
- The separating gel consists of 7.5- 15.0% acrylamide.
- This range depends on the molecular size of proteins prepared in Tris HCL buffer: pH 8.8. In the lysis of a complex mixture of proteins, the resolution is improved by the initial movement through stacking gel.
- The final bands in the separating gel are sharper and focused in a better way. After electrophoresis mixture of proteins is separated as discrete bands.

Principle

- The proteins are denatured and have a negative charge with a uniform charge to mass ratio (z / m) n treated with SDS (anionic detergent).

- Proteins migrate towards the anode at alkaline pH throughout the gel during electrophoresis.
- The smaller polypeptides moves faster followed by the larger peptides.
- Therefore, the intrinsic charge on proteins is masked in SDS-PAGE.
- Hence, the separation is based on the size.
- Mercaptoethanol reduces interpolypeptide disulfide bridges and separates the sub -units of a polymeric protein.
- Molecular weight of the separated protein is analyzed by the molecular weight of the standard protein and its mobility.

Requriments

- Ammonium per sulfate (10%), coomassie brilliant blue (0.3%).
- Destaining mixture, gel staining dish.
- Electrophoresis apparatus with supply
- Running buffer, SDS (10%) , 1.5 M Tris HCI (pH 8.8), 0.5 M Tr is HCI (pH 6.8), Laemmli .
- Stock acrylamide-bis-acrylamide solution .
- Reagents: The following reagents must be prepared in advance
- Stock acrylamide (bis-acrylamide solution: 29.2 g acrylamide, 0.8 g bis acrylamide: final volume raised to 100 ml)
- 1.5 M tris-HCI, pH 8.8: Dissolve 18.15 g of Tris in 50 ml of distilled water, adjust the n to 8.8 with HCI, make the final volume to 100 ml).
- 0.5 M Tris- HCI, pH 6.8: Dissolve 6 g, Tris, in 60 ml distilled water, adjust pH to 6.8 with HCL, make the final volume to 100 ml.
- 10% SDS: Dissolve 1g SDS in 5 ml distilled and raise the final volume equal to 10 ml.
- Gel running buffer: Dissolve 14.4 g glycine, 1 g SDS in 1 litre distilled water, adjust pH to 8.3 by adding solid Tris, make the final volume equal to 1 litre.
- Ammonium per sulfate (APS) (10%): Dissolve 500 mg of solid APS in 5 ml of distilled water. Always use freshly prepared APS

solution only.

- Coomassie Brilliant Blue R 250: Dissolve 600 mg of Coomassie brilliant 250) in 80 ml methanol, add 20 ml glacial acetic acid and make the final volume equal to 0 (CBBR- 200 ml with distilled water.
- De-staining solution: Mix 400 ml methanol, 100 ml glacial acetic acid and 500 ml distilled to get 1 liter of this solution.
- Laemmli buffer: 62.5 mM Tris-HCl, pH 6.8 (use diluted 0.5M Tris HCl pH 6.8), 10 glycerol, 5% mercaptoethanol, 2% SDS. According to the manufacturer's instructions, gel electrophoresis apparatus should be assembled protein sample should be prepared in Laemmli buffer or in the buffer supplied with the kit for the experiment.

Procedure

- Make 7.5% SDS resolving gel. The capacity of the apparatus governs the volume of the final gel solution. Accordingly, read carefully the instruction manual supplied with the electrophoresis unit before deciding the volume of the gel. A commercially available SDS-PAGE kit is also used according to instructions.
- According to the manufacturer's instruction, assemble the gel electrophoresis apparatus.
- Make the gel solution of the separating and stacking gels (find out the volume of solution needed by the apparatus). Prepare fresh solutions just before.
- APS and TEMED are added to the rest of the solution just before pouring the get solution into the 48.0 ml 768 ml 96.0 ml glass sandwich.
- Fill up the glass sandwich with a pipette upto the mark with the separating gel solution prepared (the body must not make contact with chemicals. The acrylamide solution in an unpolymerized state is neurotoxic).
- To make the gel surface straight after polymerization, overlay the acrylamide solution with water (the gel should be allowed to polymerize for 30 minutes. When the polyacrylamide layer becomes distinct below the water layer, polymerization is complete).

- Remove the distilled water by using the filter paper carefully. Thereafter, add the requisite amount of 4% stacking gel solution (freshly mixed). Then insert the comb for polymerization.
- With the polymerized gel place the glass sandwich into the electrophoretic chamber and add running buffer to level in both cathodic and anodic chambers. To prevent the sample wells from deforming, remove the comb under the buffer.
- With the help of a micropipette, load the denatured protein solution into the well (the protein sample should be denatured in Laemmli buffer by boiling for 5 minutes).
- Add standard molecular weight marker proteins in one lane. For detection by CBB dye, generally, 20 to 50 ug proteins are sufficient.
- Tightly connecting the electrodes of the apparatus with the power supply. Run the gel at a constant current of 20 mA.
- Track the mobility of the sample in the matrix with the dye (generally bromophenol blue is added to the Laemmli buffer). After completion, switch off the button and disconnect the apparatus.
- Transfer the gel to the staining tray containing the gel staining dye.
- Under shaking conditions on a rocking shaker, stain the gel for at least 2 hours or overnight. At this stage, the whole gel turns blue.
- Carefully transfer the gel de-staining solution and shake on a rocker shaker for 30 minutes. Add fresh destaining solution. Repeat these steps until the bands are clearly visible in the gel. At this stage, take a photograph of the gel.
- When the gel is clearly visible, analyze the photographed gel.

Results

Several distinct blue colored bands can be seen in the gel.
Each band represents a single or multiple bands in the lane.
Depending on the amount of the polypeptide present in the protein solution loaded in the gel, the intensity of these bands varies.

REFERENCES

1. Abollino, O., Giacomino, A., Malandrino, M., Mentasti, E., Aceto, M., Barberis, R., (2006). Assessment of metal availability in a contaminated soil by sequential extraction. Water Air Soil Pollution, 173(1-4): 315-338.
2. Agarwal, V. K., (2000). Molecular Biology. New Delhi: S. Chand.
3. Agarwal, V. K. and Verma, P. S., (2000). Concepts of Molecular Biology. New Delhi: S. Chand.
4. Ahmadpour, P., Ahmadpour, F., Mahmud, T., Abdu, A., Soleimani, M., Tayefeh, F.H., (2014). Phytoremediation of heavy metals: a green technology. In: Asrari, E. (Ed), heavy metal contamination of water and soil: Analysis, Assessment, and Remediation strategies. CRC Press, Taylor and Francis, 249.
5. Aldrich, J., and Vaker, R., (1970). Plant Dis. Reptr. 54: 446-448.
6. Alexander, M., (1977). Introduction to Soil Microbiology, John Wiley & Sons, New York.
7. Ali, H., Khan, E., Sajad, M.A., (2013). Phytoremediation of heavy metals-concepts and applications. Chemosphere, 91(7): 869-881.
8. Amirato, P.V., (1983). In Handbook of Plant Cell Culture. Vol. I, Techniques for Propagation and Breeding (Eds. D.A., Evans, W.R., Sharp, P.V., Amminrato and Y., Yamada), PP. 82-123, Macmillan., New York.
9. Anand, A., et al., (2004). Physiol. Biol. 10: 233-242.
10. Aneja, K.R., (2014). Experiments in Microbiology, Plant Patholoy and Biotechnology. Fourth Edition. New Age International (P) Limited Publisher.
11. Aneja, K.R., (2014). Laboratory Manual of Microbiology and Biotechnology. Scientific International PVT. LTD Medtech.
12. Anonymous (1980 a) Nature (London) 286:110.

13. Anonymous (1980 b). NBRI News letter 7:28.
14. Arora, P.N., and Malhan, P.K., Biostatistics.
15. Atkinson, G.F., (1961). Mushrooms, 2nd ed. Hofner Publishing Co., New York.
16. Atkins, F.C., (1983). Mushroom J., 125: 168-171.
17. Baker, R., and Nash, S.M., (1965). Phytopath. 55: 1381-1382.
18. Baker, K.F., and Snyder W.C., (1965). Ecology of Soil Borne Plant Pathogens-Prelude to Biological Control, John Murray, London, P. 571.
19. Banerjee, P.K., Problems on genetics molecular genetics and evolutionary Genetics.
20. Bhattacharya, D., and Roy, Choudury, Statistcs theory and Practices.
21. Barabasz, A., Wilkowska, A., Tracz, K., Ruszczyñska, A., Bulska, E., Mills, R.F., (2013). Expression of HvHMA2in tobacco modifies Zn-Fe-Cd homeostasis. Journal of Plant Physiology, 170: 1176-1186.
22. Barell, B.G., Air, G.M., and Hutchindon, C.A., (1976). Nature. 264: 34-41.
23. Becking, J.H., (1982). In Advances in Agricultural Microbiology, eds. N.S. Subba Rao), pp. 89-110, Oxford & IBH Publish, Co., New Delhi.
24. Bergersen, F.J., (1971). Ann. Rev. Plant Physiol. 22: 121-140.
25. Betsch, David F. "DNA Fingerprinting in Human Health and Society." Retrieved: 12th April 2006 http://www.accessexcellence.org/RC/AB/BA/DNA_Fingerprinting_Basics.html
26. Carlson, P.S., (1973). Science 180: 1366-1368.
27. Bhaskaran, S., (1987). In Biotechnology in Agriculture, (eds. Natesh, S., Chopra, V.L., and Ramachandan, S.,) PP. 101-108, Oxford &IBH Publ. Co., New Delhi.
28. Biello, David. Tobacco Plant Transformed into Plague Vaccine Factory. Retrieved 10 Jan. 2006 http://www.sciam.com/article.cfm?chanID=sa016&articleID=000E3140-EB2E-13C2-AB2E83414B7F0000
29. Binns, A.N., (1981). In Propagation of Higher Plants Through Tissue Culture: Emerging Technoloies and Strategies, (eds. Constantin, M.J., Henke, R.R., Hughes, K.W., and Conger, B.V.,) Enviorn Exp. Bot. 21: 325-332.
30. Biotechnology: A Dutch Perspectives (1981). A report by the Netherlands Study Centre for Technology Trends, (ed. J.H.F. Van Apeldoorn).
31. Boffey, S.A., (1987). In Biotechnology: The Biotechnology Principles (eds. Trevan, M.D., Boffey, S., Goulding, K.H., and Stanbury, P.,) 11-152. Open University Press: Milton Keynes, England.
32. Bouwer, E.J., and Zehnder, A.J.B., (1993). Trends Biotechnology. 11: 287-318.
33. Bouwer, E.J. et al., (1998). In Biological Treatment of Hazardous Wastes (eds. Lewandowaski, G.A., and Deflippi, L.J.) pp 237-270, John Wiley & Sons.
34. Brooks, R., Lee, J., Reeves, R.D., Jaffre, T., (1977). Detection nickeliferous rocks by analysis of herbarium specimens of indicator plants. Journal of Geochemical

Exploration, 7: 49-57.

35. Bull, A.T., Holt, G., and Lilly, M.D., (1983). Biotechnology: International Trends and Perspectives, p. 84, Oxford & IBH Publ. Co. PVT. Ltd., New Delhi.

36. Carlson, P.S., (1983). Cell and Tissue Culture Techniques for Cereal Crop Improvement, pp. 407-418, New York.

37. Christov, L., and Van Driessel, B., (2003). Indian J. Biotech. 2:444-450.

38. Clark, D.S., (1962). Biotehnol Bioeng. 4:1721.

39. Cohen, S.N., Chang, A.C.Y., Boyer, H.W., and Helling, R.B., (1973). Proc. Nat. Acad. Sci. (U.S.A) 70:3240-3244.

40. Collins, H.A.,and Watts, M., (1983). In Handbook of Plant Cell Culture: Techniques for Propagation and Breeding, Vol. I (eds. Evans, D.A., Sharp, W.T., Ammirato, P.V., and Yamada, Y.) pp. 729-747. Macmillan, New York.

41. Crueger, W., and Crueger, A., (1984). Biotechnology-A Text Book of Industrial Microbiology. Sinaver Associates Inc., Sunderland, USA, pp 23: 207-216.

42. Cummings, Michael R., and Williams S. Klug., (2004). Concepts of Genetics. New Delhi: Pearson Education.

43. David, L.N., and Michael, M.C., (2013). Lehninger principles of biochemistry 6th edition.

44. Dodds, J.H., and Roberts, L.W., (1985). Experiments in Plant Tissue Culture, 2nd Ed. Cambridge University Press, Cambridge.

45. Dubey, R.C., (2009). A textbook of Biotechnology. S.Chand and Company LTD.

46. Ferber, Dan. "The New Science of Cell Hacking." Popular Science, June 2004. pp: 43-44.

47. Frazier, W. C. and Westhoff, D.C., (1993). Food Microbiology. New Delhi: Tata McGraw-Hill.

48. Gautam, N., Verma, P.K., Verma, S., Tripathi, R.D., Trivedi, P.K., Adhikari, B., Chakrabarty, D., (2012). Genome wide identification of rice class I metallothionein gene: tissue expression patterns and induction in response to heavy metal stress. Functional and Integrative Genomics, 12: 635-647.

49. Garg, N., Garg, K.L., Mukerjii, K.G., (2013) Laboratory Manual of Food Microbiology. I.K. International Publishing House Pvt. Ltd.

50. Gibbs, W., (2001). "Cybernetic Cells." Scientific American, Aug. 54-57

51. Greipsson, S., (2011). Phytoremediation, National Education Knowledge, 3(10): 7.

52. Gritfith, A., Gelbart, W.M., Miller, S.H., and Lewontin, R.C., Modern genetic analysis.

53. Gu, C.S., Liu, L.Q., Deng, Y.M., Zhu, X.D., Huang, S.Z., Lu, X.Q., (2015). The heterologus expression of the Iris lacteal var. chinensis type 2 metallothionenin IIMT2b, gene enhances copper tolerance in Arabidopsis thaliana. The Bulletin of Environmental Contamination and Toxicology, 94: 247-253.

54. Guha, S., Maheswari, S.C., (1964). Nature (London), 204: 297.
55. Guha, S., Maheswari, S.C., (1966). Nature 212: 97-98.
56. Gupta, C.P., Sharma, A., Dubey, R.C., and Maheshwari, D.K., (1999). Cytobios. 99:183-189.
57. Gupta, C.P., Dubey, R.C., Kang, S.C., Maheshwari, D.K., Current Science. 81: 91-94.
58. Haberlandt, G., (1902). Sber Akad. Wiss. Wien. III: 69-92.
59. Hanson, Earl D., (1983) Recombinant DNA Research and the Human Prospect. Washington, DC. American Chemical Society.
60. Henshaw, G.G., (1982). In plant Tissue Culture, pp. 789-792, Tokyo.
61. Huang, G.Y., and Wang, Y.S., (2010). Expression and characterization analysis of type 2 metallothionein from grey mangrove species (Avicennia marina) in responses to metal stress. Aquatic Toxicology, 99: 86-92.
62. ICGEB. (1998). Activity Report.
63. Khoshoo, T.N., (1988). Enviornmental Concern and Strategies. II Ed. PP. 253-404. Ashish Publ. House, New Delhi.
64. King, P.J., (1980). Int. Rev. Cytol. Supplement. 11A: 25-54.
65. Kowles Richard, Solving problems in Genetics.
66. Kumar, H.D., (2003). Modern Concepts of Biotechnology. New Delhi: Vikash Publishing House.
67. Lee, S., (2014). Artifical induction of cadmium tolerance and its further enhancement via heterologus co-expression of SpHMT1 and AtPCS1 in the yeast cells. Journal of Korean Society Applied Biological Chemistry. 57: 307-310.
68. Lopez, D. A., R. M. Williams., and Michlke, K., (1994). Enzymes: The Foundation of Life. München: The Neville Press.
69. Ma, Y., Rajkumar, M., Luo, Y., Freitas, H., (2011). Inoculation of endophytic bacteria on host and non-host plants-effects on plant growth and Ni uptake. Journal of Hazard Materials, 195: 230-237.
70. Maestri, E., Marmiroli, M., Visioli, G., Marmiroli, N., (2010). Metal tolerance and hyperaccumulation: costs and trade-offs between traits and environment. Environmental and Experimental Botany, 68(1): 1-13.
71. Maheswari, S.C., Rasid, A., and Tyagi, A.K., (1983). IAPTC Newsletter 41:2-9.
72. Mandels, M., and Stenberg, D., (1976). J. Ferment. Technol. 54: 267-286.
73. Mandal, P.K., (1988). Gene Therapy. Sci. Reptr. 25: 156-157.
74. Mason, H.S., and Arntzen, C.J., (1995). Trends Biotechnol. 13: 388-392.
75. Maxam, A.M., and Gilbert, W., (1977). Proc. Natl. Acad. Sci. (U.S.A) 74: 560-564.
76. Memon, A.R., Schroder, P., (2009). Implications of metal accumulations mechanisms to phytoremediation. Environmental Science and Pollution Research, 16(2): 162-175.

77. Miller, C.O., Skoog, F., Saltz, M. and Strong, F.M., (1955). J. Am. Chem. Soc. 77: 1992.

78. "Molecular Biotechnology in Life." Author not mentioned. Retrieved: 5th April 2006 http://www.goldenessays.com/free_essays/4/technology/molecular-biotechnology-in-life.shtml.

79. Morgan, T.H., (1901). Regerantion, Macmillian: London.

80. Nitsch, C., (1974). C.R. Acad. Sci. (Paris), 278:1031-1034.

81. Murashige, T., (1974). Ann. Rev. Plant Physiol 25: 135-166.

82. Murashige, T., and Skoog, F., (1962). Physiol Plant, 15:473-497.

83. Peberdy, J.F., (1989). Mycol Res. 93:1-20.

84. Peters, P., (1993). Biotechnology- a guide to genetic engineering. Wmc. Brown Publishers, Dubuwue, P. 253.

85. Prakash, J., and Giles, K.L., (1987). Int. Rev. Cytol. 107: 273-292.

86. Prescott, L.M., Harley, J.P., and Kein, D.A., (2003). Microbiology. McGraw-Hill High Education, New York.

87. Purohit, S., (2005) Agricultural Biotechnology. India: Agrobios.

88. Purohit, S., (2004) Biotechnology: Fundamental and Applications. India: Agrobios.

89. Rajkumar, M., Ae, N., Prasad, M.N.V., Freitas, H., (2010). Potential of siderophore-producing bacteria for improve heavy metal phytoextraction. Trends Biotechnology, 28(3): 142-149.

90. Reinert, J., and Bajaj, Y.P.S., (1977). In Applied and Fundamental Apects of Plant Cell, Tissue and organ ulture (eds. Reinert, J., and Bajaj, Y.P.S) pp. 251-267, Springer-Verlag Berlin.

91. Roberts, R.J., (1983). Nuceilc Acid Res. 11: 135-168.

92. Robson, R.L., (1979). Biochem. J. 181: 569-575.

93. Ruschel, A.P., and Vose, P.B., (1983). In Hand Book of Plant Cell Culture-Techniques for Propagation and Breeding, Vol. 1 (eds. Evans, D.A; Sharp, W.R; Ammirato, pp. and Yamada, Y.) pp 828-851, Macmillian Publ. Co., New York.

94. Russel, P.J., Genetics (5th edition).

95. Sasson, A., (1984). Biotechnology: Challenges and Promises, UNESCO, Paris.

96. Schell, J., and Van Montagu, M., (1983). Biotechnology 1: 175-180.

97. Schesinger, H., (1999). "DNA Conductor." Popular Science, Aug. pp: 47.

98. Schlegel, H.S., (1986). Genera Microbiology, 6th Ed. (Tranlated by M. Kogut) Camcridge University Press, London.

99. Sharp, W.R., Sondahl, M.R., Caldas, L.S., and Maraffa, S.B., (1980). Hort. Rev. 2: 268-310.

100. Shrof, J. M., (1998). "Miracle Vaccines." US News & World Report, 23 Nov. pp: 157

101. Singh, Mahabir, and Sandeep Ahlawat. "Applied Genetics." MTBiology Today, April 2004: pp: 27-29.
102. Skoog, F., (1944). Am. J. Bot. 31: 19-24.
103. https://www.slideshare.net/syed_ismail/carbohydrates-28872311.
104. Smith, P.H., (1966). Dev. Industrial Microbiol. 7: 156-159.
105. Southern, E.M., (1975). J. Mol. Biol. 98: 503-517.
106. Snustad, D. Peter, and Simmons, M.J., Principle of genetics (4th edition).
107. Srivastava, S.K., et al., (1995). Indian J. Exp. Biol. 33: 962-966.
108. Swaminathan, M.S., (1980). Second Annual Day Lecture, NBRI, Lucknow.
109. Tangahu, B.V., Sheikh Abdullah, S.R., Basri, H., Idris, M., Anuar, N., Mukhlisin, M., (2011). A review on heavy metals (As, Pb, and Hg) uptake by plants through phytoremediation. International Journal of Chemical Engineering, 31: 939161.
110. Thomas, E and Davey, M.R., (1975). From Single Cell to plants, Wyked=ham: London.
111. White, P.R., (1934). Plant Physiol. 9: 585-600.
112. Willams, J.G., (1981). In Genetic Engineering, Vol. I, (ed. Williamson, R.,), PP. 1-59, AcademicPress London.
113. Wiseman, A., (1975). Handbook of Enzyme Technology, Halsted Press, New York.
114. WWW.scribd.com., www.slideshare.com., www.coursehero.com., www.biocyclopedia.com., www. molecular-plant-biotechnology.info., www.slideplayer.com., www.documents.mx., www.docplayer.net., www.bmb.psu.edu.,www.archive.org.,www.mafiadoc.com., www.powershow.com., www.bioscience.jbpub.com., www.outhorstream.com., www.docslide.com., www.msluay.com., www.scienceprofonlue.com., www.csus.edu., www.krishikosh.egrouth.ac.in., www.tankonyvtor.hu., www.silo.pub., www.studyblue.com., www.Preservearticles.com., www.biotechuniverse.blogspot.com.
115. Wikipedia Online Encyclopedia, Retrieved 20 Feb. 2006 http://www.wikipedia.org/wiki/Biotechnology.
116. Wikipedia Online Encyclopedia, Retrieved 20 Feb. 2006 http://www.wikipedia.org/wiki/genetic_engineering.
117. Wikipedia Online Encyclopedia, Retrieved 20 Feb. 2006 http://www.wikipedia.org/wiki/genetically_modified_food.
118. Wikipedia Online Encyclopedia, Retrieved 22 Feb. 2006 http://www.wikipedia.org/wiki/bioinformatics
119. Yamada, Y., aand Fujita, Y., (1983). In Handbook of Plant Cell Culture-Technique for Propagation and Breeding. (eds. Evans, D.A; Sharp, W.R., Ammirato and Yamada, Y) pp. 717-728. Macmillian Publ. Co., New York.

Index

D

E

F

G

FSC
www.fsc.org
MIX
Papier | Fördert gute Waldnutzung
FSC® C083411

Kolibri 360 GmbH
Ferdinand-Jühlke-Straße 7
99095 Erfurt, Deutschland
produktsicherheit@kolibri360.de